Academia Brasileira de Filosofia
Volume 2

A True Polymath
A Tribute to Francisco Antonio Doria

Volume 1
L'Imagination. Actes du 37ᵉ Congrès de l'Association des Sociétés de philosophie de langue française. Rio de Janiero, 26-31 mars 2018.
Jean-Yves Beziau and Daniel Schulthess, éditeurs

Volume 2
A True Polymath. A Tribute to Francisco Antonio Doria
J. Acacio de Barros and Décio Krause, eds.

Academia Brasileira de Filosofia collection editor
Jean-Yves Beziau   jyblogician@gmail.com

A True Polymath
A Tribute to Francisco Antonio Doria

Editors

J. Acacio de Barros

Décio Krause

© Individual authors and College Publications 2020.
All rights reserved.

ISBN 978-1-84890-351-7

College Publications
Scientific Director: Dov Gabbay
Managing Director: Jane Spurr

http://www.collegepublications.co.uk

All rights reserved. No part of this publication may be reproduced, stored in a retrieval system or transmitted in any form, or by any means, electronic, mechanical, photocopying, recording or otherwise without prior permission, in writing, from the publisher.

Foreword

Francisco Antonio Doria was born in Rio in 1945. He received a B.Sc. in chemical engineering from Rio's Federal School of Chemistry in 1968, and a Ph.D. in theoretical physics in 1977 from the Brazilian Center for Research in Physics; his advisor was Leopoldo Nachbin, a top-level Brazilian mathematician. Doria is presently a Professor of Communications, Emeritus, at Rio's Federal University.

Doria's research interests wandered from Clifford algebras in physics to the geometry of gauge fields and then to foundational questions. Clifford algebras were the subject of both his master's dissertation (under the guidance of Adel da Silveira) and his doctoral thesis, where he derived the Bargmann-Wigner high spin field equations with elementary linear algebra tools. He then became interested in the so-called gauge field copy problem, and from a question about that problem he felt he had to deal with foundational issues.

The copy phenomenon in the theory of non-Abelian gauge fields was discovered in 1975 by Nobel Prize winner Chen Ning Yang together with Tai Tsun Wu. In a brief but instigating paper they showed that the same gauge field could be derived from (at least) two gauge potentials which weren't related — not even locally — by a gauge transformation. One of their examples showed two systems with the same gauge field and different potentials; however while the first system exhibited a non-zero source, the other was a vacuum system. As no gauge transformation can map a source to zero, the two potentials cannot be gauge equivalent, not even locally.

After several efforts Doria managed to show that the gauge copy phenomenon depended on an integrability condition and on the existence of a nontrivial partial connection form with zero curvature. His first papers on the matter contained flaws, which were later corrected in two papers published in the early 1980s; the first one was very brief and discussed the copy phenomenon in General Relativity, and the second one dealt with a very general situation where one finds copy-like phenomenon.

A brief but deep personal crisis led to a halt in his research work in 1983. However he soon recovered and went on with his work. He had shown that copied gauge fields belonged to a stratified set in the space of gauge fields (which could be given a Fréchet-manifold structure). He then wondered whether different set-theoretic models for these and other theories could lead to, say, alternative physics. More succinctly: what is relevant for physics, the syntactic aspects of a theory or its semantics?

Leopoldo Nachbin had said that "he would remain Doria's advisor for life," and so Doria turned to Nachbin for advice. That proved to be a key turning point in Doria's career, as in 1985 Nachbin introduced him to Newton da Costa. Newton invited Doria to give a teach-yourself course on forcing through Boolean-valued models at the Institute of Advanced Studies at the University of São Paulo. Doria later told friends that the front row in the classroom looked like a doctoral examination committee, as it included Ofélia Alas, Edison Farah, and Newton da Costa himself... This course led to the drawing-up of a list of problems in 1987 that both Newton and Doria felt that begged for logic tools in their solution. The list included the decision problem for chaos (settled by da Costa and Doria in 1990) and the notorious P vs. NP problem — more on it later on. The problem list was a research program, on which da Costa and Doria started to work in 1987. Unsuccessful at first, Newton suggested that Doria apply for a fellowship to enjoy a sabbatical year at Pat Suppes' lab in Stanford. Doria then moved to Palo Alto in late 1989 (actually two weeks after the Loma Prieta quake) and resumed work on the list, now as a Senior Fulbright Scholar at Suppes' IMSSS. After a suggestion by Pat Suppes, he and Newton used Richardson's results on the computational handling of algebraic expressions to construct explicit examples of dynamical systems which were formally undecidable for chaos. This settled several open problems: first, there is no direct connection between Heisenberg's uncertainty principle and undecidability, something that had long been conjectured; then they exhibited an explicit expression for the halting function from computer science in the language of calculus, from which one can even derive a theory of hypercomputation for ideal analog computers. Finally they extended Rice's theorem in computer science to vast areas in mathematics. Side results, as mentioned, had to do with the possibility of hypercomputers (computers that violate Church's Thesis) and his first efforts in dealing with the P vs. NP conundrum.

The solution of the decision problem for chaos was published in 1991 and deserved a detailed comment by Ian Stewart in Nature; da Costa and Doria went on to apply the techniques they had developed, to the 1974 Arnol'd Problems in the list drawn up at the AMS Symposium on the Hilbert Problems. These problems were decision problems for the stability nature of equilibrium in polynomial dynamical systems, and their solution was published in 1994. (The 1974 AMS list of more than 100 problems, to which all lecturers in that symposium were invited to contribute, already gave Fields medals for their solution, as in the case of Mordell's Conjecture.)

The tools for the construction of undecidable sentences and problems had another, unexpected, consequence. A suggestion by Marcelo Tsuji, an economist interested in the foundations of that discipline led to the proof of the undecidability and incompleteness of game theory, which leads to the undecidability and incompleteness of Nash games and of Arrow-Debreu markets. Briefly, as chaos turned out to be undecidable — so are, in general, equilibrium prices in neoclassical economics.

Around 1992 da Costa and Doria decided to try their hand at the P vs. NP. After a few sketches which were circulated among friends, Georg Kreisel who knew da Costa, got in touch with them, and sent a message that turned out to be made into a possible research program. Two key assertions were made in the message from Kreisel and relayed to Doria by Kreisel's Austrian secretary. First, if $P < NP$ is independent of set theory, then it holds true of standard arithmetic. Second, $P < NP$ holds if and only if the counterexample function to $P = NP$ is total. Here we hit a wall: for the counterexample function is extremely fast growing in its peaks; moreover, it is noncomputable. Kreisel stated but never proved these assertions in his correspondence, but da Costa and Doria managed to prove them. Their work is still going on, and the best result they've obtained so far is a kind of relative result, which we state in full: let S be an axiomatic theory based on the classical predicate calculus and with a model that makes it arithmetically sound. Then S', the sigma-1 extension of S, proves that there is no proof of the totality of infinitely many recursive counterexample functions in S.

It's a technical result with an obscure meaning, but seems to point in a possible direction for the solution of the big mystery... The counterexample function is a fascinating object. It has fractal properties — it has copies of itself all over it; also, it includes in itself in a natural coding all possible axiomatic theories with a recursively enumerable set of theorems. These side results help us understand the difficulties which appear to block most efforts to settle the P vs NP question.

While an undergratuate student, Doria had started a career as a journalist and as a writer interested in European continental philosophy. He frankly admitted the influence of Martin Heidegger in his first books, which were published in Portuguese: these are Marcuse Vida e Obra (Marcuse Life and Works), written and published when he was 22, and O Corpo e a Existência (Body and Being), which is an essay based on Heidegger's Being and Time, and published in 1972. Rather surprisingly, despite his work on foundational problems, Doria never showed much interest in scientific philosophy. His recent work in Portuguese shows him as a memorialist; he published in 2015 Na Casa da Vovó (In Granny's Home), his remembrances of youth in his grandparent's house — his maternal grandfather was a highly influential lawyer who took active part in many episodes of Brazilian politics until 1968. He also wrote books that documented his family's history in the context of Brazilian history.

But these are side interests. His main fare is his work in the hard sciences, which he sees as a kind of artistic, rather than scientific work. Science is another way of looking for Beauty.

Newton C. A. da Costa

Preface

This book originated from a conference co-organized by one of its editors (JAB) and Professor Jean-Yves Beziau in December of 2018. The meeting, sponsored by the Brazilian Academy of Philosophy, honored Francisco Doria and celebrated his 70th birthday belatedly (see Beziau's article in this volume). We were so impressed by the large number of different scholarly areas represented in this gathering that we decided an edited book would be interesting. This book is the result of such an effort.

Here readers will find articles from many disciplines, ranging from physics to economics, passing through philosophy (both continental and analytical), mathematics, logic, and computer science. This wide disciplinary range is commensurate with Doria's interests. After all, he published in all those areas, including a book on Marcuse when he was 22 (see da Costa's Foreword). Thus, it is a tribute to a true polymath.

This volume is divided into four parts. Part I contains articles in physics, mostly on the foundations of quantum mechanics (with the notable exception of Garcia de Andrade, a former Ph.D. student of Doria, who writes about quantum field theory in gravitational fields with torsion). Part II contains articles on logic and the foundations of mathematics. In it, readers will find exciting examples of the types of mathematical problems and issues that peek Doria's interests. Part III has articles with more of a philosophical leaning, including philosophy of mind, philosophy of science, and continental philosophy. Part IV, Social Sciences, shows how Doria's mathematical constructs lead to exciting applications outside of the hard sciences. One could be tempted to think that the underlying connection between all those parts is mathematics. Still, a careful analysis reveals that this is not the case, as some articles do not use mathematics at all. Instead, the link is Doria himself.

The two editors of this volume had different starting experiences with Francisco (Chico) Doria.

I (JAB) met Doria in 1986 when I was a physics undergraduate student at the Federal University of Rio de Janeiro. One of his graduate students, Manuel Ribeiro da Silva, noticed my interest in theoretical physics and arranged for a meeting. At the time, Doria was a Professor at the School of Communications, which I found quite unusual. I remember arriving at this vast room with open cubicles where Doria's office was located. The place resembled a newspaper newsroom, with numerous desks and the frantic background noise of nonstop mechanical and electric typewriters. Those who know Doria are aware of this: he speaks very softly. So, for the thirty minutes of our meeting, Doria talked about his interest in Godel's theorem (something entirely new for me), unaware of the fact that I barely heard one out of four or five or his words. I left him extremely confused and expected never to talk to him again, as I was embarrassed to confess that I did not understand at all what he was talking about. To my surprise, about a week later, while spending the weekend in my parents' house in Petrópolis (in the mountains near Rio de Janeiro), I got a phone call from Doria. He asked whether I would be interested in coming to his house (also in Petrópolis) and chat about mathematics and physics. This time things were completely different. When I got there, he had a friend over, and we had beers and a barbecue. More importantly, there was no typewriter noise, so I could finally understand what he was talking about. This was the beginning of my work with Doria, as I became further interested in the foundations of math and eventually had Doria first as my undergraduate advisor for a CNPq scholarship (*bolsa de iniciação Científica*) and then as my supervisor for my M.Sc. and Ph.D. at the Brazilian Center for Research in Physics. From this time, I fondly remember a series of seminars that Doria organized together with Antonio Francisco Furtado do Amaral and Carlos Marcio do Amaral. Those seminars were stimulating and reminded me of why I got into this business to begin with. But in addition to getting tutored in physics and mathematics, I also had the opportunity to catch a ride with him going to Petrópolis. He commuted between Rio de Janeiro and Petrópolis, a crazy ordeal since the trip lasted, on bad days, longer than one hour. But because he commuted, I often would meet him on Friday's at the School of Communications and then go with him, since my parent's house was on his way. Those were interesting trips, not because his car was an old white VW bug, but because he would talk about all types of intriguing things that were not at all related to physics or mathematics. This experience is similar to what Décio describes below. He always had interesting things to say about current events, history, art, or reasons why his Marcuse book got him in trouble with the dictatorship by having a green and yellow cover (the Brazilian flag colors). In one of those trips, I remember him telling me that I should strive to read all of the classics by the age of 21, which I tried to, but probably failed (what is a classic text anyway?). Little did he know, though, those unfocused conversations allowed me to go beyond physics and embrace interdisciplinary research. This lesson was reinforced after my Ph.D. when Doria "convinced" me to go to Stanford and spend a year as a Capes fellow working with Patrick Suppes. So, I am glad that the first "conversation" at the School of Communications was not the end of it, but the beginning of not only a scientific collaboration but also of a long-lived friendship.

I (DK) met Doria for the first time at the University of São Paulo in 1987, when he came to work with Newton da Costa and remained for a time as a visiting scholar at the Institute for Advanced Studies. In that period, the interactions were close not only with me but also with other Ph.D. students. Doria delivered two courses at the IEA, one on Boolean models in set theory and another on classical mechanics. It was immediate for us to recognize his competence and capacity to learn, teach, and interact with beginners. A close friendship has arisen.

Even after his IEA period, we continue to interact by email, during visits to USP, in congresses, and via correspondence. He visited me at my house and the Federal University of Paraná, in Curitiba, where I lived at the time, and we have met several other times. I never lost contact with him. When I was at the University of Florence for a post-doc period with Maria Luisa Dalla Chiara (1992-93), Doria was invited by Daniele Mundici to Milan. A visit to Florence was organized, where he knew the Florence people and gave some talks. It was amazing because during the days that he stayed there, we met for several hours to visit the city, where his *antenati* lived, both the Dorias and also the Aciolis (*Accaiolis*). For instance, in other meetings in Salvador, Bahia, where we visited several churches and historical places, I identified Doria as one of those tour-guidance phones in museums that talk about the museums' collections. It seems that walking with him was similar to have one of these phones in my ear; he knew every detail about the streets, about the people who lived there, and the history of the churches (yes, we spoke with several local priests, some of them enabling us, due to Doria's convincing ability, to access parts of the places usually closed to normal visitors). It was a great time. I am sure that he will never forget the *bisteca Fiorentina* we had one day for dinner.

I read several works from him. He sends me some of his history books and writings, all of them written with precision and a good sense of humor. Nice readings indeed. It is important to acknowledge that he is also a genealogist, a topic he has dedicated to his intellectual activities. His knowledge about the subject is awe-inspiring, for instance, about the families that formed our country, Brazil. But my interests were always about science, and in this field, I found an original thinker, an entirely original one. His works with Newton da Costa have presented several significant results, such as several undecidability results in classical mechanics and the proof that chaos is undecidable and incomplete (Gödel) when adequately formalized. In several works, the duo has applied set-theoretical methods in physical theories, such as forcing, Boolean valued models, etc. A real work on foundations and philosophy of science. Most of these results are presented in their book *On the Foundations of Science*, published by the COPPE program, UFRJ in 2008.

J. Acacio de Barros, Menlo Park, CA, USA
Décio Krause, Florianópolis, SC, Brazil

Acknowledgements

This book's editors are indebted to the numerous contributors who took time from their busy schedule to write their outstanding contributions to this volume. We appreciate their hard work and their extraordinary patience while waiting for us to finish the final version. We would especially like to thank Professor Jean-Yves Beziau, editor of the Brazilian Academy of Philosophy series under College Publications. Jean-Yves offered to publish this book under his collection and helped co-organize (with Acacio de Barros) the conference in Doria's honor that jump-started this volume. This book would not have been possible without the support of Jane Spurr, from College Publications, who we wholeheartedly thank. Finally, we would like to acknowledge Francisco Doria himself, whose profound influence shaped not only our ways of thinking but also our careers and research.

Table of Contents

Foreword v

Preface ix

Acknowlegments xiii

Part I – Physics

Jonas R. Becker Arenhart and Helcio Felippe Junior
1 The Fate of Bundle and Substratum Theories Under KS Theorem . 1

J. Acacio de Barros and Adonai Sant'Anna
2 Classical Fields, Bell's Inequalities, and the Quantum Limit 23

Christian de Ronde and César Massri
3 Against 'Particles' and 'Collapses' in Quantum Entanglement 38

Luis Carlos Garcia de Andrade
4 Generation of Cosmological Magnetic Fields at Laboratory Scales in a Chiral QED Metric-Torsion Gravitational Weak Field Background . 68

Décio Krause
5 Quantum Mechanics, Ontology, and Non-Reflexive Logic 75

Part II – Logic and the Foundations of Mathematics

Felipe S. Abrahão, Klaus Wehmuth, and Artur Ziviani
6 On the existence of hidden machines in computational time hierarchies 113

Jean-Yves Beziau
7 What is an Axiom? 122

Itala M. Loffredo D'Ottaviano and Hércules de Araujo Feitosa
8 Galois Pairs with the Modal Operators of Paraconsistent Logic J_3 . 143

Newton C. A. da Costa
9 Undecidability, Incompleteness and Beyond: an adventure 168

Marcelo Finger
10 Logic in Times of Big Data 184

Adonai S. Sant'Anna
11 Epistemology of Quasi-Sets 199

Bruno Scarpellini
12 Recursively Enumerable Sets and Fourier Series 217

Apostolos Syropoulos
13 Categories of Fuzzy Structures and Fuzzy Categories 230

Part III – Philosophy

Otavio Bueno
14 The Philosophical Significance of Incompleteness 250

Gregory Chaitin
15 Consciousness and Information, classical, quantum or algorithmic? . 260

J. Acacio de Barros and Carlos Montemayor
16 Information and the Hard Problem of Consciousness 270

Marcelo Gleiser
17 What We Can't Know of the World: Science and the Limits of Knowledge 282

Muniz Sodré
18 A Question of Identity 300

Part IV – Social and Interdisciplinary Sciences

Sami Al-Suwailem
19 Gödelian Rationality 310

Francisco Caruso and Roberto Moreira Xavier

20 In what sense space dimensionality can be used to cast light into cultural anthropology?........................ 344

Maurício Vieira Kritz

21 Revisiting the systemic golden years from a contemporary organisations perspective............................. 379

Julio Michael Stern, Marcos Antonio Simplicio, Marcos Vinicius M. Silva, and Roberto A. Castellanos Pfeiffer

22 Randomization and Fair Judgment in Law and Science 399

The Fate of Bundle and Substratum Theories Under KS Theorem

Jonas R. Becker Arenhart[†] and Helcio Felippe Junior[*]

[†]Federal University of Santa Catarina, Department of Philosophy, Santa Catarina, Brazil, e-mail: jonas.becker2@gmail.com
[*]Federal University of Rio Grande do Norte, Department of Theoretical and Experimental Physics, Natal, Rio Grande do Norte, Brazil

Abstract

Basically, *bundle theories* and *substratum theories* are metaphysical accounts of the following features of concrete particulars: i) their constitution from more basic entities, ii) their individuality, and iii) their possession of properties. When connected to quantum mechanics, most discussions focus on the problem of individuality of quantum particles. In this chapter, we shall argue that irrespective of how those theories fare on *this* task, both fail on accounting for the constitution and property possession when quantum mechanics enters the stage. The Kochen-Specker theorem of quantum theory strikes directly against the account of property possession and constitution provided by the current versions of those metaphysical theories. Two major claims shall result from our investigations. First, that a revision on the nature of particular entities will have to be advanced in case one still wishes to hold that quantum entities are particulars. Second, that claims concerning metaphysical underdetermination between individuality and non-individuality will have to be revised in the light of the restrictions imposed by the Kochen-Specker theorem.

Keywords: constitution; bundle theory; substratum; Kochen-Specker theorem.

1 Introduction

Two of the most familiar approaches to the nature of particular concrete objects in metaphysics are the bundle theory and the substratum theory (see Loux

2006; French and Krause 2006, chap. 1).[1] Both theories are thought to account for three related features of such particulars:[2]

i) their constitution in terms of more basic entities,

ii) their individuality, explained in terms of their constitution and in terms of their ingredients, and

iii) the predicational nexus, accounting for how is it that a particular may be said to bear or have a property.

Roughly, *bundle theories* conceive of concrete objects as being entirely constituted by the properties they instantiate, from which it results that objects are nothing but a bundle or cluster of their co-instantiated properties. On the other hand, *substratum theories* conceive of concrete objects as being constituted by the particular's instantiated properties and by a further ingredient, a particular of non-qualitative nature, a *self-individuating substratum* or *bare particular*. Sometimes, instead of positing a particular object such as a substratum, it is also proposed that each object has only properties involved in its constitution, but that a special property of a non-qualitative nature, a *haecceity* or *individual essence*, is also present. On this account, each object has its own haecceity, which accounts for its individuality, and this makes it more similar to the substratum approach than to the bundle approach (and that is why we shall treat it so too). Both theories account for the fact that an object has a property by claiming that the property constitutes the object in some sense; their main difference concerns the individuality of particulars: is the individuality of a particular accounted solely by their qualitative features, or is a further non-qualitative ingredient needed?

Just to be sure: the problem of individuality is a metaphysical problem. It concerns explaining what is it that makes a particular object that object that it is, distinct from everything else (for a careful discussion of the terminology associated with individuality, see Krause and Arenhart 2018). Bundle theories advance the claim that properties are enough for that. Substratum theories suggest the need for a special ingredient. This is specially important, because so far, the dialectics of the debate between these theories concerns mainly the issue over the problem of individuality. For instance, substratum theorists will quickly point out an intimate relation between bundle theory and the highly controversial *Principle of the Identity of Indiscernibles* (PII), whereas bundle theorists will claim an incoherence on the very idea of an elusive substratum, lying beyond all properties, constituting objects. Either way, no consensus has been achieved and the nature of concrete particulars continues to raise important

---

[1] These are not the only theories available, of course, but we shall confine ourselves to these theories in this paper.

[2] For the sake of brevity, whenever we mention a particular in this paper, it will be understood that it is a concrete particular, unless stated otherwise.

issues within metaphysics. We believe, however, that fundamental physics is welcome in the debate and may bring some ideas to enrich and, potentially, enlighten the controversy. Our claim is that physics has important lessons for those theories on what concerns property possession, not only individuality.

This shift of focus is important, because in Quantum Mechanics (QM), the *state* of a physical system is labeled by a vector pertaining to a vector space. Traditionally, the same vector is to be regarded as a description of the physical object under study. Metaphysically speaking, this object could, *in principle*, be viewed through the lenses of both bundle and substratum theories (given that those theories should apply to every object). Nonetheless, it has been argued that, because of the commitment of bundle theories with the truth of the PII, a false instance of the latter would result in the falsehood of the former. As a matter of fact, QM is said to deal with situations in which the stronger versions of PII are demonstrably false, hence implying that bundle theories should be rejected in favor of a substratum framework in the context of quantum individuality (see French and Krause 2006, chap. 4). Yet, this is only part of the story. We will argue, following recent work done on invariance of quantum states and the Kochen-Specker (KS) theorem (de Ronde and Massri 2016), that *both* bundle and substratum theories fail to provide a satisfactory account of quantum mechanical entities. These results will drive us into further discussions regarding the nature of quantum objects and steer us towards a revision over the customary ways of understanding the metaphysics of concrete particulars. Also, this will allow us to sidestep typical discussions focusing on the role of the Identity of Indiscernibles in these discussions and extract some lessons concerning metaphysical underdetermination in quantum mechanics.

The paper is structured as follows: in section 2, we delve into the two metaphysical theories of objects under study here and how they are related to Classical Physics. In section 3, we introduce key aspects of QM, and the KS theorem enters the scene to set the stage for our arguments. Section 4 brings together the ideas fomented in the previous sections and establishes a quantum mechanical attack on both bundle and substratum theories. We finish our paper in section 5, discussing further problems, specifically aiming at the current understanding of underdetermination of metaphysics by physics in light of our findings.

2 Metaphysical theories of particulars and Classical Physics

2.1 Bundle and substratum theories

As mentioned in the introduction, we shall focus on two perspectives related to the understanding of concrete particulars. Both theories are concerned with constituting particulars from more basic ingredients. In this sense, both approaches deal with the problem of providing for the nature of a particular in terms of more basic entities that constitute the particulars; in this sense, these are reductive approaches: concrete particulars are somehow reduced to more basic ingredients that are metaphysically more fundamental, so that concrete particulars need not be part of the fundamental inventory of reality. Demirli (2010, p. 2) explains as follows the problem of constitution:

> In answering the internal constitution question, we may begin an inquiry about the various categories that go into the composition of individual substances and hope that at the end of this inquiry we will come up with a list of ingredients that constitute various individuals. Just as a certain recipe in a cook book provides us with a list of ingredients and instructions for mixing these ingredients together, we may maintain that the list or the recipe of individual substances — God's recipe book, so to say — will tell us what items from various categories are used, and how these items are combined.

Of course, unless one specifies what 'constitution' and 'composition' mean, the problem and any of its answers will remain on a very abstract and perhaps metaphorical level. Typically, 'constitution' is specified in other terms, such as set theoretic (a concrete particular is a set whose members are its ingredients) or mereologic (a concrete particular is a mereological sum of its ingredients), each with its own problems, but we shall not be concerned with specifying any particular version of constitution here (see discussion and further references in Benovsky 2016 and in Jago 2018). What is relevant for us is that the approaches with which we shall be concerned here are primarily approaches to the question of what goes on in the constitution of a particular.

Bundle and substratum theories alike conceive of objects as constituted or composed by properties instantiated by the particulars. In this sense, it is common to understand properties either as universals or as tropes, and each option will give rise to distinct versions of both bundle theories and substratum theories. We shall continue to speak in neutral terms of properties, leaving it open whether these properties should be conceptualized as universals or as tropes. Each version has its own problems and virtues, but it will not be relevant for our purposes whether one or another route is taken (for a recent account of the discussion, see Benovsky 2016). Besides agreeing that properties constitute particulars, bundle theories and substratum theories differ, however, because while bundle theorists claim that the properties exhaust the constitution of the particular, the substratum theorists posit a further underlying ingredient constituting those particulars.

That difference may be put as follows: the bundle theorist wishes a *one category metaphysics, only properties are fundamental*, and they are employed to constitute particulars. Following Loux (2006, p. 107), this is encapsulated by the principle BT:

BT Necessarily, for any concrete entity a, if for any entity, b, b is a constituent of a, then b is an attribute.

The substratum theorist is not willing to embrace this one category metaphysics, and adds a further ingredient (rejecting BT), a substratum, which is a self-individuating particular, not described in terms of qualities. Perhaps substratum theory (ST) could be defined as follows:

ST Necessarily, for any concrete entity a, if for any entity b, b is a constituent of a, then b is either an attribute or else b is a bare particular, and, if b is a bare particular, then b is unique in the constitution of a.

The main reason for adding such a further ingredient in ST concerns individuality and property bearing. It is typically thought that BT is committed to a version of the Principle of the Identity of Indiscernibles (PII):

PII Necessarily, for concrete objects a and b, if for any property P, $(P(a) \leftrightarrow P(b))$, then $a = b$.[3]

However, as argued by Rodriguez-Pereyra (2004), the bundle theory BT may live without PII, or it may even be compatible with the falsity of PII, provided that the bundles are understood as bundles of universals. In this case, one may distinguish a bundle (as a pack of universals) from its numerically distinct instances (see also Rodriguez-Pereyra 2004 for further options on defending bundle theory, as well as Demirli 2010; a further version of the bundle theory is defended in Jago 2018). We need not enter into these controversies, given that we shall not be primarily concerned with the problem of individuality of particulars. As we have mentioned, the dialectics of the debate between bundle theory and substratum theory focuses mainly on the issue of individuality. The possibility of a bundle accounting for scenarios such as Max Black's (1952) two sphere world,[4] or of quantum mechanics' indiscernible particles, are called forth, and the substratum theorist claims that a further ingredient is always needed (for the case of quantum entities, see French and Krause 2006, chap. 4; Arenhart 2017).

---

[3] One could work with at least three different versions of the PII, each of it takes the set of relevant properties to mean something slightly different. For instance, the weakest form, PII(1), states that it is impossible for two individuals to possess all properties and *relations* in common; the next strongest, PII(2), precludes spatio-temporal properties from its description; and the strongest of all, PII(3), encloses only monadic, non-relational properties. Further distinctions could be provided for stronger versions of the PII, such as restrictions to intrinsic properties, or perhaps for what are called 'pure' properties. For a discussion, see Adams (1979).

[4] In this example, recall, two isolated iron spheres, separated two miles apart from each other and indistinguishable in all their properties are, in fact, seen to be numerically distinct.

As we mentioned, we shall focus mostly on the explanation that each of these theories provide for property exemplification. Recall that both theories agree on how to account for this: a property is possessed by a concrete particular provided that the property is part of the composition of the particular (*i.e.*, is an ingredient of the particular, using Demirli's metaphor). As Jago (2018, p. 3) puts it for the case of bundle theory, the Property Possession for Bundles (PPB) is explained thus:

PPB A concrete particular a possesses property P if, and only if, P is a member of the a-bundle.

Given that 'being a member of the a-bundle' is an explanation of the very idea of constitution, this may be generalized to take into account also the case of the substratum approach. We shall call it simply the 'Property Possession' (PP):

PP A concrete particular a possesses property P if, and only if, property P constitutes a.

From PP the more specific PPB follows when one specifies that a is a bundle of properties and properties constitute such entities by being members of the bundle. Also, from PP it follows (assuming basic logical inferences) that *whenever a property does not constitute a concrete particular, it is not possessed by that particular*. In this sense, one could claim that insofar as our two target theories adhere to a form of PP, it will be metaphysically determined, for each property P, whether P is possessed or not for each particular a. That means that it is determined, for each particular, by the very nature of the particular, whether a property is possessed or not possessed by that particular. Mittlestaedt (2009, 2011) called that the 'Principle of Thoroughgoing Determination' (PTD):

PTD Given a property P and a concrete particular a, either a possesses P or else it does not posses P (that is, a possesses the complement of P, which we shall denote by $\neg P$).

In other words, if a list of every property P available for an object a could be provided, one could, at least in principle, determine for each of such properties whether a has P or does not have P.

As we have mentioned, we shall shift the focus from individuation to the bearing of properties. We shall operate on a very general level (not assuming anything about the nature of the properties — whether they are tropes or universals — and not assuming any specific account of constitution — set theoretic, mereologic, or other). Our claim will be that, insofar as these theories adhere to PP and, consequently, to PTD, they are ruled out by QM. We begin, however, by making a detour in Classical Physics, in order to illustrate how well these principles work there.

2.2 The way to Classical Physics

One could argue that the bundle approach was favored by Classical Physics because some aspects of the latter made clearer the role of properties constituting objects: its formalism centers around the main idea of physical quantities. Although defensible, such position would undermine the metaphysical details just discussed. It is uncontroversial that Classical Physics takes some form of objective (*i.e.*, preexistent) approach to physical observables, indeed making it a cornerstone to the metaphysical underpinnings of physical theories prior to the 20th century. In the classical framework, properties gain the status of definite physical quantities, that is, well-defined values on a certain interval of the real line $\mathbb{R}$ (*e.g.*, the mass m of a billiard ball, its charge q, the components x^j of its position vector $\mathbf{x}$, so forth).[5] Consequently, measurements are unproblematic and provide us with nothing but a *revelation* of the physical system's objective, actual properties. An object could then be exhausted into an amalgamation of definite, real values of physical observables through which Classical Physics asserted us the object did possess beforehand.

One encodes physical observables into the mathematical notion of real-valued functions over a space $\mathcal{S}$ of *states* of a physical system, with the understanding that, at any given time, a unique member $s \in \mathcal{S}$ can be assigned to the system. Hence, to each physical observable A there corresponds a function $f_A : \mathcal{S} \to \mathbb{R}$ such that $f_A(s)$ is the value which A possesses when the state of the system is s.[6] Also, the state s' of a system at time t' is determined *uniquely* by the state s at any earlier time $t < t'$ via dynamical maps

$$T : \mathcal{S} \to \mathcal{S}$$
$$s \mapsto s' ; \quad \forall s, s' \in \mathcal{S}, \tag{1}$$

essentially bringing determinism and the principle of causality to the classical picture. If $\mathcal{S}$ is the state space of an N-particle system, then the state is labeled by points $(q_1, q_2, \ldots, q_{3N}; p_1, p_2, \ldots, p_{3N}) \equiv (q_k, p_k)$ in a $6N$-dimensional phase space, where q_k and p_k are the respective canonical coordinates. The dynamical law is then given by the Hamiltonian $H(q_k, p_k)$ and the canonical equations

$$\frac{\mathrm{d}q_k}{\mathrm{d}t} = \frac{\partial H}{\partial p_k}, \quad \frac{\mathrm{d}p_k}{\mathrm{d}t} = -\frac{\partial H}{\partial q_k}; \quad k = 1, \ldots, 3N. \tag{2}$$

The equations in (2) are first-order differential equations and hence determine

---

[5] Note that mass and charge are both examples of *internal* or *intrinsic* physical quantities: they refer to the constitution of the thing itself; whereas position, velocity, etc, are *external* physical quantities whose values depends on a given spacetime framework.

[6] Strictly speaking, f_A is a Borel function and, depending on the observable A, it may be required to be measurable, or continuous, or smooth. Further, the set $\mathcal{S}$ (also known in classical statistical physics as the *space of microstates*) is a symplectic manifold which, in particular, corresponds to the usual phase space of a one-dimensional point particle when $\mathcal{S} = \mathbb{R}^2$ (Döring and Isham 2010).

uniquely the state (q_k, p_k) at any time t (provided we have the system's initial conditions), therefore satisfying the requirements of the dynamical maps in (1).

The objectivity feature of properties, that is, the observer-independent character of Classical Physics is captured nicely by the notion of *invariants*: physical quantities having the same value for any reference frame. The transformations between different frames of reference have the mathematical property of constituting a *group*. Since the laws of both classical mechanics and special relativity theory are invariant against the transformations of the Galilei and Poincaré group, respectively, it follows that, in Classical Physics, both *nomological* (static) and *dynamical* properties are consistently translated between distinct frames of reference and, additionally, it allows one to speak meaningfully of independent, preexistent physical observables: an actual state of affairs (ASA).[7]

Let f_A be the value of an observable A in the state $s \in \mathcal{S}$ and $f_A \in \Delta$, where $\Delta \subset \mathbb{R}$ is a Borel subset that represents the real-valued interval of an observable's values. Then, equipped with the usual logical operators $\wedge$, $\vee$, $\neg$ and $\preceq$ for material implication, it is instructive to construct the mapping

$$\Phi : \mathcal{B}(\mathbb{R}) \to L_C \qquad (3)$$

from the Borel subset $\mathcal{B}(\mathbb{R})$ onto the Boolean lattice L_C of propositions $P_i \in L_C$ such that $P_i = 0$ or $P_i = 1$, effectively creating a whole structure of definite yes-no (true- or false-valued) propositions corresponding to the presence or absence, respectively, of physical observables' values (properties) associated with a system in a state s. In other words, *a version of the PTD is justified*!

Here is where classical logical structure (*i.e.*, Boolean logic) gets incorporated into the mathematical structure of Classical Physics, and one can actually talk of a given proposition P_i as an assignable property. Further, one is now able to coherently define an *individual object* by means of the dynamical maps $T(s)$ and the invariance of definite properties P_i (nomological and dynamical): any state $s \in \mathcal{S}$ labeling a collection of properties P_i is uniquely identified, and continually re-identified through its dynamical trajectory in the classical state space $\mathcal{S}$. Moreover, the notion of an individual persists even when one is dealing with indistinguishable physical systems, only this time the argument relies on either a principle of impenetrability (non-overlapping of spatio-temporal trajectories) or particle permutation via Maxwell-Boltzmann statistics.[8] Based on that, it becomes clear that one could make sense of the individuality of classical particles within the background of either a weak version of the PII (allowing spatio-temporal properties to account for the qualitative distinction) or through

---

[7] We are adopting the terminology advanced by de Ronde and Massri (2016).

[8] Physical systems, such as particles, are said to be indistinguishable in that they possess the same state independent (intrinsic) properties (mass, charge and spin — the nomological properties in de Ronde and Massri 2016). Given two particles A and B, and two *microstates* 1 and 2, particle permutation on the states yields different outcomes regardless of A and B being indistinguishable, hence, A and B are individuals (French and Rickles 2003).

something that goes beyond all properties, a substratum. So, although it is true that Classical Physics concentrates around the idea that physical observables are objectively determined for classical systems, it is neither obvious nor compelling to assert that its formalism is committed to bundle theory. Admittedly, Classical Physics seems merely to somehow suggest the individuality of physical systems, independently of whether that individuality is achieved through bundle or substratum theories of objects. Indeed, the present situation constitutes a dilemma known as the *underdetermination of metaphysics by physics*, in which a physical theory formalism is compatible with two or more *metaphysical theories*; in our case, a classical physical system could be individuated by both bundle and substratum theories of objects, but no *physical argument* can be made in favor or against one or the other.

2.3 Bundle and substratum theories (revisited)

On what concerns some attempt to look for help from physics on deciding which is best, bundle or substratum, it appears we are back to square one: Classical Physics does not have the resources to decide between bundle or substratum frameworks. It does satisfy the PTD, which both theories do endorse, and even more: on what regards the individuality of classical indistinguishable systems, one is left between the weakened form of PII allowing for spatial properties to account for individuality or the substratum's elusiveness. The choice for which is better or more appropriate requires a digression into controversial metaphysical issues, and not more investigation into physics. The latter is silent about the relevant issues, and that is why the situation we face is called 'the underdetermination of metaphysics by the physics'. One could try to break the underdetermination by claiming that bundle theory, allowing for spatial properties to account for individuality, is a far more economic proposal, avoiding substratum (this is suggested in French and Krause 2006, chap. 2). However, that kind of claim does not come form a necessity of physics itself, and as such, will not move the substratum theorist, which has her own battery of *metaphysical* arguments against the bundle theory.

It should be clear by now that both bundle and substratum theories are *distinct* metaphysical positions regarding the nature of concrete particulars; but it would be a mistake to fully embrace their divergences and, in the process, neglect the common ground they stand. Here is why: prior to their farewell, bundle and substratum theorists would concede in mutual agreement on which properties a given concrete particular instantiate; they would always converge on which properties to assign to a certain object (*i.e.*, which corresponding propositions P_i are well-defined and attached to a system). The bifurcation only happens when one of them stays true to the cluster of properties assigned to an object, whereas the other decides to go beyond the mere qualities observed and proposes a new particular ingredient gluing them all together, establishing, at last, an *individual* through the very existence of that ingredient. It is at this moment that one can say with certainty that the bundle and substratum theorists are each speaking their own language. Yet, no matter how foreigner one's dialect may sound to the other, both share the same proto-language: each of them is committed to the basic operation of, initially, assigning a collection of definite properties to a given system; a set of well-defined propositions P_i (this is the PTD at work, of course).

In fact, such fairly overlooked move contains the key to our main argument against bundle and substratum theories; an argument that goes beyond the usual controversies surrounding the PII or the substratum's transcendence by looking at the conceptual frameworks both theories are founded on, namely, the PTD. However, to materialize such endeavor, we shall bring quantum theory and, in particular, the KS theorem into the discussion. Therefore, we now turn our attention to QM.

3 Quantum theory and the KS theorem

3.1 The way to QM

Quantum theory was born out of a body of empirical results and assumptions that were gathered in the early 20th century and systematically organized by the end of the century's first quarter. The founding fathers of QM laid out a mathematical structure that, more than a hundred years later, is still able to match theory and experiment with the utmost precision. But, despite experimental success, on what regards our understanding of the theory, a hundred years have elapsed with no scientific nor philosophical consensus on what exactly quantum theory is talking about; that is, its ontology and its metaphysical counterpart are still an open question.

The general structure of QM may be approached through the lenses of a 'minimal interpretation', a pragmatic approach that takes quantum theory as an algorithm for predicting the probabilistic distributions of *measurement* outcomes done on suitably prepared copies of a given physical system. Although not stressed as fundamental, the probabilities are interpreted as the *relative frequencies* of possible outcomes if corresponding measurements were to be repeated a sufficiently large number of times. In sharp contrast to Classical Physics, nothing is said about whether a system possesses values for a physical observable prior to its measurement.

Traditionally, the minimal interpretation is mathematically translated via complex Hilbert spaces that contain all possible quantum states of a system. It may be a finite- or an infinite-dimensional space, but for our purposes we will be mainly dealing with the former structure. To a given physical system we associate a *separable* Hilbert space $\mathcal{H}$, such that *normalized* vectors

$|\psi\rangle \in \mathcal{H}$ correspond to the states of the system.[9] To compose multiple quantum mechanical systems, say N particles, we employ the tensor product between all the respective Hilbert spaces of each system, that is, $\mathcal{H} = \mathcal{H}_1 \otimes \mathcal{H}_2 \otimes \cdots \otimes \mathcal{H}_N$. Accordingly, the corresponding composite quantum state $|\psi\rangle \in \mathcal{H}$ is the tensor product of all $|\psi_i\rangle \in \mathcal{H}_i$,

$$|\psi\rangle = \bigotimes_{i=1}^{N} |\psi_i\rangle. \qquad (4)$$

Physical observables are represented by *self-adjoint* linear operators on $\mathcal{H}$. The expected result of measuring an observable A in a state $|\psi\rangle \in \mathcal{H}$ is given by $\langle \psi | \hat{A} | \psi \rangle$, where $\hat{A}$ is the corresponding operator of A. The *spectral theorem* guarantees that, to any self-adjoint operator $\hat{A}$ in an n-dimensional Hilbert space $\mathcal{H}$, there exists an *orthonormal* basis $\{|a_1\rangle, |a_2\rangle, \ldots, |a_n\rangle\} \subseteq \mathcal{H}$ consisting of *eigenvectors* of $\hat{A}$ such that

$$\hat{A} = \sum_{i=1}^{n} a_i |a_i\rangle\langle a_i| \,; \quad a_i \in \mathbb{R}. \qquad (5)$$

The values a_i constitute the *spectrum* of the *bounded* operator $\hat{A}$ and each one of them corresponds to the possible result of a (sharp) measurement of A. Given a general state $|\psi\rangle$, the probability of obtaining a_m as the measurement outcome of A is determined by the Born rule $|\langle \psi | a_m \rangle|^2$. To that effect, one can benefit from the *expansion theorem* which states that any vector $|\psi\rangle \in \mathcal{H}$ has a unique expansion

$$|\psi\rangle = \sum_{i=1}^{n} c_i |a_i\rangle \,; \quad c_i = \langle a_i | \psi \rangle \in \mathbb{C}, \qquad (6)$$

to which it follows from Born's rule that the probability of obtaining a_m is $|c_m|^2$. In the absence of external influences (*i.e.*, in a *closed* system) the dynamical evolution of quantum systems is determined by the self-adjoint Hamiltonian operator $\hat{H} : \mathcal{H} \to \mathcal{H}$, such that the time development of any state $|\psi(t)\rangle$ is given by the time-dependent Schrödinger equation

$$i\hbar \frac{\mathrm{d}}{\mathrm{d}t} |\psi(t)\rangle = \hat{H} |\psi(t)\rangle \,; \quad \forall \, |\psi(t)\rangle \in \mathcal{H}. \qquad (7)$$

---

[9]Technically, *pure* states as opposed to the *mixed* states

$$\hat{\rho} = \sum_i p_i |\psi_i\rangle\langle\psi_i| \,; \quad 0 \leq p_i \leq 1,$$

of the more general 'density matrix' formalism of QM. Also, two vectors $|\psi\rangle, |\phi\rangle \in \mathcal{H}$ which differ by a complex factor, that is, $|\psi\rangle = \alpha |\phi\rangle \,; \alpha \in \mathbb{C}$, correspond to the same (pure) state since global phases are immaterial in QM.

Note that, as with the classical dynamical maps $T : \mathcal{S} \to \mathcal{S}$ and the canonical equations in (2), the dynamical law governing closed quantum mechanical systems is a first-order differential equation; therefore, given proper initial conditions, the state $|\psi(t)\rangle$ at any later time is uniquely determined by solving (7). In this sense, QM is as deterministic as Classical Physics. However, in general, the Principle of Thoroughgoing Determination (PTD) fails in the quantum regime, so that there is no way to guarantee a positive or negative stance regarding the proposition (representing property attribution) P_i in QM, for any P_i. To see this, let us take a closer look at the structure of physical observables in quantum theory.

Starting out with equation (5), we define the *projection* operator $\hat{P}_i$ such that

$$\hat{A} = \sum_{i=1}^{n} a_i |a_i\rangle\langle a_i| \equiv \sum_{i=1}^{n} a_i \hat{P}_i. \tag{8}$$

Projection operators $\hat{P}$ are operators that project onto some subspace of the Hilbert space. They satisfy the properties of being self-adjoint and idempotent, that is, $\hat{P}^\dagger = \hat{P}$ and $\hat{P}^2 = \hat{P}$, respectively. As such, their only eigenvalues are manifestly 0 and 1; therefore, they can be understood as propositions about properties. For instance, the operator $\hat{P}_m$ represents an observable whose numerical value is defined to equal unity if the result a_m is obtained whenever A is measured (and defined to equal zero otherwise). Thus, P_m is a positive proposition stating the value a_m to A and, accordingly, $\neg P_m$ is its negation. In Classical Physics, the Boolean lattice L_C guarantees that either P_m or $\neg P_m$ applies (the PTD is in action). Now, let $\hat{Q}_j = |b_j\rangle\langle b_j|$ be a projection operator of the spectral decomposition of a self-adjoint operator $\hat{B}$ corresponding to a physical observable B. Projection operators $\hat{P}_i$ and $\hat{Q}_j$ are said to be *orthogonal* if their commutator amounts to zero, that is, $[\hat{P}_i, \hat{Q}_j] \equiv \hat{P}_i \hat{Q}_j - \hat{Q}_j \hat{P}_i = 0$ for all $|\psi\rangle \in \mathcal{H}$; equivalently, the subspaces onto which they project are orthogonal. In QM, commuting operators correspond to *compatible* observables, that is, properties that can be measured simultaneously; thus, commuting projectors are associated with propositions that can be simultaneously asserted in some sense. But, in general, quantum mechanical operators do not commute, so that their corresponding physical observables are said to be incompatible. In particular, if $[\hat{P}_i, \hat{Q}_j] \neq 0$ then their corresponding propositions P_i and Q_j cannot be jointly asserted in any meaningful way. Equipped with logical operators, the quantum mechanical analog of the map Φ in equation (3) then becomes a mapping from the Borel subset $\mathcal{B}(\mathbb{R})$ onto a complete orthomodular lattice L_Q of *quantum logic*, a structure of propositions in which the distributive law from classical propositional calculus no longer holds. As it stands, it becomes difficult to define a quantum mechanical object the same way it was done in Classical Physics. Since a thoroughgoing determination is no longer possible, physical systems can only — if ever — be constituted *incompletely* by means of the restricted set of their objective properties (Mittelstaedt 2009). Roughly speaking, at any given time, *only one half of the classical phase space properties can be meaningfully assigned to a quantum system*.[10] Finally, it will be useful to observe that the corresponding projectors $\hat{P}_1, \hat{P}_2, \ldots, \hat{P}_n$ of a self-adjoint operator $\hat{A}$ are pairwise orthogonal so that equation (6) can be rewritten as

$$|\psi\rangle = \sum_{i=1}^{n} |a_i\rangle\langle a_i|\psi\rangle = \hat{\mathbb{1}}|\psi\rangle, \qquad (9)$$

that is, the sum of all projection operators $|a_i\rangle\langle a_i|$ constitutes a resolution of the identity in $\mathcal{H}$.

3.2 KS theorem and contextuality

Because the quantum mechanical formalism vastly disagrees with the picture of Classical Physics on what concerns property attribution, one could ask whether there is a way to go beyond the statistical predictions of measurement outcomes[11] and, in a sense, *complete* quantum theory. Historically, 'completion approaches' have been called *hidden-variable* theories because they posit a set of observably occult parameters that would assure definite (preexistent) properties to quantum systems. The question, then, is whether or not physical observables can be interpreted in terms of definite, albeit unknown, actual values pertaining to a quantum system. This was answered negatively by Kochen and Specker in 1967. Let us pave the way to their *no-go theorem* by further developing the posed question.

Given any quantum state of a system with observables A, B and C, is there a way to respectively assign numerical values $v(A)$, $v(B)$ and $v(C)$ to those observables?[12] Classically, there are no conceptual problems in constructing such *valuation functions* since, as we saw in section 2, to each physical observable A there corresponds a function $f_A : \mathcal{S} \to \mathbb{R}$ such that the value of A in the state $s \in \mathcal{S}$ is just the value of f_A at s, that is, $f_A(s) \equiv v_s(A)$, where we defined v_s as being the classical valuation function. Further, let $h : \mathbb{R} \to \mathbb{R}$ be a real-valued function and define a new physical observable $h(A)$ by requiring that its corresponding function $f_{h(A)}(s) \equiv h(f_A(s))$ for all s.[13] Employing the

---

[10] What we have in mind with such an assertion is the canonical commutation relation $[\hat{q}_j, \hat{p}_k] = i\hbar\, \delta_{jk}$.

[11] If the state of a system is an eigenstate of the operator to which its observable is being measured, then clearly the associated eigenvalue is the uniquely predicted result. But, in general, quantum mechanical probabilities less than unity prevail.

[12] The tacit assumption here is that of *non-contextuality*: if A is compatible with both B and C, but B and C are incompatible observables, the value assigned to A will not depend on whether A is being jointly measured with B or C (Mermim 1993).

[13] We have $f_{h(A)} \equiv (h \circ f_A) : \mathcal{S} \to \mathbb{R}$. Thus, $h(A)$ is *defined* by saying that its value in any state s is the result of applying the function h to the value of A (Isham and Butterfield 1998).

definition of v_s, it follows that both physical observables $h(A)$ and A satisfy the *functional composition principle (FUNC)*

$$v_s(h(A)) = h(v_s(A)); \quad \forall s \in \mathcal{S}. \tag{10}$$

Equation (10) is reasonably telling us that the value of a function of a physical observable is equal to the function evaluated on the value of that observable. Going back to QM, apart from the special case where a system is in an eigenstate of an operator, there is not an obvious way to construct a quantum valuation function since, within the minimal interpretation adopted here, it is not assumed that an observable has a value before it is measured.[14] However, there are some reasonable conditions that should be applied in constructing a *global* valuation v on the set of all bounded, self-adjoint operators on $\mathcal{H}$, namely:

(i) the *value-rule* holds, that is, the valuation $v(\hat{A}) \in \mathbb{R}$ belongs to the spectrum of the operator $\hat{A}$;

(ii) the *FUNC* principle holds, that is, $v(h(\hat{A})) = h(v(\hat{A}))$ for any real-valued function h.

These requirements suggest the following. First, if two operators $\hat{A}$ and $\hat{B}$ commute, then the global valuation is additive in the sense that, for all $|\psi\rangle \in \mathcal{H}$,

$$v(\hat{A} + \hat{B}) = v(\hat{A}) + v(\hat{B}). \tag{11}$$

Second, if again $[\hat{A}, \hat{B}] = 0$ for all $|\psi\rangle \in \mathcal{H}$ then

$$v(\hat{A}\hat{B}) = v(\hat{A})v(\hat{B}). \tag{12}$$

From this last equation, let $\hat{A} = \hat{\mathbb{1}}$, then $v(\hat{B}) = v(\hat{\mathbb{1}})v(\hat{B})$, thus, $v(\hat{\mathbb{1}}) = 1$ (provided that $v(\hat{B}) \neq 0$). This becomes interesting if the valuation function is applied to projection operators. Recall that projection operators play the role of propositions in QM and, as such, their eigenvalues are either 0 or 1. Hence, the global valuation of any projection operator $\hat{P}$ is either $v(\hat{P}) = 0$ or $v(\hat{P}) = 1$. Moreover, we saw that a set $\{\hat{P}_1, \hat{P}_2, \ldots, \hat{P}_n\}$ containing the projectors of the spectral decomposition of a self-adjoint operator $\hat{A}$ forms a resolution of the identity,

$$\sum_{i=1}^{n} \hat{P}_i = \hat{\mathbb{1}}. \tag{13}$$

It implies at once that

$$v(\hat{\mathbb{1}}) = v\left(\sum_{i=1}^{n} \hat{P}_i\right) = \sum_{i=1}^{n} v(\hat{P}_i) = 1, \tag{14}$$

---

[14] Because *quantization schemes* are still an open problem, it is not even obvious if the quantum valuation function should be over the physical quantity A or the operator $\hat{A}$ (see Isham 1995). Here, we take the latter approach.

which means that, whenever we have a collection of n pairwise orthogonal projectors $\hat{P}_i = |a_i\rangle\langle a_i|$, there is a single $\hat{P}_i$ for which $v(\hat{P}_i) = 1$, while $v(\hat{P}_j) = 0$ for all $i \neq j$. There are n different ways of associating the value 1 with only one of these projectors (that is, with one of the vectors $|a_i\rangle$). If we consider other distinct orthogonal bases in $\mathcal{H}$ and assume that the value (1 or 0) associated with that vector is the same, irrespective of the choice of the other basis vectors, then we are led to a contradiction and hence our primary hypothesis, the global valuation function, must be contested. This is, in fact, the KS theorem:

KS theorem. *There is no global valuation function if the Hilbert space $\mathcal{H}$ is such that* $\dim(\mathcal{H}) > 2$.

The KS theorem precludes the existence of global valuation functions whenever the dimension of the Hilbert space $\dim(\mathcal{H})$ is greater than 2. To answer our original question, an interpretation of physical observables in terms of definite, actual values pertaining to a quantum system is problematic since no global valuation exists in order to *globally* assign definite, actual values to projection operators. Instead, a value ascribed to an observable A must depend on some specific *context* from which A is to be considered. In algebraic terms, the valuation function $v[\mathcal{A}]$ over the *algebra* $\mathcal{A}$ is (once again) a global valuation if $\mathcal{A}$ is the set of all bounded, self-adjoint operators on $\mathcal{H}$. We say that $\mathcal{C}$ is a context if $\mathcal{C} \subset \mathcal{A}$ is a *commutative subalgebra* generated by the set $\{\hat{A}_1, \hat{A}_2, \ldots, \hat{A}_n\}$ of bounded, self-adjoint operators on $\mathcal{H}$. At last, we define a *local* valuation $v[\mathcal{C}]$ as a valuation function over a context $\mathcal{C}$. Yet again, the KS theorem precludes the existence of global valuation functions such as $v[\mathcal{A}]$, so that Hilbert spaces with dimension greater than 2 admits local valuations only. This quantum mechanical feature is known as *contextuality*, that is, one can have a set of projectors that commute in a given context, *i.e.*, propositions (or properties) that can be simultaneously asserted (or measured), but that no truth-value can be globally assigned to them in the totality of contexts (see de Barros, Holik, and Krause 2017); hence, quantum theory is said to be contextual. However, whereas an *epistemic* reading of contextuality says that measurement outcomes of an observable A depend on whether another set of physical observables are being jointly measured with A (Peres 2002), an *ontic* reading focuses on the orthodox structure of Hilbert spaces to claim that, regardless of measurements, contexts are bases (or *complete sets of commuting operators*) to which their projectors cannot be interpreted as preexistent properties possessing definite values (de Ronde 2019).

We encourage the reader to consult both the original (Kochen and Specker 1967) and more recent proofs of the theorem. Numerous discussions surrounding contextuality and its implications are amalgamated in the reference (*e.g.*, Isham 1995; Mermim 1993; Amaral and Cunha 2018). In particular, we will now follow the work done by de Ronde and Massri (2016) on invariance of quantum states as it relates to the KS theorem.

3.3 Invariance of quantum states: a corollary to KS

To begin with, let us note that the algebra of observables in Classical Physics is commutative, hence the classical local valuation function always coincides with a global valuation (reflecting the Boolean structure of the lattice L_C and the Principle of Thoroughgoing Determination embedded in it). Moreover, the laws of both classical mechanics and relativity theory are invariant under their respective group of transformations; in particular, nomological and dynamical properties are consistently translated between different frames of reference. As we saw, the fulfillment of these conditions is what underpins classical objectivity and allows one to speak of the state $s \in \mathcal{S}$ as being an actual state of affairs — *ASA*, as it was called. Algebraically, this corresponds to the mathematical feature of consistently pasting together multiple contexts of local valuations into a single global valuation function, to which we call such feature as *value invariance* (*VI*). One can be motivated to specify a *VI* with respect to nomological and dynamical properties, *VINP* and *VIDP*, respectively. If both *VINP* and *VIDP* are satisfied (and hence we get *VI*) then we have an *ASA*. In reality, the invariance of the valuation of both the sets of nomological and dynamical properties of a state $s \in \mathcal{S}$ is, effectively, what reifies an actual state of affairs in the classical picture. In quantum theory, however, the algebra of observables is non-commutative, so that the KS theorem prohibits global valuations from existing and imposes restrictions over a *VI*: even though the *state-vector* $|\psi\rangle \in \mathcal{H}$ is *defined* to be invariant under rotations, it is not a mathematical invariant of the kind needed to provide an interpretation in terms of objects possessing definite physical properties. More specifically, although the invariance of nomological properties (*VINP*) is respected in the formalism, valuations of dynamical magnitudes are not preserved under rotations (failure of *VIDP*) and thus we do not have, in general, *VI* in QM. Therefore, within the quantum mechanical formalism, one cannot reify the vector $|\psi\rangle$ as an *ASA*. This was shown by de Ronde and Massri (2016) through a corollary to the KS theorem which we reproduce next.

Corollary. *If* $\dim(\mathcal{H}) > 2$, *then the* VIDP *of a vector* $|\psi\rangle \in \mathcal{H}$ *is precluded.*

Proof. We refer the reader to the proof in de Ronde and Massri (2016), but the basic idea consists of submitting a vector in Hilbert space to a sequence of rotations. The authors showed that this, in turn, exhausts the local valuation function defined over a *maximal* context.[15] The failure of *VI* leads to the stated result. □

In their work, de Ronde and Massri pointed out that, since invariance of nomological properties is a *necessary* but not a *sufficient* condition to characterize a physical system, one cannot interpret a vector as being an actual

---

[15] We say that a context $\mathcal{A}$ is maximal if, given a self-adjoint operator $\hat{B}$ such that $\hat{B}\hat{A} = \hat{A}\hat{B}$ for all $\hat{A} \in \mathcal{A}$, then $\hat{B} \in \mathcal{A}$.

individual entity. This is our cue to resume the discussion regarding bundle and substratum theories of objects — we are finally ready to make our case against them.

4 Bundle and substratum theories (reconsidered)

When we left the discussion around bundle and substratum theories in order to explore the machinery of quantum mechanics, we saw that, despite all the problems surrounding the metaphysical dispute, both theories were equally successful in accounting for property possession and for individuality of concrete particulars, agreeing with our pre-theoretic intuition that such entities are individuals and, as such, with the Boolean structure of Classical Physics. This, in turn, created what was called an underdetermination problem of metaphysics by physics, inasmuch as our physical theories were completely silent with respect to their preference of bundle over substratum frameworks, and vice versa. As it happens, the situation in QM is even more dramatic. While it is mostly considered that Classical Physics is at least committed to the notion of individuality of physical systems, objects in quantum theory can either be interpreted in terms of individuals or, unprecedentedly, as *non-individuals* (see again French and Krause 2006, chap. 4).

To see that, recall the PII: it states that there are no numerically distinct indistinguishable objects. Classically, only a weak version of the PII, such as PII(1) (in which intrinsic properties and relations are assignable to an object), can survive the subtleties of classical physical situations. In quantum theory, however, a textbook-example such as two electrons in a box serves to decimate the PII in just a back-of-the-envelope calculation: compose the Hilbert space of two electrons in accordance to equation (4) and then apply the expectation value to *any* observable; the result obtained is invariant under particle permutation. There are no physical means whatsoever to tell which electron is which. Therefore, it is said that both electrons lost their individuality in some sense. Accordingly, physical systems are no longer subject to the Maxwell-Boltzmann statistics; the latter becomes obsolete in QM, and one needs to comply with the formalism by adopting the so-called Fermi-Dirac and Bose-Einstein statistics. To the founding fathers of QM, such as Schrödinger, this was enough to abandon the individuality of quantum mechanical entities and adopt the framework of non-individuals (French and Krause 2006, chap. 4; see also Arenhart 2017). But then, again, that would undermine the metaphysical details discussed in section 2 in the following sense: the failure of PII implies the failure of bundle theories in accounting for the individuality of concrete particulars, but nothing related can be said about the substratum approach to individuality (or perhaps other weaker versions of bundle theories such as the one presented by Rodriguez-Pereyra 2004). Actually, one could even argue that QM exposes the limitations within the strategy of exhausting an object solely into its constitutive properties, vindicating the need for a substratum in order to constitute and individualize concrete particulars (a related claim may be found in French and Krause 2006, chap. 4). Thus, quantum theory takes the underdetermination of metaphysics by physics to an upper level. The space of metaphysical discourse is expanded in a way to cover both metaphysics of individual systems (through substratum theory) and non-individual systems. If Classical Physics made us turn all the way 'back to square one', QM surpassed the rules and changed the game itself. We believe, however, that the new rules are in our favor.

It was stated that, amid the well-known disparities between bundle and substratum theories, both frameworks were inherently committed with the basic operation of property assignment via PTD, that is, given a property P and a particular a, either a possesses P or else it possesses $\neg P$. In the physicist jargon, we were then able to show that such attribute assignment is equivalent to physical theories to which the notion of physical observables (properties) are 'isomorphic' to propositions about physical systems. Hence, we can say that the instantiation of properties by concrete particulars is on equal footing with the assignment of truth-valued propositions describing physical systems. In particular, it was shown that Classical Physics is a fertile soil for both bundle and substratum theories, since the classical formalism contains in it a Boolean structure of definite-valued propositions suited enough to accommodate the PTD. However, as we have seen all throughout this work, such feature is simply another description of what is completely untenable in quantum theory: as a result of the algebra of non-commutative observables, the KS theorem follows from the quantum mechanical formalism itself to preclude a global assignment of preexistent values to projection operators (properties) of physical systems. One cannot assert values to properties of a quantum system because there are no definite values to be asserted in the first place. Our approach aims at the conceptual foundations shared by both bundle and substratum theories, to which one finds the PTD as the common ground they stand. Albeit a solid bedrock in Classical Physics, the PTD, and hence bundle and substratum theories, does not survive the restrictions imposed by QM and, in particular, the KS theorem.

In other words, instead of concentrating on whether distinct versions of the bundle theory could be cooked up to account for indiscernible quantum entities, and whether substratum theories are up to the job, we have advanced an argument to the effect that a fairly neglected principle of both theories (the PTD) fails in quantum theory. That brings both metaphysical theories down in a single stroke, showing clearly that quantum mechanics may be profitably used as a test field for metaphysical theories (a claim advanced, for instance, in Arenhart 2012, Arroyo 2020). Obviously, one may attempt to weaken the PTD in order to account for the indeterminacy of quantum theory, but then, such an account still has to be worked out in details, and it is not clear that what will result will still have the attractions that the original theories had to begin with.

Instead of speculating on how such theories could be modified to resist the KS argument, it is more interesting in this moment to check how deep the result goes. It affects even recent modifications of bundle theory, without having to delve into the details of which version of PII is valid or not. For instance, Rodriguez-Pereyra (2004) made a compelling argument in favor of bundle theory without the need of any PII version; indeed, the author was able to refute the PII within his version of bundle theory. Nevertheless, that's immaterial to how the approach deals with property possession, and the KS kind of argument presented here can be applied to generate trouble to Rodriguez-Pereyra's bundle theory, since it makes use of the very account of attribute assignment that does not survive the results of quantum theory. Equivalently, Jago's *essential bundle theory* (Jago 2018), which attempts to circumvent problems of distinguishing between essential and accidental property attribution in quantum theory will also have problems with the KS argument. In order to account for the distinct modal status of some properties, one still needs to provide for a consistent thorough distribution of truth values for the propositions attributing properties to entities (the PTD), while making a distinction (that Jago is willing to ground) between essential ones and non-essential ones. The fact that some properties are essential while others are not essential does nothing to prevent the argument above from running. Furthermore, Jago's approach requires that spatio-temporal location is a definite property constituting every bundle. Given that spatio-temporal location may be undefined in some contexts in QM, this approach too faces the consequences of the argument developed here.

Notice that this is a general argument against versions of bundle theory and substratum theory that are willing to account for property attribution and composition. The simplicity and explanatory power of such theories make them appealing at first; however, the KS theorem makes for such theories unable to account for quantum entities. This test of such theories needs not discuss the issue of individuality. This, it seems to us, is an advantage, given that the rod for these theories is blocked independently of the problem of individuality, already much discussed in the literature.

5 Conclusion

In this work, we attempted to enrich and potentially enlighten an old discussion regarding the nature of concrete particulars by welcoming into the conversation quantum theory and recent developments on invariance of quantum states as it relates to the Kochen-Specker theorem (as developed by de Ronde and Massri 2016). We have argued that current versions of both bundle theory and substratum theory all adopt a version of what we have called (following Mittelstaedt) the Principle of Thoroughgoing Determination, which requires that a particular either instantiates or does not instantiate any given property P. This principle is vindicated by bundle and substratum theories by the very approach they offer on the constitution of a concrete particular. We have shown that this principle is untenable when QM enters the stage, so that these approaches to particulars are not consistent with quantum theory. Of course, this opens the door for structuralist accounts of quantum entities, but that was not the topic of our discussion.

A further consequence of the result being advanced here concerns metaphysical underdetermination. It was argued, for instance, by French and Krause (2006, chap. 4), that QM underdetermines its metaphysics of particular objects. That is, as we have discussed, quantum theory, when understood as dealing with objects (*i.e.*, as providing for an object-oriented ontology), may be interpreted as dealing with individual objects (whose principle of individuality is provided by substratum theory, and not by bundle theory) or else with non-individual objects (not individuated, of course).[16] By taking into account only the issue of individuality, French and Krause were able to argue that a substratum theory may live with indiscernible quantum entities, given that individuality is provided for by the substratum. However, this focus on the problem of individuality somehow blinded them to the fact that substratum theories (and bundle theories without the PII too, if those were taken into account) are clearly *incompatible* with quantum theory due to the KS theorem. As a result, substratum theories are also not really an option to account for those entities' individuality; that is, those theories are not an option to account for the very objecthood of quantum entities if those items are understood as concrete particulars. If that is correct, then, only the non-individuals interpretation is left as a legitimate metaphysics of concrete particulars. But is it? Well, it all depends on how property possession for non-individuals is accounted for, and non-individuals have a nebulous metaphysics, to say the least. But that is an issue for another discussion.

References

[1] Adams, R. M. (1979). Primitive Thisness and Primitive Identity. *The Journal of Philosophy* 76(1): 5-26.

[2] Amaral, B., Cunha, M. T. (2018). *On Graph Approaches to Contextuality and their Role in Quantum Theory.* Springer International Publishing. https://doi.org/10.1007/978-3-319-93827-1

[3] Arenhart, J. R. B. (2012). Ontological frameworks for scientific theories. *Foundations of Science* 17: 339-56.

---

[16] French and Krause (2006) argued that bundle theories fail because the strong versions of the PII fail in QM; they did not take into account neither Rodriguez-Pereyra's (2004) version of the bundle theory, nor any other versions of the bundle theory that may live without PII.

[4] Arenhart, J. R. B. (2017). The received view on quantum non-individuality: formal and metaphysical analysis. *Synthese* 194: 1323–47.

[5] Arroyo, R. W. (2020). *Discussions on physics, metaphysics and metametaphysics: Interpreting quantum mechanics.* PhD Thesis in Philosophy. Florianópolis: Universidade Federal de Santa Catarina.

[6] Benovsky, J. (2016). *Metametaphysics. On Metaphysical Equivalence, Primitiveness, and Theory Choice.* Switzerland: Springer.

[7] Black, M. (1952). The Identity of Indiscernibles. *Mind* 61: 153-64.

[8] de Barros, J. A., Holik, F., Krause, D. (2017). Contextuality and Indistinguishability. *Entropy* 19, 435.

[9] Demirli, S. (2010). Indiscernibility and bundles in a structure. *Philosophical studies* 151: 1-18.

[10] de Ronde, C., Massri, C. (2016). Kochen-Specker Theorem, Physical Invariance and Quantum Individuality. *Cadernos de História e Filosofia da Ciência* 2(1): 107-30.

[11] de Ronde, C. (2019). Unscrambling the Omelette of Quantum Contextuality (Part I): Preexistent Properties or Measurement Outcomes?. *Foundations of Science.* https://doi.org/10.1007/s10699-019-09578-8

[12] Döring A., Isham C. (2010). 'What is a Thing?': Topos Theory in the Foundations of Physics. *New Structures for Physics* 813: 753-937.

[13] French, S., Rickles, D. (2003). Understanding permutation symmetry. *Symmetries in Physics: Philosophical Reflections.* Cambridge University Press: 212-38.

[14] French, S., Krause, D. (2006). *Identity in Physics: A Historical, Philosophical, and Formal Analysis.* Oxford: Oxford University Press.

[15] Isham, C. J. (1995). *Lectures On Quantum Theory: Mathematical and Structural Foundations.* London: Imperial College Press.

[16] Isham, C. J., Butterfield, J. (1998). Topos Perspective on the Kochen-Specker Theorem: I. Quantum States as Generalized Valuations. *International Journal of Theoretical Physics* 37(11): 2669-733.

[17] Jago, M. (2018). Essential bundle theory and modality. *Synthese.* https://doi.org/10.1007/s11229-018-1819-3

[18] Kochen, S.; Specker, E. (1967). On the problem of hidden variables in quantum mechanics. *Journal of Mathematics and Mechanics* 17(1): 59-87.

[19] Krause, D.; Arenhart, J. R. B. (2018). Quantum Non-Individuality: Background Concepts and Possibilities. In: S. Wuppuluri; F. A. Doria (eds.) *The Map and the territory. Exploring the foundations of science, thought and reality*, pp.281-305. Cham: Springer.

[20] Loux, M. J. (2006). *Metaphysics: A contemporary introduction*. 3rd ed. New York, NY: Routledge.

[21] Mermim, N. D. (1993). Hidden Variables and the Two Theorems of John Bell. *Reviews of Modern Physics* 65: 803-15.

[22] Mittelstaedt, P. (2009). Cognition versus Constitution of Objects: From Kant to Modern Physics. *Foundations of Physics* 39(7): 847-59.

[23] Mittelstaedt, P. (2011). The Problem of Interpretation of Modern Physics. *Foundations of Physics* 41(11): 1667-76.

[24] Peres, A. (2002). *Quantum Theory: Concepts and Methods*. Dordrecht: Kluwer Academic Publishers.

[25] Rodriguez-Pereyra, G. (2004). The Bundle Theory is compatible with distinct but indiscernible particulars. *Analysis* 64(1): 72-81.

Classical Fields, Bell's inequalities, and the quantum limit

J. Acacio de Barros*, Adonai S. Sant'Anna[†]

*School of Humanities and Liberal Studies, San Francisco State University, San Francisco, CA 94312. E-mail: barros@sfsu.edu

[†]Mathematics Department, Federal University at Paraná, CP 19081, 81530-900, Curitiba, PR, Brazil. E-mail: adonaisantanna@gmail.com

Abstract

In this paper, we use a homodyne detection of a classical field to violate Bell's inequalities. This violation is achieved with local random variables that are continuous, which does not preclude the existence of a joint probability distribution. This result and meanings are discussed.

We dedicate this article to Professor Francisco Doria (FAD) on the occasion of his 70th birthday. We believe it is apt to honor him with this article for several reasons. First, this paper is an update of a previous article co-authored with Pat Suppes (available on arXiv:quant-ph/9606019). Pat and Doria were good friends, and Pat would have been delighted to contribute to this volume. Second, Doria spent a one-year sabbatical leave in the '90s as a Fulbright scholar visiting Pat at Stanford University's Institute for Mathematical Studies in the Social Sciences (IMSSS), which Pat founded and directed. This contact between Pat and Doria led to ASS and JAB spending some time at the IMSSS, where this paper's ideas germinated. Third, this paper exemplifies the type of research that Doria does: foundation issues in physics and mathematics that lead to different insights. Finally, both ASS and JAB owe their passion for science and mathematics to Doria's guidance and tutoring. So, this work would not have been written without his early support and involvement.

Most of the outcomes of this paper are not new. As mentioned above, the main result of a violation of Bell's inequalities with classical fields was present in our '96 manuscript with Pat Suppes. However, that article was never published. Furthermore, in recent years much interest appeared in connection with violations of quantum inequalities with classical systems (see [26] and references therein). In a certain sense, our paper came to this area too early when few worked on it. Thus, we believe it is as relevant today as it was more than 20 years ago.

1 Introduction

Quantum theory is bizarre. This strangeness comes mostly from assigning properties to a quantum system, famously exemplified by the Einstein-Bohr dialogues. Bohr believed that a property of physical systems only existed if an observer made an actual measurement of this property. To Bohr, talking about properties that were not measured was nonsensical. Einstein, on the other hand, believed that properties had a reality independent of the observer. To him, the reality of a system's property did not require performing an actual measurement. Since quantum theory did not provide a way to talk about unobserved quantities, the observer-independent reality, Einstein concluded that quantum theory was incomplete and that a complete theory should be developed. Such theories that extended quantum mechanics are known as *hidden-variable theories*.

Perhaps the most persuasive argument in favor of a hidden variable theory was that of Einstein, Podolsky, and Rosen (EPR) [19]. Consider two spin-1/2 particles are produced in the singlet state[1]

$$|\psi\rangle = \frac{1}{\sqrt{2}} \left(|+-\rangle - |-+\rangle \right),$$

where $|+\rangle$ and $|-\rangle$ are the eigenvectors of the spin operator (say, in direction **z**). One of the particles is sent to Alice's laboratory (A) and the other to Bob's lab (B). EPR noticed that if Alice measured the spin to be "+," she could immediately infer that Bob's measurement of the spin would be "−." To EPR, this meant that we could infer the value of the particle's spin at Bob's without interacting with it, i.e., without disturbing it, since we should rule out non-local interactions between particles in Alice and Bob's labs, as those labs can be placed as far as we want. Thus, EPR concluded that quantum theory was incomplete, as there were elements of reality (the value of spin) that could be inferred without any measurement. To complete quantum mechanics, one would need a hidden variable theory.

There are several obstacles to developing hidden-variable theories. For example, without resorting to EPR's argument, a question remains as to whether we can simultaneously assign values to complementary properties, such as momentum and position. In a famous paper, Kochen and Specker [23] showed that if we try to do so, we reach a contradiction, unless we assume that a property changes with the experimental context. For example, imagine a quantum system with four properties, A, B, C, and D. Further assume that

---

[1] The version of the argument presented here is Bohm's, and not EPR's [9]. EPR used momentum and position as correlated variables instead of spin.

we can create experiments where we can only measure the following variables simultaneously: A and C; A and D; B and C; and B and D. In this example, A shows up in two measuring contexts: A in the context of C and A in the context of D. Given that an experimenter can choose at will which context to measure, for the property A to be independent of the experimenter's capricious choices, it needs to be independent of the experimental context. However, Kochen and Specker proved that this idea that a property, in this case, A, is independent of those two contexts, leads to contradictions. In other words, Kochen and Specker showed that any hidden-variable theory has to be contextual.

Another important obstacle to hidden-variable theories was shown in 1963 by John Bell [6, 7]. In his paper, Bell discussed Bohm's setup for the Einstein-Podolsky-Rosen's (EPR) Gedankenexperiment [9, 19]. By assuming EPR's criteria of realism and locality, Bell derived a set of inequalities that any local hidden-variable theory had to satisfy, known as Bell's inequalities. Bell then proceeded to show that, for certain situations, quantum mechanics violated his inequalities. In other words, if EPR was right that quantum theory was incomplete and should be substituted by a local hidden-variable theory, then Bell determined that certain predictions of quantum mechanics had to be wrong. In 1982 Alain Aspect and collaborators confirmed that correlated quantum systems indeed violate Bell's inequalities. Thus, local hidden-variable theory are wrong [4, 3, 2].

Bell's assumptions are considered equivalent to an underlying physical reality, with added locality conditions. As such, they are equivalent to the existence of a joint probability distribution for all possible outcomes of an experiment in all possible contexts [32, 20]. What this means is that, for particular conjunction of properties, one cannot assign a probability value consistent with the observed marginals[2]. This is equivalent to saying that properties are contextual (see [15] and references therein).

Because quantum non-locality and contextuality are among the most disturbing aspects of the theory, it was commonly believed that any classical system satisfies Bell's inequalities. In this paper, we show that classical fields *do not* satisfy Bell's inequalities. Hence classical fields, e.g., electromagnetic fields, are not Bell-type hidden variables. We do this by using a simple experimental setup proposed by Tan *et al.* [33, 34, 35] for single photons. We then reinterpret this setup for classical electromagnetic fields with randomized phase. For this proposed experiment, we derive from the classical field properties a violation of Bell's inequalities [6, 7, 8, 32], with, at the same time, locality being preserved in a sense to be made precise.

Most of the outcomes of this paper are not new. As we mentioned above, the main result of a violation of Bell's inequalities with classical fields was present in our '96 paper with Pat Suppes. However, that paper was never

---

[2]An alternative is to use extended probabilities. See [30, 31, 17, 18, 16, 15] for examples using non-monotonic upper probabilities as well as (signed) negative probabilities.

published. Furthermore, in recent years much interest appeared in connection with violations of quantum inequalities with classical systems [24, 1, 11, 22, 28, 5, 27, 26, 25]. In a certain sense, our paper came to this area too early when nobody was working on it. Thus, we believe it is as relevant today as it was more than 20 years ago.

2 Experimental Setup

As mentioned above, we use a similar experimental scheme to that of Tan, Walls, and Collett [34]. In their experiment, Tan et al. used a single-photon source and a beam splitter to create an entanglement between a single photon and the vacuum in two separate beams. Each beam's phase was measured via two homodyne detections [14] with same frequency but fixed phases θ_1 and θ_2. They showed that for two different homodyne phases θ_1 and θ_2, the detection statistics showed correlations between each detector that violated Bell's inequalities. Equally important, according to Tan et al., a classical coherent field would not violate Bell's inequalities. Here we argue that their conclusion about classical fields is not correct, as they assumed a fixed phase for the coherent state. In what follows, we show that a uniformly distributed phase in the interval $[0, 2\pi]$ yields a correlation between detectors that violates Bell's inequalities.

Our experimental scheme uses three *classical* coherent sources: $\alpha_1(\theta_1)$, with fixed phase θ_1; $\alpha_2(\theta_2)$, with fixed phase θ_2; and $u(\theta)$, with a unknown stochastic phase θ. The geometry of the setup is shown in Figure 1. The experimental configuration has two homodyne detections, (D_1, D_2) being one and (D_3, D_4) the other, such that the measurements are sensitive to phase changes in $u(\theta)$. Similarly to Tan et al.'s experiment, in Figure 1, BS, BS_1, and BS_2 are beam splitters that reflect 50% of the incident electromagnetic field and let 50% of it pass. When the electromagnetic field is reflected, the mirrors add a phase of $\pi/2$ to the field, while no phase is added when the field passes through BS, BS_1, or BS_2. We will look for correlations between the pairs of detectors (D_1, D_2) and (D_3, D_4).

To compute the correlation functions, we first define the *continuous* random variables in terms of which we derive the Bell-type correlations. On this matter we shall be as explicit as possible. Associated to the source $u(\theta)$ at D_1 is the random variable $U_1(t)$, whose value at t is just the value of the classical field at D_1, namely,

$$U_1(t) = \frac{1}{4}\beta \cos(\omega t + \theta + \pi/2), \tag{1}$$

where β is the amplitude of the field at the source, θ is the unknown phase and $\pi/2$ is a phase gained when u is reflected at BS_2.

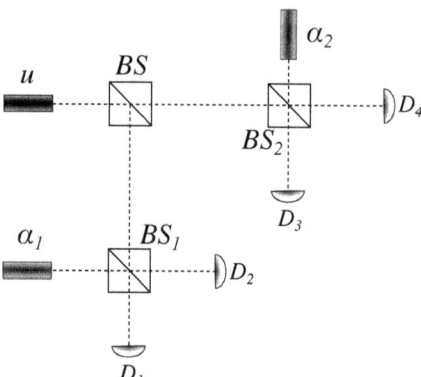

Figure 1: Experimental configuration. A laser u with stable phase θ impinges on a 50:50 beam splitter (BS). The two equal-intensity beams then impinge on two additional 50:50 beam splitters, BS_1 and BS_2. At BS_1 and BS_2, the beam from u is combined with lasers α_1 and α_2 (with half the intensity of u), respectively, with phases θ_1 and θ_2. Detectors D_1, ..., D_4 register the intensity of the fields at each arm of the beam splitters.

Probability enters initially by using the time average to compute the expectation of $U_1(t)^2$

$$U_1^2 = \langle U_1(t)^2 \rangle = \langle [\frac{1}{4}\beta \cos(\omega t + \theta + \pi/2)]^2 \rangle,$$

which is just the standard intensity, but here we treat it probabilistically. In the above expression, $\langle f(t) \rangle$ is defined as the temporal expectation given by

$$\langle f(t) \rangle = \lim_{T \to \infty} \frac{1}{T} \int_0^T f(t) dt. \qquad (2)$$

We emphasize that (2) is expressed in terms of a limit as T goes to infinity for mathematical simplicity, but in practice it suffices that T is large enough to stabilize the expectation values (e.g., if $T \gg 1/\omega$). In a similar fashion, associated to the source $\alpha_1(\theta_1)$ at D_1 is the random variable $A_1(t)$,

$$A_1(t) = \frac{1}{2}\alpha \cos(\omega t + \theta_1 + \pi/2)$$

and thus

$$A_1^2 = \langle A_1(t)^2 \rangle = \left\langle \left[\frac{1}{2}\alpha \cos(\omega t + \theta_1 + \pi/2)\right]^2 \right\rangle.$$

At D_1, the total field is the random variable $F_1(t) = U_1(t) + A_1(t)$. So, the intensity of the total field at D_1 is just the second moment of $F_1(t)$, i.e.,

$$\begin{aligned} I_1(\theta) &= F_1^2 = \langle F_1(t)^2 \rangle = \langle (U_1(t) + A_1(t))^2 \rangle \\ &= \langle U_1(t)^2 \rangle + 2\langle (U_1(t) A_1(t)) \rangle + \langle A_1(t)^2 \rangle, \end{aligned}$$

where we used θ as an argument for I_1 to make it explicit that it depends on θ. We can see that the cross moment in the expression above is the classical interference term.

We can compute I_1 directly from the expression for $U_1(t)$ and $A_1(t)$ in the following way

$$I_1(\theta) = \lim_{T \to \infty} \frac{1}{T} \int_0^T \left[\frac{1}{2} \alpha \cos(\omega t + \theta_1 + \pi/2) + \frac{1}{4} \beta \cos(\omega t + \theta + \pi/2) \right]^2 dt,$$

which yields

$$I_1(\theta) = \frac{1}{32} \beta^2 + \frac{1}{8} \alpha \beta \cos(\theta - \theta_1) + \frac{1}{8} \alpha^2. \tag{3}$$

In similar fashion, we can compute for the other three detectors,

$$I_2(\theta) = \frac{1}{32} \beta^2 - \frac{1}{8} \alpha \beta \cos(\theta - \theta_1) + \frac{1}{8} \alpha^2, \tag{4}$$

$$I_3(\theta) = \frac{1}{32} \beta^2 - \frac{1}{8} \alpha \beta \sin(\theta - \theta_2) + \frac{1}{8} \alpha^2, \tag{5}$$

and

$$I_4(\theta) = \frac{1}{32} \beta^2 + \frac{1}{8} \alpha \beta \sin(\theta - \theta_2) + \frac{1}{8} \alpha^2. \tag{6}$$

The intensities obtained are conditional on θ. To obtain the unconditional intensities we assume a uniform distribution for θ and integrate the expressions for all possible values of θ. Not only is θ unknown, but the phase would vary randomly for repeated runs of the experiment. If θ were a coherent source with fixed θ, Bell's inequalities would not be violated [33].

The unconditional intensities I_1, I_2, I_3, and I_4 for the detectors D_1, D_2, D_3, and D_4 are

$$I_1 = \frac{1}{32} \beta^2 + \frac{1}{8} \alpha^2, \tag{7}$$

$$I_2 = \frac{1}{32} \beta^2 + \frac{1}{8} \alpha^2, \tag{8}$$

$$I_3 = \frac{1}{32} \beta^2 + \frac{1}{8} \alpha^2, \tag{9}$$

$$I_4 = \frac{1}{32}\beta^2 + \frac{1}{8}\alpha^2. \tag{10}$$

We can see from (7)–(10) that the intensities are the same for all detectors, and are similar to those given by Walls and Milburn [35] in the case of a classical source. This is expected, since (7)–(10) express the concept that the intensity at the detectors is the sum of the intensities from the two sources.

We now start computing the covariance between intensities in the homodyne detectors. The covariance we are interested in is between $(I_1 - I_2)$ and $(I_3 - I_4)$.

$$\begin{aligned}\text{Cov}(I_1 - I_2, I_3 - I_4) &= \frac{1}{2\pi}\int_0^{2\pi}[(I_1(\theta) - I_2(\theta)) \times (I_3(\theta) - I_4(\theta))]d\theta \\ &\quad - \frac{1}{2\pi}\int_0^{2\pi}(I_1(\theta) - I_2(\theta))\,d\theta \\ &\quad \times \frac{1}{2\pi}\int_0^{2\pi}(I_3(\theta) - I_4(\theta))\,d\theta.\end{aligned} \tag{11}$$

It is straightforward to show from (3)–(6) and (11) that

$$\text{Cov}(I_1 - I_2, I_3 - I_4) = -\frac{1}{32}\beta^2\alpha^2 \sin(\theta_1 - \theta_2).$$

In order to compute the correlation we have to know the variance of the random variables $(I_1 - I_2)$ and $(I_3 - I_4)$, which are defined in a standard way as

$$\begin{aligned}\text{Var}(I_1 - I_2) &= \frac{1}{2\pi}\int_0^{2\pi}(I_1(\theta) - I_2(\theta))^2 d\theta - \left[\frac{1}{2\pi}\int_0^{2\pi}(I_1(\theta) - I_2(\theta))d\theta\right]^2 \\ &= \frac{1}{32}\beta^2\alpha^2\end{aligned}$$

and

$$\begin{aligned}\text{Var}(I_3 - I_4) &= \frac{1}{2\pi}\int_0^{2\pi}(I_3(\theta) - I_4(\theta))^2 d\theta - \left[\frac{1}{2\pi}\int_0^{2\pi}(I_3(\theta) - I_4(\theta))d\theta\right]^2 \\ &= \frac{1}{32}\beta^2\alpha^2.\end{aligned}$$

Finally, we are in a position to compute the correlation between the two random variables $A \equiv (I_1 - I_2)$ and $B \equiv (I_3 - I_4)$. This is done in a standard way by just dividing the covariance by the square-root of the variances [29]:

$$\rho(A, B) = \frac{\text{Cov}(A, B)}{\sqrt{\text{Var}(A)\,\text{Var}(B)}},$$

and we have the following expression for the correlation

$$\rho(A, B) = -\sin(\theta_1 - \theta_2),$$

which we may rewrite as

$$\rho(A(\theta_1), B(\theta_2)) = -\sin(\theta_1 - \theta_2), \tag{12}$$

to make explicit the dependency of A and B on homodyning phase angles θ_1 and θ_2. This correlation is the same as the well-known correlations for Bell's setup [6].

So, we are now in a position to show that we can violate Bell-type inequalities. We may now choose angles θ_1, θ_2, θ'_1, and θ'_2 such that we obtain at once, for the four correlations $\rho(\theta_1, \theta_2)$, $\rho(\theta_1, \theta'_2)$, $\rho(\theta'_1, \theta_2)$ and $\rho(\theta'_1, \theta'_2)$ a violation of Bell's inequalities in the form due to Clauser, Horne, Shimony, and Holt (CHSH) [12, 13], by choosing the four angles such that

$$\theta_1 - \theta_2 = \theta'_1 - \theta'_2 = 60°,$$

$$\theta_1 - \theta'_2 = 30°,$$

$$\theta'_1 - \theta_2 = 90°.$$

In particular,

$$\rho(\theta_1, \theta_2) - \rho(\theta_1, \theta'_2) + \rho(\theta'_1, \theta_2) + \rho(\theta'_1, \theta'_2) = -\frac{\sqrt{3}}{2} + \frac{1}{2} - 1 - \frac{\sqrt{3}}{2} < -2.$$

However, in the case of continuous random variables, which is what we have in the present context for intensity, or differences of intensity, failure to satisfy Bell's inequalities in the CHSH form does not imply that there can be no joint distribution of the four random variables compatible with the four given correlations. It is easy to show that for selected values of the two missing correlations, there does, for this example, exist a joint probability of the four random variables compatible with the four given correlations.

3 Measurement and Photon Counts

Because classical field theory is a deterministic theory, our introduction of expectations and probabilities might be questioned. Our response is that the strength of a classical field at a space-time point cannot be measured, as was emphasized long ago by Bohr and Rosenfeld [10]. As they pointed out, classical field strength cannot be represented by true point functions, but by average values over space-time regions. This is exactly what we have done in introducing random variables and their expectations. The casual reader might claim that we should do an analysis of coincidence counts with photocounters. This makes no sense in the case of classical fields, where the number of photons arriving at the same time at each detector is incredibly large. What makes sense is not discrete but continuous measurement of intensity.

Despite that, we are going to use the previous result to model discrete photon counts, and show that this time it does not in such a way that they violate Bell's inequalities. For this, we define two new discrete random variables $X = \pm 1$ and $Y = \pm 1$. These random variables correspond to nearly simultaneous correlated counts at the detectors, and are defined in the following way.

$$X = \begin{cases} +1 & \text{if detector } D_1 \text{ triggers a count} \\ -1 & \text{if detector } D_2 \text{ triggers a count} \end{cases}$$

$$Y = \begin{cases} +1 & \text{if detector } D_3 \text{ triggers a count} \\ -1 & \text{if detector } D_4 \text{ triggers a count.} \end{cases}$$

To compute the expectation of X and Y we use the stationarity of the process and do the following. First, let us note that

$$I_1 - I_2 = N_X \cdot P(X = 1) - N_X \cdot P(X = -1),$$

where N_X is the expected total number of photon counts at D_1 and D_2 and $P(X = \pm 1)$ is the probability that the random variable X has values ± 1. The same relation holds for

$$I_3 - I_4 = N_Y \cdot P(Y = 1) - N_Y \cdot P(Y = -1).$$

To simplify we put as a symmetry condition that $N_X = N_Y = N$, i.e., the expected number of photon counts at each homodyne detector is the same. But we know that

$$I_1 + I_2 = N \cdot P(X = 1) + N \cdot P(X = -1) = N,$$

and

$$I_3 + I_4 = N \cdot P(X = 1) + N \cdot P(X = -1) = N.$$

Then we can conclude from equations (3)–(6), assuming maximum visibility, that

$$E_d(X|\theta) = \frac{I_1 - I_2}{I_1 + I_2} = \cos(\theta - \theta_i),$$

$$E_d(Y|\theta) = \frac{I_3 - I_4}{I_3 + I_4} = \sin(\theta - \theta_j),$$

where E_d represents the expected value of the counting random variable. It is clear that if θ is uniformly distributed we have at once:

$$E(X) = E_\theta(E_d(X|\theta)) = 0, \qquad (13)$$

$$E(Y) = E_\theta(E_d(X|\theta)) = 0. \qquad (14)$$

We can now compute $\text{Cov}(X,Y)$. Note that

$$\begin{aligned}\text{Cov}(X,Y) &= E(XY) - E(X)E(Y) \\ &= E_\theta(E_d(XY|\theta)) - E_\theta(E_d(X|\theta))E_\theta(E_d(Y|\theta))\end{aligned}$$

and so

$$\begin{aligned}\text{Cov}(X,Y) &= \frac{1}{2\pi}\int_0^{2\pi} E_d(XY|\theta)d\theta \\ &\quad -\frac{1}{2\pi}\int_0^{2\pi} E_d(X|\theta)d\theta \times \frac{1}{2\pi}\int_0^{2\pi} E_d(Y|\theta)d\theta.\end{aligned}$$

In order to compute the covariance, we also use the conditional independence of X and Y given θ, which is our locality condition:

$$E_d(XY|\theta) = E_d(X|\theta)E_d(Y|\theta),$$

because given θ, the expectation of X depends only on θ_i, and of Y only on θ_j. Then, it is easy to see that

$$\rho(X,Y) = \text{Cov}(X,Y) = -\sin(\theta_i - \theta_j). \quad (15)$$

The correlation equals the covariance, since X and Y are discrete ± 1-valued random variables with zero mean, as shown in (13) and (14), and so $\text{Var}(X) = \text{Var}(Y) = 1$. It follows at once from (15) that for a given set of θ_i's and θ_j's Bell's inequalities are violated.

However, there is an important detail underlying our computations above. When we are looking for correlations, we are only considering the cases where an "observation" is made on both detectors, i.e., we have coincidence counts for X and Y. However, to have a coincidence count, there are two assumptions that were tacitly introduced. First, for the given observation time window, the assumption is that we have a detection in one of the photodetectors. For the hidden-variable model that associates with field intensity a probability of photo detection, there should be a non-zero probability that no "detection" happens during a time window. This was not included in our model. Second, even if we accounted the no-detection event, and conditioned on coincidence detections, we would have to use a non-enhancement hypothesis [21] to compute the correlations in the way we are doing.

4 Final Remarks

The experiment proposed in [34] supposes a single photon source that is split into the two homodyne detectors. Tan *et al.* also analyze the classical case and get no violation of Bell's inequalities. However, they assume a weak coherent source with randomized phase as the classical analogue of their single photon source. This would be equivalent to having a classical thermal source, where coherence would not be a strong feature. In our experiment we suppose that this source is not only classical, i.e., with high intensity, but also that it is coherent with the phase unobservable and varying randomly on repeated runs. The different source used here, as opposed to that used in [34] implies that the expectations given by (3)–(10) are computed in a different way than in [34]. Here we first integrate with respect to t and then θ. It is easy to supply a source that would fit our requirements. This would be, for example, a radio source, a microwave, or a laser source, all with unstabilized phases. To realize this experiment, one must also use two additional coherent sources with stable known phases and with the same frequency as the nonstabilized source. If a data table is then built that keeps track of all the measured values on the detectors, we can compute the correlations and see a violation of Bell's inequalities.

There are several other remarks that we must add in order to clarify some points. First, when using classical fields the number of photons is overwhelmingly large. For that reason, we would not need to compute any photon count correlation. What we measure is intensity. On the other hand, Bell's inequalities are not enough to show that we do not have a joint probability distribution for classical fields, because they assume a continuous range of values. That is why we computed the correlation matrix, showing that for this case a joint probability distribution does indeed not exist.

Another point is that intensity of classical fields does not satisfy the basic assumption made by Bell, because it can take an infinite range of values; Bell considered spin measurements that can take only two possible values. To show that this does not present any constraint, we did an analysis of photon counts, which can only take, as in Bell's assumptions, two discrete values. However, to violate Bell's inequalities here, we must use an enhancement strategy. This strategy requires communication between the two different homodyne detectors, and it would not be non-local, as opposed to the true quantum example.

Finally, the last point. One can argue that if classical fields violate Bell's inequalities, then, since they are classical, Bell's theorem must be wrong, and we must show why it is wrong. First, we did not show that Bell's theorem is wrong; we just showed that a classical field approximates the quantum correlations and that Bell's inequalities are violated for them. Second, we emphasized that classical fields are continuous, and Bell's inequalities are derived from dichotomous random variables. Therefore, the conclusion that there does not exist a joint probability distribution for the intensities does not follow from violations of Bell's inequalities[3].

---

[3]In our arXiv:quant-ph/9606019 paper, we claimed that the correlation matrix for the intensities had negative and positive eigenvalues, which implied the non-existence of a joint probability distribution. This claim is incorrect, as our claim was based in a specific assumption for the hidden-variable model. However, as our example shows, a joint probability does exist,

Acknowledgements. The authors thank Pawel Kurzynski, José Huguenin, Walter Balthazar, and Gary Oas for discussions and suggestions. JAB was partially supported by the Patrick Suppes Gift Fund for the Suppes Brain Lab. He is also thankful to John Perry for his hospitality while visiting CSLI, Stanford University, where part of this work was conducted. This article is the continuation of our work with Patrick Suppes, who passed away in November 2014. Though Pat was a significant contributor to its main ideas, which appeared in a version on the ArXiv's (arXiv:quant-ph/9606019), he passed away before we discussed the current version, which changed many of our original points and conclusions. Therefore, any mistakes or problems are the sole responsibility of JAB and AS.

References

[1] A. Aiello, G. Puentes, and J. P. Woerdman. Linear optics and quantum maps. *Physical Review A*, 76(3):032323, September 2007. Publisher: American Physical Society.

[2] Alain Aspect, Jean Dalibard, and Gã©rard Roger. Experimental Test of Bell's Inequalities Using Time- Varying Analyzers. *Physical Review Letters*, 49(25):1804–1807, December 1982.

[3] Alain Aspect, Philippe Grangier, and G\'erard Roger. Experimental Realization of Einstein-Podolsky-Rosen-Bohm Gedankenexperiment: A New Violation of Bell's Inequalities. *Phys. Rev. Lett.*, 49(2):91–94, 1982.

[4] Alain Aspect, Philippe Grangier, and Gã©rard Roger. Experimental Tests of Realistic Local Theories via Bell's Theorem. *Physical Review Letters*, 47(7):460–463, August 1981.

[5] W. F. Balthazar, C. E. R. Souza, D. P. Caetano, E. F. GalvÃ£o, J. a. O. Huguenin, and A. Z. Khoury. Tripartite nonseparability in classical optics. *Optics Letters*, 41(24):5797–5800, December 2016. Publisher: Optical Society of America.

[6] J.S. Bell. On the Einstein-Podolsky-Rosen paradox. *Physics*, 1(3):195–200, 1964.

[7] J.S. Bell. On the Problem of Hidden Variables in Quantum Mechanics. *Rev. Mod. Phys.*, 38(3):447–452, 1966.

---

and can be constructed from the field itself. A quick examination reveals that the field-intensity model does not satisfy the assumptions of the hidden-variable model considered in our quant-ph paper.

[8] J.S. Bell. *Speakable and Unspeakable in Quantum Mechanics: Collected Papers on Quantum Philosophy*. Cambridge University Press, June 2004.

[9] David Bohm. *Quantum Theory*. Courier Corporation, April 2012.

[10] N. Bohr and L. Rosenfeld. Field and Charge Measurements in Quantum Electrodynamics. *Phys. Rev.*, 78(6):794–798, 1950.

[11] C. V. S. Borges, M. Hor-Meyll, J. A. O. Huguenin, and A. Z. Khoury. Bell-like inequality for the spin-orbit separability of a laser beam. *Physical Review A*, 82(3):033833, September 2010. Publisher: American Physical Society.

[12] J F Clauser and A. Shimony. Bell's theorem. Experimental tests and implications. *Reports on Progress in Physics*, 41(12):1881–1927, 1978.

[13] J.F. Clauser, M.A. Horne, A. Shimony, and R.A. Holt. Proposed Experiment to Test Local Hidden-Variable Theories. *Physical Review Letters*, 23(15):880–884, October 1969.

[14] M. J. Collett, R. Loudon, and C. W. Gardiner. Quantum Theory of Optical Homodyne and Heterodyne Detection. *Journal of Modern Optics*, 34(6-7):881–902, June 1987. Publisher: Taylor & Francis _eprint: https://doi.org/10.1080/09500348714550811.

[15] J. Acacio de Barros, Janne V. Kujala, and Gary Oas. Negative probabilities and contextuality. *Journal of Mathematical Psychology*, 74:34–45, October 2016.

[16] J. Acacio de Barros, Gary Oas, and Patrick Suppes. Negative probabilities and Counterfactual Reasoning on the double-slit Experiment. In J.-Y. Beziau, D. Krause, and J.B. Arenhart, editors, *Conceptual Clarification: Tributes to Patrick Suppes (1992-2014)*. College Publications, London, 2015.

[17] J. Acacio de Barros and P. Suppes. Probabilistic results for six detectors in a three-particle GHZ experiment. In Jean Bricmont, D. Durr, Maria Carla Galavotti, G. Ghirardi, F. Petruccione, and N. Zanghi, editors, *Chance in Physics*, volume 574 of *Lectures Notes in Physics*, page 213, 2001.

[18] J. Acacio de Barros and P. Suppes. Probabilistic Inequalities and Upper Probabilities in Quantum Mechanical Entanglement. *Manuscrito*, 33(1):55–71, 2010.

[19] A. Einstein, B. Podolsky, and N. Rosen. Can Quantum-Mechanical Description of Physical Reality Be Considered Complete? *Physical Review*, 47(10):777–780, May 1935.

[20] Arthur Fine. Joint distributions, quantum correlations, and commuting observables. *Journal of Mathematical Physics*, 23(7):1306–1310, July 1982.

[21] Susana F. Huelga, Miguel Ferrero, and Emilio Santos. Loophole-free test of the Bell inequality. *Physical Review A*, 51(6):5008–5011, June 1995. Publisher: American Physical Society.

[22] Kumel H. Kagalwala, Giovanni Di Giuseppe, Ayman F. Abouraddy, and Bahaa E. A. Saleh. Bell's measure in classical optical coherence. *Nature Photonics*, 7(1):72–78, January 2013.

[23] Simon Kochen and E. P. Specker. The Problem of Hidden Variables in Quantum Mechanics. *Journal of Mathematics and Mechanics*, 17:59–87, 1967.

[24] K. F. Lee and J. E. Thomas. Experimental Simulation of Two-Particle Quantum Entanglement using Classical Fields. *Physical Review Letters*, 88(9):097902, February 2002. Publisher: American Physical Society.

[25] Marcin Markiewicz, Dagomir Kaszlikowski, P. Kurzynski, and Antoni Wojcik. From contextuality of a single photon to realism of an electromagnetic wave. *npj Quantum Information*, 5(1):5, January 2019.

[26] M. H. M. Passos, W. F. Balthazar, J. Acacio de Barros, C. E. R. Souza, A. Z. Khoury, and J. A. O. Huguenin. Classical analog of quantum contextuality in spin-orbit laser modes. *Physical Review A*, 98(6):062116, December 2018.

[27] M. H. M. Passos, W. F. Balthazar, A. Z. Khoury, M. Hor-Meyll, L. Davidovich, and J. A. O. Huguenin. Experimental investigation of environment-induced entanglement using an all-optical setup. *Physical Review A*, 97(2):022321, February 2018. Publisher: American Physical Society.

[28] Xiao-Feng Qian and J. H. Eberly. Entanglement and classical polarization states. *Optics Letters*, 36(20):4110–4112, October 2011. Publisher: Optical Society of America.

[29] L. J Savage. *The foundations of statistics*. Dover Publications Inc., Mineola, New York, 2nd edition, 1972.

[30] P. Suppes and M. Zanotti. Existence of hidden variables having only upper probabilities. *Foundations of Physics*, 21(12):1479–1499, 1991.

[31] Patrick Suppes, J. Acacio de Barros, and Gary Oas. A Collection of Probabilistic Hidden-Variable Theorems and Counterexamples. In Riccardo Pratesi and L. Ronchi, editors, *Waves, Information, and Foundations of Physics: a tribute to Giuliano Toraldo di Francia on his 80th birthday*, Florence, Italy, October 1996. Italian Physical Society.

[32] Patrick Suppes and Mario Zanotti. When are probabilistic explanations possible? *Synthese*, 48(2):191–199, 1981.

[33] S. M Tan, M. J Holland, and D. F Walls. Bell's inequality for systems with quadrature phase coherence. *Optics Communications*, 77(4):285–291, 1990.

[34] S. M Tan, D. F Walls, and M. J Collett. Nonlocality of a single photon. *Phys. Rev. Lett.*, 66(3):252–255, 1991.

[35] D. F. Walls and G. J. Milburn. *Quantum Optics*. Springer-Verlag, New York, 1994.

Against 'Particles' and 'Collapses' in Quantum Entanglement

Christian de Ronde[1,2,3] and César Massri[4,5]

1. Philosophy Institute Dr. A. Korn, University of Buenos Aires - CONICET
2. Center Leo Apostel for Interdisciplinary Studies
   Foundations of the Exact Sciences - Vrije Universiteit Brussel
3. Institute of Engineering - National University Arturo Jauretche.
4. Institute of Mathematical Investigations Luis A. Santaló, UBA - CONICET
5. University CAECE

Abstract

The basis of what is known today as Standard Quantum Mechanics (SQM) was established by Paul Dirac and John von Neumann during the early 1930s. According to this view, QM talks about a microscopic realm constituted by elementary particles represented by quantum superpositions of many states which "collapse" when observed in the act of measurement. One of the major attacks against SQM was produced during the year 1935 when Albert Einstein and Erwin Schrödinger discussed what they called "entanglement" (*Verschränkung*, in German). However, mainly due to what was considered the triumph of Niels Bohr, their notion together with their critical arguments remained buried for almost half a century. This was until, due to Alain Aspect's experiments, physicists begun to recognize the relevance of quantum entanglement for information processing. During the 1990s, with the rise of *Foundations of QM* and *Quantum Information*, entanglement became to be known as just a new chapter of SQM itself. In this work we attempt to address some of the difficulties present in such contemporary understanding of entanglement specifically related to the existence of *quantum particles* and their *separability*. We will then turn our attention to two recent approaches which might offer a way out of this conundrum by dropping the notion of 'quantum system'. While the first, called device-independent approach, proposes an (anti-metaphysical) operational-linguistic scheme which reminds us of Bohr's interpretation; the second approach which might be linked to the works of Einstein, Heisenberg and Pauli, takes an essentially creative metaphysical path which seeks to develop a new (non-classical) representation of entanglement grounded on the potential coding of *intensive* and *effective* relations.

Keywords: *quantum entanglement, collapse, pure state, particle metaphysics.*

1 The Critical Origin of Quantum Entanglement

Critical thought implies the possibility of analysis of the foundation of thought itself. The analysis of the conditions under which thinking becomes possible. By digging deeply into the basic components of thinking, one is able to understand the preconditions and presuppositions which support the architecture of argumentation itself. In this work we attempt to provide a critical analysis of the definition of quantum entanglement which goes back to the —today famous— EPR *Gedankenexperiment* presented by Einstein, Podolsky and Rosen in 1935 [40]. Many analysis of this thought-experiment have been already provided within the philosophical and foundational literature (e.g., see [2, 3, 52]). In the following, we attempt to extend this analysis paying special attention to the notion of *separability* and to the famous definition of *element of physical reality* in order to approach —later on— the presuppositions involved within the definition of quantum entanglement itself.

Even though Albert Einstein was certainly a revolutionary in many aspects of his research, he was also a classicist when considering the preconditions of physical theories themselves. His dream to create a unified field theory was grounded in his modern belief that physical theories, above all, must always discuss in terms of specific situations happening within space and time. In this respect, the influence of transcendental philosophy in Einstein's thought cannot be underestimated [54]. That space and time are the *forms of intuition* that allow us to discuss about objects of experience was one of the most basic *a priori* dictums of Kantian metaphysics, difficult to escape even for one of the main creators of relativity theory. In a letter to Max Born dated 5 April, 1948, Einstein writes:

> "If one asks what, irrespective of quantum mechanics, is characteristic of the world of ideas of physics, one is first stuck by the following: the concepts of physics relate to a real outside world, that is, ideas are established relating to things such as bodies, fields, etc., which claim a 'real existence' that is independent of the perceiving subject —ideas which, on the other hand, have been brought into as secure a relationship as possible with the sense-data. It is further characteristic of these physical objects that they are thought of as arranged in a space-time continuum. An essential aspect of this arrangement of things in physics is that they lay claim, at a certain time, to an existence independent of one another, provided these objects 'are situated in different parts of space'. Unless one makes this kind of assumption about the independence of the existence

(the 'being-thus') of objects which are far apart from one another in space
—which stems in the first place in everyday thinking— physical thinking
in the familiar sense would not be possible. It is also hard to see any way
of formulating and testing the laws of physics unless one makes a clear
distinction of this kind." [12, p. 170]

This precondition regarding objects situated in different parts of space can
be expressed, following Howard [53, p. 226], as a principle of spatio-temporal
separability:

Separability Principle: *The contents of any two regions of space separated
by a non-vanishing spatio-temporal interval constitute separable physical systems,
in the sense that (1) each possesses its own, distinct physical state, and (2) the
joint state of the two systems is wholly determined by these separated states.*

In other words, the presence of a non-vanishing spatio-temporal interval is a
sufficient condition for the individuation of physical systems and their associated
states. Everything must "live" within space-time; and consequently, the characterization of every system should be discussed in terms of *yes-no questions*
about physical properties. But, contrary to many, Einstein knew very well the
difference between a conceptual presupposition of thought and the conditions
implied by mathematical formalisms. In this respect, he also understood that
his principle of separability was *only for him* a necessary metaphysical condition
for doing physics. More importantly, he was aware of the fact there was no
logical inconsistency in dropping the separability principle in the context of
QM. At the end of the same letter to Born he points out the following:

"There seems to me no doubt that those physicists who regard the descriptive methods of quantum mechanics as definite in principle would
react to this line of thought in the following way: they would drop the
requirement for the independent existence of the physical reality present
in different parts of space; they would be justified in pointing out that
the quantum theory nowhere makes explicit use of this requirement." [12,
p. 172]

This famous passage shows that Einstein was completely aware of the fact
that QM is not necessarily committed to the metaphysical presupposition of
space-time separability.

Also kernel to the EPR argument was the introduction of a "reality criterion"
which would stipulate a sufficient condition for considering an element of physical
reality:[1]

Element of Physical Reality: *If, without in any way disturbing a system,
we can predict with certainty (i.e., with probability equal to unity) the value of a*

---

[1] We are thankful to Prof. Don Howard for pointing us the specificity of the reality criterion.

physical quantity, then there exists an element of reality corresponding to that quantity.

This definition introduced a co-relation between, on the one hand, a *certain prediction*, and on the other, the value of a physical quantity (or property) of a system. Following the orthodox criteria imposed by Paul Dirac, *certainty* was understood as probability equal to unity. Notice that this remark is crucial in order to filter the predictions provided by QM. Only those predictions which could provide a *yes-no answer* to an experimental situation could be considered to be related to physical reality. This means, implicitly, that the rest of the quantum mechanical probabilistic predictions which were not equal to one or zero —namely, those which pertain the open interval $(0,1)$— were not considered as being truly real. Of course, given a quantum state, Ψ, there is only one meaningful operational statement (or property) that can be predicted with certainty. The rest of quantum properties are considered as being *indeterminate*. This has lead some researchers to conclude that for each quantum system there is only one single property which can be regarded as *actual* (or real).[2] The important point is that the "non-certain" predictions —namely, those with probability $p \in (0,1)$— are not directly related to physical reality. Unlike real actual properties, indeterminate properties are considered as being only "possible" or "potential" properties; i.e., properties that might become actual in a future instant of time (see for a detailed analysis [68]). Until these properties are not actualized they remain in a sort of limbo. As Heisenberg [48, p. 42] explains, such properties stand "in the middle between the idea of an event and the actual event, a strange kind of physical reality just in the middle between possibility and reality." The filtering of indeterminate properties —something which, at least from an operational perspective, seems completely unjustified—, is directly related to the actualist spatio-temporal (metaphysical) understanding of physical reality which Einstein so willingly wanted to retain. As he made the point [55]: "that which we conceive as existing ('actual') should somehow be localized in time and space. That is, the real in one part of space, A, should (in theory) somehow 'exist' independently of that which is thought of as real in another part of space, B."[3] But, as noticed by Bohr himself in his famous reply to EPR [9], it is the first part of the definition which introduces a serious "ambiguity". Indeed, the previous specification, *"If, without in any way disturbing a system,"* refers explicitly to the possibility of measuring the system in question. It thus involves an improper scrambling between ontology and epistemology, between physical reality and measurement. A scrambling —let us stress—, completely foreign to all classical physics. This scrambling, might

---

[2]This idea has been strongly endorsed by Dennis Dieks within his neo-Bohrian modal interpretation.

[3]As we discussed in detail in [28], it is not difficult to see that this actualist understanding of existence is grounded in the classical representation of physics provided in terms of an *actual state of affairs* and *binary valuations*.

be regarded as one between the many "quantum omelettes" created during the early debates of the founding fathers [56, p. 381]. However, it might be also interesting to notice that the EPR criteria goes against Einstein's [32, p. 175] own characterization of physical theories: "[...] it is the purpose of theoretical physics to achieve understanding of physical reality which exists independently of the observer, and for which the distinction between 'direct observable' and 'not directly observable' has no ontological significance". This is of course, even though "the only decisive factor for the question whether or not to accept a particular physical theory is its empirical success." According to Einstein the physical representation of a physical theory is always *prior* to the possibility of epistemic inquiry of which 'measurement' is obviously one of its main ingredients. As he would also remark to a young Heisenberg: it is only the theory which decides what can be observed —not the other way around.[4]

The collapse of the quantum wave function was added by Dirac to his axiomatic formulation of the theory in order to evade the theoretical reference to quantum superpositions and account for the observation of single measurement outcomes (see for a detailed analysis: [24]). Since, in a positivist fashion, it was assumed that physical theories should be able to describe observations, the famous *projection postulate* became a necessary ingredient of a theory that would secure the observation of single outcomes. This idea was presented by Paul Dirac and John von Neumann in their famous books [36, 69] at the beginning of the 1930s. Dirac [36, p. 36] was the first to explicitly introduce this new "quantum jump" — already present in Born's probabilistic rule— between superpositions and single measurement outcomes: "When we measure a real dynamical variable ξ, the disturbance involved in the act of measurement causes a jump in the state of the dynamical system. From physical continuity, if we make a second measurement of the same dynamical variable ξ immediately after the first, the result of the second measurement must be the same as that of the first." In turn, von Neumann [69, p. 214] also addressed the need of such measurement collapses and famously argued that: "[I]f the system is initially found in a state in which the values of $\mathcal{R}$ cannot be predicted with certainty, then this state is transformed by a measurement M of $\mathcal{R}$ into another state: namely, into one in which the value of $\mathcal{R}$ is uniquely determined. Moreover, the new state, in which M places the system, depends not only on the arrangement of M, but also on the result of M (which could not be predicted causally in the original state) —because the value of $\mathcal{R}$ in the new state must actually be equal to this M-result." As we mentioned above, the only ones who seemed worried about the introduction of these "invisible collapses" were Einstein and Schrödinger. As a matter of fact, their analysis attempted to show the inconsistencies of assuming such subjectively induced process. In particular, the EPR *Gedeankenexperiment*

---

[4]This might be one of the main reasons why Einstein did not like the reality criteria presented in the EPR paper. For a detailed analysis of the disagreements of Einstein with the EPR paper see: [52].

makes explicit use of the collapse in order to show the strange non-local influence that appears when measuring one of the particles on the other distant (entangled) partner. Indeed, if one accepts the orthodox interpretation of QM according to which the measurement of a quantum superposition induces a "collapse" to only one of its terms, Einstein, Podolsky and Rosen then show that there seems to exist a super-luminous collapse activity between the particles. Once the entangled particles are separated, all their properties remain *indeterminate*. But, the moment we perform a measurement of an observable in one of the particles we also learn instantaneously what is the value of the distant partner —of course, in case we would choose to measure the same observable. Every time we measure an observable in one of the particles, the other particle —as predicted by QM— will be found to possess a strictly correlated value.[5] Thus, the (real) "collapse" of one of the particles also produces a (real) "collapse" in her entangled companion. Einstein was of course clearly mortified by this seemingly non-local "quantum effect" which he ironically called *spukhafte Fernwirkung*, translated later as "spooky action at a distance". That same year, continuing the EPR's critical reflections, Erwin Schrödinger introduced in an explicit manner the notion of notion of *entanglement*.

> "When two systems, of which we know the states by their respective representatives, enter into temporary physical interaction due to known forces between them, and when after a time of mutual influence the systems separate again, then they can no longer be described in the same way as before, viz. by endowing each of them with a representative of its own. I would not call that one but rather the characteristic trait of quantum mechanics, the one that enforces its entire departure from classical lines of thought. By the interaction the two representatives (or ψ-functions) have become entangled." [65, p. 555]

Making explicit reference to the EPR paper, Schrödinger remarked that:

> "Attention has recently [40] been called to the obvious but very disconcerting fact that even though we restrict the disentangling measurements to one system, the representative obtained for the other system is by no means independent of the particular choice of observations which we select for that purpose and which by the way are entirely arbitrary. It is rather discomforting that the theory should allow a system to be steered or piloted into one or the other type of state at the experimenter's mercy in spite of his having no access to it." [65, pp. 555-556]

Following the reality criteria proposed in the EPR paper, Schrödinger also assumed —implicitly— that *maximal knowledge* had to be understood as *certain*

---

[5]Let us remark that observability is used in this case a sufficient condition to define reality itself. There is involved here a two sided definition of what accounts for physical reality, either in terms of computing the certainty of an outcome (= real) or by observing an outcome which was uncertain but became actual (= real).

knowledge; i.e., as knowledge involving probability equal to unity. As he critically remarks, the astonishing aspect of QM is that when two systems get entangled through a known interaction, the knowledge we have of the parts might anyhow decrease.

> "If two separated bodies, each by itself known maximally, enter a situation in which they influence each other, and separate again, then there occurs regularly that which I have just called *entanglement* of our knowledge of the two bodies. The combined expectation-catalog consists initially of a logical sum of the individual catalogs; during the process it develops causally in accord with known law (there is no question of measurement here). The knowledge remains maximal, but at the end, if the two bodies have again separated, it is not again split into a logical sum of knowledges about the individual bodies. What still remains *of that* may have become less than maximal, even very strongly so.—One notes the great difference over against the classical model theory, where of course from known initial states and with known interaction the individual states would be exactly known." [64, p. 161]

It is the projection postulate interpreted as a "real collapse" of the quantum wave function which ends up scrambling —just like in the case of the measurement process— the objective theoretical representation provided by the mathematical formalism and the subjective observation of a particular 'click' in the lab. The entanglement of systems and outcomes within the same representation determines then the scrambling of the objective knowledge related to the representation provided by the theory, and the subjective knowledge related to the purely epistemic process of measurement. Thus, Einstein's criticism against the "spooky action", was in fact a criticism against the addition of a subjectively produced collapse. As he would write to Schrödinger in a letter dated December 22, 1950:

> "You are the only contemporary physicist, besides Laue, who sees that one cannot get around the assumption of reality, if only one is honest. Most of them simply do not see what sort of risky game they are playing with reality —reality as something independent of what is experimentally established. They somehow believe that the quantum theory provides a description of reality, and even a complete description; this interpretation is, however, refuted most elegantly by your system of radioactive atom + Geiger counter + amplifier + charge of gun powder + cat in a box, in which the Ψ-function of the system contains both the cat alive and blown to bits. Nobody really doubts that the presence or absence of the cat is something independent of the act of observation." [41, p. 39]

In this same respect, Einstein is quoted by Everett [60, p. 7] to have said that he "could not believe that a mouse could bring about drastic changes in the universe simply by looking at it". Schrödinger would also make fun of the existence of such induced collapses:

> "But jokes apart, I shall not waste the time by tritely ridiculing the attitude that the state-vector (or wave function) undergoes an abrupt change, when 'I' choose to inspect a registering tape. (Another person does not inspect it, hence for him no change occurs.) The orthodox school wards off such insulting smiles by calling us to order: would we at last take notice of the fact that according to them the wave function does not indicate the state of the physical object but its relation to the subject; this relation depends on the knowledge the subject has acquired, which may differ for different subjects, and so must the wave function." [60, p. 9]

2 The Contemporary Definition of Quantum Entanglement

Mainly due to the deep anti-metaphysical and anti-realist influence of the Bohrian-positivist approach to physics, the questioning introduced by Einstein and Schrödinger regarding the meaning and representation of physical reality in QM remained silenced for almost half a century. As remarked by Jeffrey Bub [14], "[...] it was not until the 1980s that physicists, computer scientists, and cryptographers began to regard the non-local correlations of entangled quantum states as a new kind of non-classical resource that could be exploited, rather than an embarrassment to be explained away." The reason behind this change in attitude is interesting: "Most physicists attributed the puzzling features of entangled quantum states to Einstein's inappropriate 'detached observer' view of physical theory, and regarded Bohr's reply to the EPR argument (Bohr, 1935) as vindicating the Copenhagen interpretation." Indeed, the notion of entanglement was erased from debates by a community of physicists who under the Bohrian spell were turning into the instrumentalist side. In 1964, John Stewart Bell, an Irish physicist working at the Conseil Européen pour la Recherche Nucléaire (CERN), would rediscover the forbidden text. In his spear time, during the weekends, Bell would take the time to write a paper titled: *On the Einstein-Podolsky-Rosen Paradox* [6] were he was able to derive from the rules of classical probability theory the same set of inequalities that Boole had found almost exactly one century before in 1882.[6] Boole-Bell's statistical inequality gave an upper bound to the classical correlations derived from any classical local-binary realistic theory. This work, which had opened

---

[6]The relation between the famous Bell's inequalities from 1964 and Boole's less known inequalities from 1862, derived almost exactly one century before, has been explicitly considered by Itamar Pitowsky in [62].

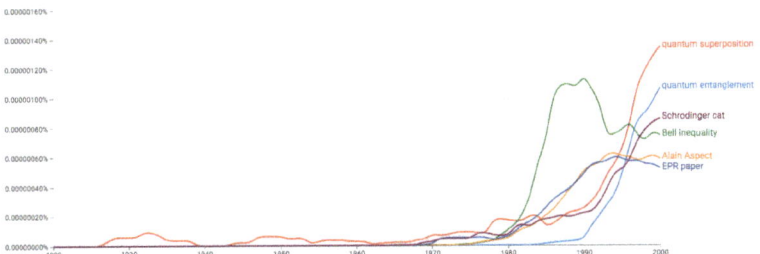

Figure 1: The image from Google Ngram Viewer shows the relevance of the terms 'quantum entanglement', 'quantum superposition', 'Bell inequality', 'Alain Aspect', 'EPR paper', 'Schrödinger cat' present in books between the years 1920 and 2000.

the possibility of testing EPR's reasoning, would also remain in the dark for about two more decades. It was only at the beginning of the 1980s that —as described by Bub— Alain Aspect, Philippe Grangier and Gerard Roger [5] would finally test in Orsay the correlations found in an EPR situation (with pairs of entangled spin "particles") and exhibit a violation of the Boole-Bell inequality.[7] The violation showed —against Bell's physical intuitions— that a classical local-binary valued[8] theory would not be able to describe the (non-classical) correlations of Aspect's experiment, showing that quantum entanglement had to be taken very seriously. It took still one more decade for physicists to realize that *entanglement* could be used as a resource for information processing. Finally, during the 1990s entanglement would rapidly begin to populate the journals, labs and research projects of all Universities and research centers around the world. The technological era of quantum information processing had woken up from its almost half century hibernation. An hibernation —let us not forget— caused by the uncritical attitude of a physics community captivated by Bohr's preaching about complementarity, correspondence, waves and particles.

With the advent of the new millennia the era of quantum information processing became one of the kernel lines of research and technology worldwide. It became then necessary for physicists to reach a consensus regarding the definition and meaning of quantum entanglement. This implied a very deep problem which most physicists were unwilling to recognize. Schrödinger's definition, given on the basis of *separability* and the *purity* of states, was essentially designed to go against SQM —in particular, against the reference to "elementary particles" and the introduction of "collapses" [66]. But regardless of this historical fact, physicists —still uncomfortable with what they considered

---

[7] This test has been repeated countless times up to the present [7, 51].

[8] Even though the original term is "realistic", we prefer to add "binary valued" for reasons that will become evident in the forgoing part of the paper.

to be "philosophical debates"— simply added the definition as just a new chapter of SQM.[9] Ever since, entanglement has been uncritically understood as correlating the distant collapses of particles. A good example is the explanation of entanglement provided by Mintert et al.:

> "Composite quantum systems are systems that naturally decompose into two or more subsystems, where each subsystem itself is a proper quantum system. Referring to a decomposition as 'natural' implies that it is given in an obvious fashion due to the physical situation. Most frequently, the individual susbsystems are characterized by their mutual distance that is larger than the size of a subsystem. A typical example is a string of ions, where each ion is a subsystem, and the entire string is the composite system. Formally, the Hilbert space $\mathcal{H}$ associated with a composite, or multipartite system, is given by the tensor product $\mathcal{H}_1 \otimes ... \otimes \mathcal{H}_N$ of the spaces corresponding to each of the subsystems." [59, p. 61]

This introduction to QM is an explicit exposure of standard ideas. First, it implies that Hilbert spaces can adequately represent 'physical systems'; i.e., small elementary particles such as ions. And secondly, it also implies that such particles inhabit space-time, that one can make reference to distances and that the subspaces —which are considered as 'parts' of the original Hilbert space— describe 'subsystems'. The problem is that none of these physical concepts can be adequately represented within this mathematical framework.

Assuming right from the start the metaphysics of particles as a "common sense" *given* of physical representation, the story of entanglement is told in the following manner. In general, the Hilbert space associated with a composite system is given by the tensor product $\mathcal{H}_1 \otimes ... \otimes \mathcal{H}_n$ of the spaces corresponding to each of the subsystems. As explained by Dennis Dieks and Andrea Lubberdink [35, p. 1052]: "The standard quantum mechanical treatment of particles starts simply enough, with the uncontroversial case of one particle described by a state in a single Hilbert space. In the case of two or more particles the tensor product of such individual Hilbert spaces is formed, $\mathcal{H}_1 \otimes \mathcal{H}_2 \otimes \mathcal{H}_3 \otimes ...$. The natural interpretation, especially with the classical case in mind, is that in such formulas $\mathcal{H}_i$ is the state space of particle i." From this standpoint, the idea is that we should focus on a finite dimensional bipartite quantum system described by the Hilbert space $\mathcal{H} = \mathcal{H}_1 \otimes \mathcal{H}_2$. In order to define *separability*, another essential element enters the scene, namely, the notion of *pure state* which rests in the following operational definition: 'If a quantum system is prepared in such way that one can devise a maximal test yielding with *certainty* (i.e., probability

---

[9]One might seriously wonder about the general comprehension of the community of physicists and philosophers regarding the critical analysis of entanglement as well as the argument presented in the EPR paper. As shown by Stancy Blake in [8]: "Notwithstanding its great influence in modern physics, the EPR thought-experiment has been explained incorrectly a surprising number of times."

equal to unity) a particular outcome, then it is said that the quantum system is in a *pure state*'. A pure state of a quantum system is described by a unit vector in a Hilbert space which in Dirac's notation is denoted by $|\psi\rangle$.[10] Assume now that each subsystem is prepared in the following pure states $|\psi\rangle$ and $|\psi'\rangle$. The state of the composite system is then $|\psi\rangle \otimes |\psi'\rangle$. Suppose that one had access to only one of the subsystems at a time. Then, after a measurement of any local observable $A \otimes \mathbb{I}$ on the first subsystem, (where A is a hermitian operator acting on $\mathcal{H}_1$, and $\mathbb{I}$ is the identity acting on $\mathcal{H}_2$), the state of the first subsystem will be projected onto an eigenstate of A, but the state of the second subsystem will remain unchanged. If later on, one performs a second local measurement, now on the second subsystem, it will yield a result that is completely independent of the result of the first measurement pertaining to the first subsystem. Hence, the measurement outcomes on the two subsystems are uncorrelated between each other and only depend on their own subsystem states. Furthermore, unlike with superposed states, the knowledge of the subsystems grant the complete knowledge of the whole they compose. However, in general, depending on the basis, a pure state in $\mathcal{H}$ is given by a superposition of pure states, $|\varphi\rangle = \sum a_i |\psi\rangle_i \otimes |\psi'\rangle_i$. For a local operator on the first subsystem, the expected value is

$$\text{Tr}(A \otimes \mathbb{I} |\varphi\rangle\langle\varphi|) = \text{Tr}_1(A\rho_1), \quad \rho_1 := \text{Tr}_2(|\varphi\rangle\langle\varphi|),$$

where Tr_1 and Tr_2 are the partial traces over the first and second subsystem and ρ_1 is the reduced density matrix of the first subsystem. Then, one can conclude that the state of the first subsystem is given by ρ_1 and the state of the second subsystem by ρ_2 (where $\rho_2 := \text{Tr}_1(|\varphi\rangle\langle\varphi|)$). However, the state of the composite system is different from $\rho_1 \otimes \rho_2$. Moreover, if one performs a local measurement on one subsystem, this leads to a state reduction of the entire system state, not only of the subsystem on which the measurement had been performed. Therefore, according to orthodoxy, the probabilities for an outcome of a measurement on one subsystem are influenced by the measurements on the other distant subsystem.

Definition 2.1 *States that can be written as a product of pure states are called* product *or* separable states. *In contradistinction, the states which are not separable are then defined as* entangled states.

As it is explicit from this definition the notions of *purity* and *separability* play an essential role within the orthodox understanding of quantum entanglement. Both notions are intrinsically related to features present within the classical representation of systems. While purity is related to certain knowledge and

---

[10] As discussed in [19, 30] this definition is ambiguous due to the non-explicit reference to the basis in which the vector is written. It is this ambiguity which, in turn, mixes the notion of 'state of a system' and 'property of a system'.

actual properties, separability is linked to the independent characterization of systems and their classical composition.

3 Is Particle Metaphysics Adequate for Quantum Theory?

As pointed out by Werner Heisenberg [16, p. 218]: "The strongest influence on the physics and chemistry of the past [19th] century undoubtedly came from the atomism of Democritos. This view allows an intuitive description of chemical processes on a small scale. Atoms can be compared with the mass points of Newtonian mechanics, and from this a satisfactory statistical theory of heat was developed. [...] the electron, the proton, and possibly the neutron could, it seemed, be considered as the genuine atoms, the indivisible building blocks, of matter." The funny thing is that even though QM was born from the most radical and explicit departure from the modern space-time representation of nature, physicists and philosophers would very soon return to metaphysical atomism and claim that QM made —obviously— reference to a microscopic realm constituted by elementary particles. In his famous paper: *What is an elementary particle?* Schrödinger criticized in depth this situation: "Atomism in its latest form is called quantum mechanics. [...] In the present form of the theory the 'atoms' are electrons, protons, photons, mesons, etc. The generic name is elementary particle, or merely particle. The term 'atom' has very wisely been retained for chemical atoms, though it has become a misnomer." And yet, more that half a century from Schrödinger's critical reflections, the reference of the theory of quanta to atoms has remained untouched. Advocated as dogma, still today most physicists and philosophers assume uncritically metaphysical atomism as an unescapable language for representing QM. But apart from its being "weird" and "quantum", no one seems to understand what a "quantum particle" really is. When asked about such microscopic quantum realm, both physicists and philosophers escape any meaningful reply by confirming to their audience that this was "just a way of talking". Of course, even if they don't recognize it, atomism is much more than that. It implies a way of thinking. As a matter of fact, its application in the context of QM has imposed the introduction of the notions of pure state and separability. Even though all the scientific research in the last century points explicitly to the fact that we do not understand what QM is talking about, orthodox textbooks used in Universities all around the world keep teaching students that the theory of quanta talks about "quantum particles" which can be characterized in terms of actual properties and their separability.

In SQM, pure states guarantee the existence of an observable which can be predicted with *certainty* (probability equal to 1) in a particular basis. In turn, this observable can be interpreted as an *actual property*.[11] At the opposite corner, superposed states of more than one term do not describe observables which, when measured, will be obtained with certainty. The properties constituting superposed states[12] of more than one term are referred to in the literature as being *indeterminate* or *potential* properties. Indeterminate properties might, or might not become actualized in a future instant of time and thus, cannot be considered as elements of physical reality (in the EPR sense). It is at this point that, since Dirac, the empiricist-positivist understanding of physical theories as a formal schemes capable of describing observations has imposed the need of a 'projection postulate' that would allow to transform superpositions into single outcomes —which is, as argued by orthodoxy, what we actually observe in the lab. Just like in the case of particles, this projection is sometimes interpreted as a real "collapse" process and sometimes —also— as "just a way of talking". However, as remarked by Dennis Dieks [33, p. 120]: "Collapses constitute a process of evolution that conflicts with the evolution governed by the Schrödinger equation. And this raises the question of exactly when during the measurement process such a collapse could take place or, in other words, of when the Schrödinger equation is suspended. This question has become very urgent in the last couple of decades, during which sophisticated experiments have clearly demonstrated that in interaction processes on the sub-microscopic, microscopic and mesoscopic scales collapses are never encountered." In the last decades, the experimental research seems to confirm there is nothing like a "real collapse process" suddenly happening when measurement takes place [47]. Unfortunately, as Dieks [34] also acknowledges: "The evidence against collapses has not yet affected the textbook tradition, which has not questioned the status of collapses as a mechanism of evolution alongside unitary Schrödinger dynamics."

It is at this point that it might be wise to recall that physics is not a program which attempts to justify our (metaphysical) prejudices about reality. In fact, it is exactly the opposite. Physics begins with the humble acceptance of the unknown and it is only from this standpoint that theories attempt to provide understanding of experience through the creation of unified consistent and coherent representations. Within this process, critical thought is certainly essential. The addition of fictitious inadequate concepts —such as quantum

---

[11]Only pure states allow to interpret observables as *elements of physical reality* (in the EPR sense). The acceptance of EPR's reality criteria goes in line with the operational quantum logic approach proposed by Constantin Piron [61] and subsequently developed by Aerts himself [1]. As Aerts and Massimiliano Sassoli de Bianchi [4, p. 20] —both students of Piron— explain with great clarity: "the notion of 'element of reality' is exactly what was meant by Einstein, Podolsky and Rosen, in their famous 1935 article. An element of reality is a state of prediction: a property of an entity that we know is actual, in the sense that, should we decide to observe it (i.e., to test its actuality), the outcome of the observation would be certainly successful."

[12]See [24] for an explicit definition of quantum superposition.

particles, collapses, pure sates, separability, etc.— in order to address the reference of the theory is not the way out of the labyrinth, it is the entrance. It is the scrambling of particle metaphysics and invisible collapses with the quantum formalism which is responsible for creating a numerous set of (pseudo)problems that are —still today— being discussed in the specialized literature. The fact that the notion of 'system' is inadequate to explain the formalism of QM — something which is exposed in an extreme manner by the superposition principle and Kochen-Specker theorem (see [24, 27])— or the fact that "collapses" have no empirical support nor play any essential role within the operational application of the theory, has not stoped contemporary physicists from repeating their mantra: "QM talks about elementary particles which collapse each time we observe a 'click' in a detector."

In the following sections we will discuss two recent approaches to QM which attempt to understand the orthodox formalism escaping —right from the start— the reference to "tiny (separable) particles". While the first, taking as a standpoint a neo-Bohrian linguistic understanding of physics goes directly against the addition of a metaphysical picture; the second approach, following the footsteps left by Einstein, Heisenberg and Pauli, presents a new (non-classical) conceptual representation which attempts to provide an objective-invariant formal-conceptual representation of the theory of quanta.

4 The Device-Independent Approach: Beyond Particle Metaphysics?

Operational axiomatic approaches to QM have a long history going back to Dirac's and von Neumann's formulation of the theory during the 1930s. Later on, in the 1960s and 70s the Geneva School commanded by Josef-Maria Jauch and Constantin Piron kept developing these ideas further in the context of quantum logic. As explained by Sonja Smets:

> "In the language provided by the Geneva approach, a physical property is called *actual* if and only if the DEP's [Definite Experimental Project] which test it are certain and is *potential* otherwise. When a property is actual or not, depends on the state in which one considers the system to be. The Geneva approach adopts here a realistic stance towards physical properties. The underlying assumption is that the physical properties have an extension in reality, can be described and characterized by physicists and are considered to be measurable. In particular the EPR-'criterion of reality' (see [40]) is explicitly adopted and explains why measurability is an important ingredient. Indeed, an 'actual property' is closely linked to the notion of 'element of reality' introduced in [40]." [67, p. 47]

But while the Geneva approach adopted a realist viewpoint with respect to 'properties' and 'systems' (or 'entities', as they prefer to call them), the more recent device-independent approach to QM —even though continues the operational trend of thought— remains at safe distance from realist claims. As remarked by Alexei Grinbaum [46]: "Operational axiomatic approaches to quantum mechanics focus on the inputs and outputs of the observer: a 'box' picture. The postulates that successfully constrain the box to behave according to the rules of quantum theory become our best candidates for fundamental principles of Nature. In a device-independent approach, such postulates are also at work: they are the only content of physical theory along with the inputs and the outputs of the parties." From this standpoint, the device-independent approach is then ready to drop the notion of 'system'. The reason is simple: "If a theory contains no notion of system, there is no reason to picture reality as comprised of physical entities." Grinbaum continues to explain: "Systems in the device-independent approach are unnecessary not only for the purposes of interpretation, but also on the theoretical side. They cannot correspond to objective reality because they are absent from the theory. Both in the philosophy of physics and in its mathematics systems are no more a requirement." Of course, the idea that the formalism of QM should be regarded beyond a direct reference to an intuitive conceptual representation is not new. This idea goes obviously back to Bohr himself who denied explicitly the possibility of representing QM beyond classical concepts. Bohr stressed the reference to (classical) language, and in particular, he argued that "[...] the unambiguous interpretation of any measurement must be essentially framed in terms of classical physical theories, and we may say that in this sense the language of Newton and Maxwell will remain the language of physicists for all time." Furthermore, he claimed [10] that: "Physics is to be regarded not so much as the study of something a priori given, but rather as the development of methods of ordering and surveying human experience. In this respect our task must be to account for such experience in a manner independent of individual subjective judgement and therefor objective in the sense that it can be unambiguously communicated in ordinary human language." And it is exactly the role of played by language which Grinbaum takes into consideration by following Wheeler, one of Bohr's famous advocates:

> "Wheeler's methodology provides guidance: at the beginning, 'take a formal system.' In a rebuttal of his position, one could argue that a semantic view of device-independent models is necessary if the theory is to be physical rather than purely mathematical: physical, hence somehow relating to nature. A minimal requirement would be to link with empirical data. Wheeler's 'not about anything' can be circumvented by keeping the preposition 'about' while clearing oneself of 'any thing': in the absence of semantics, the only available interpretation is 'about languages.' If strings are not 'about' some elements of reality, they can be said to be 'about' languages from which they are formed. These languages are the only

remaining theoretical constituents, making the 'about-languages' reading a minimal interpretation one level below 'about-systems.' The emphasis on languages rather than physical entities is not new in the philosophy of physics." [46]

Indeed, this primacy of language can be easily linked not only to Wheeler but also to his mentor, Niels Bohr (see [21]). Grinbaum advances in the same direction: "the propositions are themselves elements of reality and [...] they do not need to refer to any entities whatsoever, whether empirical or theoretical. Device-independent models proceed on a similar view replacing Wheeler's 'undecidable propositions' by an ensemble of operationally defined inputs and outputs." It is from this standpoint that the device-independent approach engages in the problem of the *reference*. Grinbaum attempt to answer the following question: If the theory is not a theory about physical systems, what would it be a theory of?

> "One finds a tentative answer in a definition using only the strictly necessary concepts: *For Alice (respectively for Bob), an experiment is a process or black box to which she feeds an input x from the alphabet X and from which she receives an output a from the alphabet A. Alphabets X; Y; A; B are of finite cardinality.* [60] On this view, physical theory is about languages: it is defined by a choice of alphabets for the inputs and the outputs and by the conditions imposed on this algebraic structure. Strings, or words in such alphabets, form a common mathematical background of device-independent approaches." [46]

It might be noticed that there exist interesting connections between the device-independent approach presented by Grinbaum, Bohr's interpretation of QM, and the information theoretic proposals by Jeffrey Bub and Chris Fuchs (see e.g., [13, 15, 43, 45]). The analysis of such relations exceed the scope of the present paper.

The device independent approach is critical about the notion of 'system' when applied to QM. But even though Grinbaum shows that 'systems' are nowhere to be found within the theory of quanta, he still takes for granted the quantum informational contemporary definition of entanglement which —as we have stressed— is directly linked to the separability of (quantum) systems. Thus, there seems to exist a problem, for while the notion of 'system' is dropped the orthodox notion of entanglement is retained without any further reconsideration. As we have pointed out above, even though entanglement was defined by both Einstein and Schrödinger within a critical reconsideration of Standard QM, its use and application within the field of quantum information has not been revisited. This fact might be related to the widespread contemporary instrumentalist understanding of physics as a discipline which only needs to predict measurement outcomes. In fact, many physicists and —even— some philosophers tend to believe that the reference to quantum particles is "just a way of talking". However, from the present debate we might argue that the reference to particles, collapses and separability is much more than that, it implies a way of thinking. It is at this point that we must address the role played by metaphysics within physical theories. A role which might be also regarded as much more essential than "just a way of talking".

5 The Role of Metaphysics Within Physics

According to the standard view theories are mainly understood as sets of mathematical models developed in order to account for empirical observations. As explained by Miklos Redei [63]: "physics carries out precision measurements aiming at determining values of operationally defined physical quantities [and] sets up mathematical models of physical phenomena that make explicit the functional relationships among the measured quantities." In particular, Redei [63] has characterized the role played by mathematics within physics in terms of what he calls the 'supermarket picture': "mathematics is like a supermarket and physics is a customer. That is to say, it is tacitly assumed that when a physicist needs a mathematical concept, a mathematical structure, or any mathematical tool to formulate the mathematical model of a physical phenomenon, then (s)he just goes to the mathematics-supermarket, looks at the shelves [and] takes off the product needed." The 'supermarket picture' discussed by Redei is also perfect for characterizing the standard *praxis* of philosophers of physics when dealing with the 'interpretation' of theories. Indeed, when a philosopher needs a (metaphysical) concept in order to build up an 'interpretation', then she just goes to the 'concept-supermarket', looks in the shelves and takes the notions (s)he likes. But there is an essential distinction, while the 'mathematics supermarket' is essential for the survival of theories, the 'concept-supermarket' is not. According to the standard view, the latter is only used by fancy customers —such as realist philosophers— who are interested in creating a metaphysical 'picture' of what the world is like would the theory be true [70]. Of course, more down to earth customers who realize that they already possess an empirically adequate theory and actually do care about their money[13] do not need to venture into the suburbs in order to find these expensive metaphysical chains full of ornamental ideas.

This standard scheme is intrinsically linked to the 20th Century Bohrian-positivist re-foundation of physics which diminished the role of metaphysics within physics to that of a narrative or interpreatation. However, there is also a completely different understanding of the role that metaphysics should play within physical theories which goes back to the ideas shared by many of the

---

[13] As argued by Fuchs in [42]: "The issue remains, when will we ever stop burdening the taxpayer with conferences devoted to the quantum foundations?"

founding fathers of QM —certainly by Einstein, Heisenberg and Pauli. From this perspective metaphysics should be essentially regarded as the systematic creation of an inter-related net of concepts. Concepts do not make reference to 'things'. Concepts are always related to other concepts weaving a net which allows us to conceive and capture a specific type of experience. There is no way to discuss about experience without presupposing a conceptual representation for it is only the theory which decides what can be observed. A very good example of how a metaphysical system works is provided by Aristotle's hylomorphic scheme in which the notion of *entity* was characterized in terms of two modes of existence: the potential and the actual. In particular, the actual mode of existence —which is the only one that survived after Newton's actualist metaphysical choices [23]— characterized entities in terms of three logical and ontological principles: existence, non-contradiction and identity (see for a detailed analysis [25]). These principles, which —as David Hume himself would remark about causality— are not to be found in the empirical world, allow us to make sense of phenomena. They allow us to predicate existence of 'something', to claim that the existent has non-contradictory properties, and furthermore, that this non-contradictory existent remains identical to itself thorough time. Acting together —and only together— these principles allow us to think about an object of experience in a systematic relational manner. As remarked by Einstein:

> "From Hume Kant had learned that there are concepts (as, for example, that of causal connection), which play a dominating role in our thinking, and which, nevertheless, can not be deduced by means of a logical process from the empirically given (a fact which several empiricists recognize, it is true, but seem always again to forget). What justifies the use of such concepts? Suppose he had replied in this sense: Thinking is necessary in order to understand the empirically given, *and concepts and 'categories' are necessary as indispensable elements of thinking.*" [39, p. 678] (emphasis in the original)

Any scientific discourse must always presuppose a conceptual representation of what is meant by a 'state of affairs'. This is not —at least for the metaphysician— something "self evident" nor part of the "common sense" of the layman but the very precondition for understanding phenomena in a scientific manner. It is the recognition of the essential role played by metaphysical representation which allows science to be critical about its own foundation. For willingly or not, we physicists, are always producing our praxis *within* a specific representation. Representation is always first, experience and perception are necessarily second. Paraphrasing Wittgenstein's famous remark regarding language, the physical representation we inhabit presents the limits of the physical world we understand.[14] This marks a point of departure with respect to naive empiricism and

---

[14]Let us notice, firstly, that "physical" should not be understood as a *given* "material reality",

positivism. A point that was also stressed by Einstein:

"I dislike the basic positivistic attitude, which from my point of view is untenable, and which seems to me to come to the same thing as Berkeley's principle, *esse est percipi*. 'Being' is always something which is mentally constructed by us, that is, something which we freely posit (in the logical sense). The justification of such constructs does not lie in their derivation from what is given by the senses. Such a type of derivation (in the sense of logical deducibility) is nowhere to be had, not even in the domain of pre-scientific thinking. The justification of the constructs, which represent 'reality' for us, lies alone in their quality of making intelligible what is sensorily given." [39, p. 669]

It is only adequate conceptual nets which allow us to capture a specific field of experience for as remarked by Heisenberg [50, p. 264]: "The history of physics is not only a sequence of experimental discoveries and observations, followed by their mathematical description; *it is also a history of concepts*. For an understanding of the phenomena the first condition is the introduction of adequate concepts. *Only with the help of correct concepts can we really know what has been observed.*" But the creation of new physical concepts is not an easy task, it is a difficult process which always requires breaking the chains of old representations that might have become —as Hume remarked— through habit part of our "common sense". In this respect it is important to recall the warning of Einstein:

"Concepts that have proven useful in ordering things easily achieve such an authority over us that we forget their earthly origins and accept them as unalterable givens. Thus they come to be stamped as 'necessities of thought,' 'a priori givens,' etc. The path of scientific advance is often made impossible for a long time through such errors. For that reason, it is by no means an idle game if we become practiced in analyzing the long common place concepts and exhibiting those circumstances upon which their justification and usefulness depend, how they have grown up, individually, out of the givens of experience. By this means, their all-too-great authority will be broken. They will be removed if they cannot be properly legitimated, corrected if their correlation with given things be far too superfluous, replaced by others if a new system can be established that we prefer for whatever reason." [38, p. 102]

---

but rather as a procedure for representing reality in theoretical —both formal and conceptual— terms. And secondly, that the relation between such physical representation and reality is not something "self evident". The naive realist account according to which representation "discovers" an already "fixed" reality is not the only possibility that can be considered. A one-to-one correspondence relation between theory and reality is a very naive solution to the deep problem of relating theory and *physis*.

Of course, we do not believe that the structural relationship between the formal and the conceptual levels of a theory are developed in a linear or straightforward manner. On the contrary, this is in general an entangled interrelated process that goes back and forth between the creation of new mathematics, conceptual schemes and technical achievements. One level helps the other in an interrelated process of development. Unfortunately, the orthodox attempt rather than discussing this relation and developing new adequate concepts has been focused —following Bohr's correspondence principle— in "bridging the gap" between the mathematical formalism and our manifest (classical) image of the world [37]. Instead, taking the mathematical formalism as a standpoint and advancing beyond Bohr's prohibitions, one can also attempt to develop a new adequate non-classical conceptual framework. This requires a careful analysis of the conditions implied by the mathematical formalism of the theory. Following such type of analysis, as we argued in [25], the notion of *entity* ('system' or 'object') even though is essential for classical mechanics, becomes in the context of quantum theory an *epistemological obstruction*.[15]

If we accept that there must exist an adequate structural relationship between mathematical formalisms and conceptual frameworks it is not difficult to see that the quantum formalism cannot represent 'separable systems'. While the equation of motion in classical mechanics can be expressed in $\mathcal{R}^3$ allowing an interpretation in 3-dimensional Euclidean space, QM works in a configuration space. The difference is essential when attempting to consider existents within space. Classically, if we add two systems, the properties are summed. Given two systems with a number of properties, R and R', respectively; their joint consideration is just the sum of the properties of each system, namely, $R + R$. A paradigmatic example is the completely inelastic crash of two systems. While before the crash the two particles are separated and their mass are m and m', and their velocities are v and v', respectively; after the crash they become a (non-separable) single system of mass $m+m'$ with a common velocity v_f. The essential property characterizing the two systems —namely, their mass— becomes nothing else than the sum of masses. In this case, there is nothing created after the crash that was not already there before the crash. On the contrary, as it is we known, in QM there is an essential difference when considering the addition of (quantum) systems mathematically represented by vector spaces, $\mathcal{H} = \mathcal{H}_1 \otimes \mathcal{H}_2$. Let us consider two rays which intersect each other. In the case of classical set theory, their addition or union would be just the sum of the two rays. But when considered as vector spaces the addition of the two rays (now considered as subspaces) will be more than just their sum: the two rays $\mathcal{H}_1$ and $\mathcal{H}_2$ *generate* the whole plane $\mathcal{H}$ (see also the analysis provided by Rob Griffiths in [44, Sect. 2]). An excellent example of this problematic situation within

---

[15]The fact that the notion of *entity* is not adequate in order to interpret the quantum formalism is explicit from the Kochen-Specker theorem [23, 27, 28], the existence of quantum superpositions [24] and the non-separability theorem [1, 3].

orthodox QM is discussed by Rob Clifton in [17, 18] where he analyzes from a classical perspective the inconsistency present in QM when attempting to provide a common valuation to the properties pertaining to a 'system' and its 'subsystems'.[16] In QM subspaces do not relate in the way that systems do. So why should we even talk about "systems" in QM? While the logic of classical mechanics follows that of Boolean sets, the logic of quantum mechanics is non-distributive and the *joint* is clearly non-classical. In the context of quantum logic, Diederik Aerts has even derived a non-separability theorem which shows that quantum systems are essentially non-separable [1, 3]. All these different results might be regarded as —what Wolfgang Pauli would call— "road signs" that all point in the same direction, namely, the necessity to go beyond the classical notion of separability imposed by particle metaphysics. Extending Grinbaum's remark: If a theory contains no notion of separability, there is no reason to picture reality as comprised of separable entities.

6 Intensive Relational Metaphysics and Entanglement

Today, the Standard version assumes that QM describes (quantum) 'systems' in an algorithmic fashion allowing us to predict observable measurement outcomes. This understanding of physics restricted by the classical paradigm has blocked the possibility to advance in the development of a new conceptual scheme. Niels Bohr was explicit regarding this point arguing repeatedly that: "it would be a misconception to believe that the difficulties of the atomic theory may be evaded by eventually replacing the concepts of classical physics by new conceptual forms." Exactly this type of warning, is what David Deutsch [31] has rightly characterized as "bad philosophy", namely, "[a] philosophy that is not merely false, but actively prevents the growth of other knowledge." Breaking the Bohrian law, the logos approach to QM attempts to develop a new metaphysical

---

[16]Clifton developed an example in which the violations of *Property Composition* and *Property Decomposition* seem to show implications which seem at least incompatible with the everyday description of reality. In the example Clifton takes a Boeing 747 which has a possibly wrapped left-hand wing: α is the left-hand wing and $\alpha\beta$ is the airplane as a whole. Q_α represents the property of being wrapped and $Q_{\alpha \otimes I_\beta}$ represents the plane property of the left wing being wrapped. In such an example a violation of *Property Composition* ($[Q_\alpha] = 1, [Q_{\alpha \otimes I_\beta}] \neq 1$) leads, according to Clifton [18, p. 385], to the following situation: "a pilot could still be confident flying in the 747despite the fault in the left hand wing". If, on the other hand, *Property Decomposition* fails ($[Q_{\alpha \otimes I_\beta}] = 1, [Q_\alpha] \neq 1$) the implication reads "no one would fly in the 747; but, then again, a mechanic would be hard-pressed to locate any flaw in its left-hand wing". The situation gets even stranger when the pilot notices that the plane as a whole has the property $[Q_{\alpha \otimes I_\beta}] = 1$ and concludes (incorrectly) following *Property Decomposition* that the left-hand wing is wrapped, that is, that $[Q_\alpha] = 1$. The mechanic is then sent to fix the left hand-side wing but according to Clifton cannot locate the flaw because the wing does not possess the property Q_α

scheme with specially suited non-classical concepts that are able to explain in an intuitive manner what QM is really talking about. In fact, by carefully studying the orthodox quantum formalism it is possible to derive an important set of consequences which have been always there, at plain sight. To take the formalism seriously means for us to seek for an objective set of concepts which are grounded on the mathematical structure of the formalism itself. In particular, as we have argued in [27], the key to understand the objective aspect of the mathematical formalism of QM is not something related to observations, it is exposed in the invariant mathematical structure of the theory. As Max Born himself reflected: [11]: "the idea of invariant is the clue to a rational concept of reality, not only in physics but in every aspect of the world." In physics, invariants are quantities having the same value for any reference frame. The transformations that allow us to consider the physical magnitudes from different frames of reference have the property of forming a group. It is this feature which allows us to determine what, according to a mathematical formalism, can be considered to be *the same* independently of any particular viewpoint or observation. While in classical mechanics invariance is provided via the Galilei transformations and in relativity theory we find the Lorentz transformations, in QM the invariance is given by the Unitary group.

Born Rule: *Given a vector Ψ in a Hilbert space, the following rule allows us to predict the average value of (any) observable P.*

$$\langle \Psi|P|\Psi \rangle = \langle P \rangle$$

This prediction is independent of the choice of any particular basis.

This rule, which provides the invariant operational foundation of the theory, points implicitly to the way in which physical reality should be conceived according to the theory of quanta. Taking distance from the famous Bohrian prohibition to consider physical reality beyond the theories of Newton and Maxwell, we have proposed the following extended definition of what can be naturally considered —by simply taking into account the mathematical invariance of the Hilbert formalism— as a generalized element of (quantum) physical reality (see [22]).

Generalized Element of Physical Reality: *If we can predict in any way (i.e., both probabilistically or with certainty) the value of a physical quantity, then there exists an element of reality corresponding to that quantity.*

This redefinition implies a deep reconfiguration of the way in which the quantum formalism must be addressed, the type of predictions it provides and even the way in which data must be analyzed. It also allows us to understand Born's probabilistic rule in a new light; not as providing information about a (subjectively observed) measurement result, but instead, as providing (objective) information of a theoretically described (potential) state of affairs (see [26]).

Objective probability does not mean that particles behave in an intrinsically random manner. Objective probability means that probability characterizes a feature of the conceptual representation itself which can be understood —in invariant terms— independently of any subjective observation. This account of probability allows us to restore a representation in which the state of affairs is completely detached from the observer's choices to measure (or not) a particular property —just like Einstein requested. Consequently, the Born rule always provides complete knowledge of the quantum mechanically described state of affairs; in cases where the probability is equal to 1 and also in cases in which probability is different to 1. Any vector or matrix, independently of the context (or basis), provides *maximal knowledge* of the represented (quantum) state of affairs. Since there is no essential mathematical distinction between any matrix (of any rank), both pure states and mixtures have to be equally considered; none of them can be regarded as being "less real" or "less well defined" than the others. Thus, it is not necessary at all to distinguish between pure states and mixed states.[17] In turn, through the strict application of the Born rule in order to define *intensive valuations* we have also been capable to derive a non-contextuality intensive theorem which bypasses the Kochen-Specker theorem in a natural manner and allows us to restore a global objective reference to all —observed or not— projection operators [28].

Unlike many approaches which assume a classical metaphysical standpoint when analyzing QM —introducing implicitly or explicitly classical notions within the theory—, the logos approach attempts to stay close to the quantum formalism in the most strict manner. This implies for us, a suspicious attitude towards the (classical) notions of 'system', 'state' and 'property' which —as a matter of fact— do not seem adequate to represent the orthodox formalism. Taking their place, we have created new (non-classical) concepts which attempt to satisfy the features of the quantum formalism —and not the other way around. According to the logos approach, QM talks about a potential realm which is independent of actuality. There is never a "collapse" from a quantum superposed state to a measurement outcome, simply because physics does not describe the observations of agents. Physics represents in a formal-conceptual manner states of affairs and their evolution. Measurement is always outside of the picture of the theory. Following this line of reasoning, we have argued that QM talks about a potential state of affairs constituted by immanent powers with definite potentia. From this standpoint, we have shown how through the aid of these newly introduced notions we are able to explain the distance between the objective representation provided by the theory and the subjective measurements taking place *hic et nunc* in a lab [28] —dissolving in this way the infamous measurement riddle. Forced by the need to replace particle metaphysics and collapses with a new adequate metaphysical scheme, in [29]

---

[17]This point has been already addressed by David Mermin in [58, Sect. VII].

we have also derived a new objective definition of entanglement in terms of the potential coding of intensive and effective relations.

EFFECTIVE RELATIONS: *The relations determined by a difference of possible actual effectuations. Effective relations discuss the possibility of an actualist definite coding. It involves the path from intensive relations to definite correlated (or anti-correlated) outcomes. They are determined by a binary valuation of the whole graph in which only one node is considered as true, while the rest are considered as false.*

INTENSIVE RELATIONS: *The relations determined by the intensity of different powers. Intensive relations imply the possibility of a potential intensive coding. They are determined by the correlation of intensive valuations.*

These relations provide an intuitive grasp of what can be done in a lab and what type of relations are at play. The following definitions provide a new account of entanglement which rests on the analysis of relational intensive and effective correlations.

Definition 6.1 (Quantum Entanglement) *Given Ψ_1 and Ψ_2 two PSAs, if Ψ_1 and Ψ_2 are related intensively and effectively we say there exists* quantum entanglement *between Ψ_1 and Ψ_2.*

According to this definition entanglement relates to the potential coding of intensive and effective relations between two distant measuring set-ups. We also have the possibility to provide an intuitive non-spatial definition of separability which relates to the lack of correlations between two distant screens.

Definition 6.2 (Relational Separability) *Given Ψ_1 and Ψ_2 two PSAs, if Ψ_1 and Ψ_2 are not related intensively nor effectively we say there is* relational separability *between Ψ_1 and Ψ_2.*

It is interesting to notice that our definitions of potential coding in terms of intensive and effective relations allows us to address a third possibility which considers the cases in which there are only intensive relations involved but not effective ones.

Definition 6.3 (Intensive Correlation) *Given Ψ_1 and Ψ_2 two PSAs, if Ψ_1 and Ψ_2 are related intensively but not effectively we say there exists an* intensive relation *between Ψ_1 and Ψ_2.*

This new approach shows how metaphysical considerations are essential for the analysis of operational data. In fact, the analysis of intensive relations has been completely bypassed within the orthodox definition of entanglement exclusively focused on the collapse of invisible particles characterized in terms of properties and outcomes described in binary terms. From this viewpoint, what needs to be stressed is that since the notion of entanglement is grounded explicitly on both 'particle metaphysics' and the 'collapse' of the quantum wave function, the rejection of these notions —also present within the orthodox axiomatic Dirac-von Neumann formulation of the theory— implies the rejection of the contemporary definition of quantum entanglement. Our proposed redefinition of the notion of entanglement beyond classical notions and *ad hoc* rules hopes to open the debate about such a possibility.

7 Conclusion

In this paper we have provided arguments against the orthodox definition of quantum entanglement grounded on particle metaphysics and the existence of unobserved "collapses". We have discussed and analyzed two different approaches which attempt to go beyond the notion of 'system' in QM. While the first, so called, device-independent approach retains the orthodox definition of entanglement, the latter logos approach to QM presents a new definition which going back to Heisenberg's matrix formulation reintroduces the consideration of intensive relations.

Acknowledgements

We want to thank an anonymous referee for his comments and remarks which have allowed us to change the manuscript substantially. C. de Ronde would like to thank Don Howard for historical references. He would also like to thank Diederik Aerts and Massimiliano Sassoli de Bianchi for related discussions. This work was partially supported by the following grants: the Project PIO CONICET-UNAJ (15520150100008CO) "Quantum Superpositions in Quantum Information Processing", UNAJ INVESTIGA 80020170100058UJ.

References

[1] Aerts, D., 1981, *The one and the many: towards a unification of the quantum a classical description of one and many physical entities*, Doctoral dissertation, Brussels Free University, Brussels.

[2] Aerts, D., 1984, "The missing elements of reality in the description of quantum mechanics of the EPR paradox situation", *Helvetica Physica Acta*, **57**, 421-428.

[3] Aerts, D., 1984, "How do we have to change quantum mechanics in order to describe separated systems", in *The Wave-Particle Dualism* pp. 419-431, S. Diner, D. Fargue, G. Lochak and F. Selerri (Eds.), Springer, Dordrecht.

[4] Aerts, D. & Sassoli di Bianchi, M., 2015, "Many-Measurements or Many-Worlds? A Dialogue", *Foundations of Science*, **20**, 399-427.

[5] Aspect A., Grangier, P. & Roger, G., 1981, "Experimental Tests on Realistic Local Theories via Bell's Theorem", *Physical Review Letters*, **47**, 725-729.

[6] Bell, J.S., 1964, "On the Einstein-Podolsky-Rosen paradox", *Physics*, **1**, 195-200.

[7] Bernien, H., Hensen, B., Pfaff, W., Koolstra, G., Blok, M.S., Robledo, L., Taminiau, T.H. Markham, M. Twitchen, D.J. Childress, L. & Hanson, R., 2013, "Heralded entanglement between solid-state qubits separated by three meters", *Nature*, 497, 86-90

[8] Blake C.S., 2018, "Misreading EPR: Variations on an Incorrect Theme", preprint. (quant-ph:1809.01751).

[9] Bohr, N., 1935, "Can Quantum Mechanical Description of Physical Reality be Considered Complete?", *Physical Review*, **48**, 696-702.

[10] Bohr, N., 1960, *The Unity of Human Knowledge*, in *Philosophical writings of Neils Bohr*, vol. 3., Ox Bow Press, Woodbridge.

[11] Born, M., 1953, "Physical Reality", *Philosophical Quarterly*, **3**, 139-149.

[12] Born, M., 1971, *The Born-Einstein Letters*, Walker and Company, New York.

[13] Bub, J., 2004, "Why the quantum?", *Studies in the History and Philosophy of Modern Physics*, **35**, 241-266.

[14] Bub, J., 2017, "Quantum Entanglement and Information", *The Stanford Encyclopedia of Philosophy (Spring 2017 Edition)*, E.N. Zalta (ed.), URL = https://plato.stanford.edu/archives/spr2017/entries/qt-entangle/.

[15] Bub, J., 2017, "Why Bohr Was (Mostly) Right", preprint. (quant-ph:1711.01604)

[16] Castellani, E., 1998, *Interpreting Bodies. Classical and Quantum Objects in Modern Physics*, Princeton University Press, Princeton.

[17] Clifton, R.K., 1995, "Why Modal Interpretations of Quantum Mechanics must Abandon Classical Reasoning About Physical Properties", *International Journal of Theoretical Physics*, **34**, 1302-1312.

[18] Clifton, R.K., 1996, "The Properties of Modal Interpretations of Quantum Mechanics", *British Journal for the Philosophy of Science*, **47**, 371-398.

[19] da Costa, N. & de Ronde, C., 2016, "Revisiting the Applicability of Metaphysical Identity in Quantum Mechanics", in *Probing the Meaning of Quantum Mechanics*, D. Aerts, J. Arenhart, C. de Ronde and G. Sergioli (Eds.), World Scientific, Singapore, in press.

[20] de la Torre, A.C., Goyeneche, D. & Leitao, L., 2010, "Entanglement for all quantum states", *European Journal of Physics*, **31**, 325-332.

[21] de Ronde, C., 2015, "Epistemological and Ontological Paraconsistency in Quantum Mechanics: For and Against Bohrian Philosophy." in *The Road to Universal Logic. Studies in Universal Logic*. A. Koslow and A. Buchsbaum (Eds) Birkhäuser, Cham.

[22] de Ronde, C., 2016, "Probabilistic Knowledge as Objective Knowledge in Quantum Mechanics: Potential Immanent Powers instead of Actual Properties", in *Probing the Meaning of Quantum Mechanics: Superpositions, Semantics, Dynamics and Identity*, pp. 141-178, D. Aerts, C. de Ronde, H. Freytes and R. Giuntini (Eds.), World Scientific, Singapore.

[23] de Ronde, C., 2017, "Causality and the Modeling of the Measurement Process in Quantum Theory", *Disputatio*, **9**, 657-690.

[24] de Ronde, C., 2018, "Quantum Superpositions and the Representation of Physical Reality Beyond Measurement Outcomes and Mathematical Structures", *Foundations of Science*, **23**, 621-648.

[25] de Ronde, C. & Bontems, V., 2011, "La notion d'entité en tant qu'obstacle épistémologique: Bachelard, la mécanique quantique et la logique", *Bulletin des Amis de Gaston Bachelard*, **13**, 12-38.

[26] de Ronde, C., Freytes, H. & Sergioli, G., 2019, "Quantum Probability: a reliable tool for an agent or a reliable source of reality?", *Synthese*, DOI: 10.1007/s11229-019-02177-x. (quant-ph/arXiv:1903.03863)

[27] de Ronde, C. & Massri, C., 2017, "Kochen-Specker Theorem, Physical Invariance and Quantum Individuality", *Cadernos da Filosofia da Ciencia*, **2**, 107-130.

[28] de Ronde, C. & Massri, C., 2018, "The Logos Categorical Approach to Quantum Mechanics: I. Kochen-Specker Contextuality and Global Intensive Valuations.", *International Journal of Theoretical Physics*, DOI: 10.1007/s10773-018-3914-0. (quant-ph:1801.00446)

[29] de Ronde, C. & Massri, C., 2019, "A New Objective Definition of Quantum Entanglement as Potential Coding of Intensive and Effective Relations.", *Synthese*, DOI: 10.1007/s11229-019-02482-5. (quant-ph:1807.08344)

[30] de Ronde, C. & Massri, C., 2019, "Against the Tyranny of Pure States in Quantum Theory", submitted. (quant-ph:1902.01667)

[31] Deutsch, D., 2004, *The Beginning of Infinity. Explanations that Transform the World*, Viking, Ontario.

[32] Dieks, D., 1988, "The Formalism of Quantum Theory: An Objective Description of Reality", *Annalen der Physik*, **7**, 174-190.

[33] Dieks, D., 2010, "Quantum Mechanics, Chance and Modality", *Philosophica*, **83**, 117-137.

[34] Dieks, D., 2019, "Quantum Mechanics and Perspectivalism", in *What is Quantum Information?*, pp. 51-70, O. Lombardi, S. Fortin, F. Holik and C. López (eds.), Cambridge University Press, Cambridge.

[35] Dieks, D. & Lubberdink, A., 2011, "How Classical Particles Emerge From the Quantum World", *Foundations of Physics*, **41**, 1051-1064.

[36] Dirac, P. A. M., 1974, *The Principles of Quantum Mechanics* (4th Edition), Oxford University Press, London.

[37] Dorato, M., 2015, "Events and the Ontology of Quantum Mechanics", *Topoi*, **34**, 369-378.

[38] Einstein, A., 1916, "Ernst Mach", *Physikalische*, **17**, 101-104.

[39] Einstein, A., 1949, "Remarks concerning the essays brought together in this co-operative volume", in *Albert Einstein. Philosopher-Scientist*, P.A. Schlipp (Eds.), pp. 665-689, MJF Books, New York.

[40] Einstein, A., Podolsky, B. & Rosen, N., 1935, "Can Quantum-Mechanical Description be Considered Complete?", *Physical Review*, **47**, 777-780.

[41] Einstein, A., Schrödinger, E., Planck, M. & Lorentz H.A., 1967, *Letters on Wave Mechanics*. K. Przibram K (Ed.), Philosophical Library, New York.

[42] Fuchs, C.A., 2002, "Quantum mechanics as quantum information (and only a little more)" in *Quantum Theory: Reconsideration of Foundations*, pp. 463-543, A. Khrennikov (Ed.) Växjö University Press, Växjö. (quant-ph:0205039)

[43] Fuchs C.A., 2017, "On Participatory Realism", in *Information and Interaction* The Frontiers Collection, pp. 113-134, Durham I., Rickles D. (eds), Springer, Cham.

[44] Griffiths, R., 2002, "Probabilities and Quantum Reality: Are There Correlata?", *Foundations of Physics*, **33**, 1423-1459. (quant-ph:0209116)

[45] Grinbaum, A., 2015, "Quantum theory as a critical regime of language dynamics", *Foundations of Physics*, **45**, 1341-1350.

[46] Grinbaum, A., 2017, "How device-independent approaches change the meaning of physical theory", *Studies in History and Philosophy of Science Part B*, **58**, 22-30.

[47] Gurtovoi, V.L., Il'in A.I. & Nikulov, A.V., 2020, "Experimental investigations of the problem of the quantum jump with the help of superconductor nanostructures", *Physics Letters A*, **26**.

[48] Heisenberg, W., 1958, *Physics and Philosophy*, World perspectives, George Allen and Unwin Ltd., London.

[49] Heisenberg, W., 1971, *Physics and Beyond*, Harper & Row, New York.

[50] Heisenberg, W., 1973, "Development of Concepts in the History of Quantum Theory", in *The Physicist's Conception of Nature*, pp. 264-275, J. Mehra (Ed.), Reidel, Dordrecht.

[51] Hensen, B., Bernien, H., Dréau, A.E., Reiserer, A., Kalb, N., Blok, M.S., Ruitenberg, J., Vermeulen, R.F.L., Schouten, R.N., Abellán, C., Amaya, W., Pruneri, V., Mitchell, M.W., Markham, M., Twitchen, D.J., Elkouss, D., Wehner, S., Taminiau T.H. & Hanson, R., 2015, "Loophole-free Bell inequality violation using electron spins separated by 1.3 kilometres", *Nature*, **526**, 682-686.

[52] Howard, D., 1985, "Einstein on Locality and Separability", *Studies in History and Philosophy of Science*, **16**, 171-201.

[53] Howard, D., 1989, "Holism, Separability and the Metaphysical implications of the Bell inequalities", in |it Philosophical Consequences of Quantum Theory: Reflections on Bell?s Theorem, pp. 224-253, Cushing and McMullin (Eds.), University of Notre Dame Press, Indiana.

[54] Howard, D., 1994, "Einstein, Kant, and the Origins of Logical Empiricism", in *Logic, Language, and the Structure of Scientific Theories: Proceedings of the Carnap-Reichenbach Centennial, University of Konstanz, 21-24 May 1991*, 45-105, W. Salmon and G. Wolters (Eds.), University of Pittsburgh Press, Pittsburgh.

[55] Howard, D, 2010, "Einstein's Philosophy of Science", *The Stanford Encyclopedia of Philosophy (Summer 2010 Edition)*, E. N. Zalta (Ed.), URL: http://plato.stanford.edu/archives/sum2010/entries/einstein-philscience/.

[56] Jaynes, E.T., 1990, *Complexity, Entropy, and the Physics of Information*, W. H. Zurek (Ed.), Addison-Wesley.

[57] Li, J.-L. & Qiao, C.-F., 2018, "A Necessary and Sufficient Criterion for the Separability of Quantum State", *Scientific Reports*, **8**, 1442.

[58] Mermin, D., 1998, "What is quantum mechanics trying to tell us?", *American Journal of Physics*, **66**, 753-767.

[59] Mintert, F., Viviescas, C. & Buchleitner, A., 2009, "Basic Concepts of Entangled States" in *Lecture Notes in Physics*, **768**, 61-86.

[60] Osnaghi, S, Freitas, F. & Freire, O., 2009, "The origin of the Everettian heresy" *Studies in History and Philosophy of Modern Physics*, **40**, 97-123.

[61] Piron, C., 1976, *Foundations of Quantum Physics*, W.A. Benjamin Inc., Massachusetts.

[62] Pitowsky, I., 1994, "George Boole's 'Conditions of Possible Experience' and the Quantum Puzzle", *The British Journal for the Philosophy of Science*, **45**, 95-125.

[63] Redei, M., 2019, "On the tension between physics and mathematics", preprint.

[64] Schrödinger, E., 1935, "The Present Situation in Quantum Mechanics", *Naturwiss*, **23**, 807-812. Translated to english in *Quantum Theory and Measurement*, J. A. Wheeler and W. H. Zurek (Eds.), 1983, Princeton University Press, Princeton.

[65] Schrödinger, E., 1935, "Discussion of Probability Relations between Separated Systems", *Mathematical Proceedings of the Cambridge Philosophical Society*, **31**, 555-563.

[66] Schrödinger, E., 1950, "What is an elementary particle?", *Endeavor*, VolIX, N35, July 1950.

[67] Smets, S., 2005, "The Modes of Physical Properties in the Logical Foundations of Physics", *Logic and Logical Philosophy*, **14**, 37-53.

[68] Sudbery, A., 2016, "Time, Chance and Quantum Theory", in *Probing the Meaning and Structure of Quantum Mechanics: Superpositions, Semantics, Dynamics and Identity*, pp. 324-339, D. Aerts, C. de Ronde, H. Freytes and R. Giuntini (Eds.), World Scientific, Singapore.

[69] Von Neumann, J., 1996, *Mathematical Foundations of Quantum Mechanics* (12th. Edition), Princeton University Press, Princeton.

[70] Van Fraassen, B., 2009, "Rovelli's World", *Foundations of Physics*, *40*, 390-417.

[71] Vedral, "Quantum Entanglement", *Nature Physics*, **10**, 256-258.

Generation of Cosmological Magnetic fields at laboratory scales in a chiral QED metric-torsion gravitational weak field background

By L.C. Garcia de Andrade[1]

Abstract

Previously Landsteiner et al have investigated the chiral anomalies in the framework of gravito-magnetism where the magnetic field is in certain sense given by vorticity. To investigate the role of the truly magnetic fields in the case of laboratory or earth space experiments for gravitational anomalies we make use here of the weak field approximation of the gravitational field. Phenomena of transport are not investigated here, but instead we start from a QED set of equations on a metric-torsion background of non-Riemannian space. It is shown in the present paper that presence of a long range gravitational potential in the metric presents a possibility of investigate cosmology in the lab since we obtain a magnetic field of the order $10^{-13}G$ which is of the order of the cosmological magnetic fields. This computation is obtained from geophysical data for mass and radius of the earth and for an estimated torsion of $10^{-31}GeV$ from the Hughes-Drever experiment. Gianluca Grigori and his team have recently obtained a cosmological metric of FRW induced in the laser plasma lab to produce axion particles from quantization with spinless particles. When only torsion effects are taken into account a chiral dynamo effect due to torsion amplify magnetic fields of the order of $10^{-21}G$ up to $10^{-16}G$.

---
[1]Departamento de Física Teórica - IF - UERJ - Rua São Francisco Xavier 524, Rio de Janeiro, RJ, Maracanã, CEP:20550. e-mail:garcia@dft.if.uerj.br

1 Introduction

Earlier de Sabbata and Gasperini [1] have obtained a QED electrodynamics with torsion derived from the polarization of vacuum and electron positron pairs. In this electrodynamics torsion axial vector appears naturally in terms of Dirac spinors ψ and gamma matrices γ where torsion is then given by $Q^\mu = -3\pi\bar{\psi}\gamma_5\gamma^\mu\psi$. Here we adopt the cgs systems of units and the fine structure constant is $\alpha = e^2$ where e is the unit of electric charge of electron. More recently [2] we use this electrodynamics in the case where torsion vector is a constant background and show that this expression which reduces then to Chern-Simons electrodynamics (CS) gives rise to torsionful chiral electrodynamics and chiral dynamos ar electrodynamics. In analogous idea B P Dolan [3] associates torsion from spin-spin interaction of Einstein-Cartan gravity [1] in terms of spinor fields where torsion would obey the Cartan equation in differential forms language as before. In this paper following similar ideas of Lansteiner and his group [4] where they computed the gravitational mixed with chiral anomalies as

$$A \sim \int F(\wedge F + R \wedge R \tag{1}$$

where $\mu = 0, 1, 2, 3$ are orthonormal indices and $F = dA$ is the Maxwell 2-curvature differential form and R is the Riemannian differential 2-form. Landsteineir computed the anomalies coefficients in the background of the AdS black hole metric. In this paper instead we shall compute instead of the gravito-magnetic field given by the vorticity of the fluid $B_g = \nabla \times \mathbf{v}$ we use the real magnetic field induction $\mathbf{B}$ and actually the only place where there is a point of contact between the two papers is that we may use the Larmor frequency relation between real magnetic fields and vorticity to compute Flaschi Fukushima Riemannian chiral currents extended to non-Riemannian spaces. It is also shown that by using this metric-torsion spacetime that the magnetic field instability at present universe is extremely weak and of the order of $\frac{\delta B}{B} \sim 10^{-24}$. Certainly at the early university this instability should grow. The plan of the paper is as follows: In section 2 one presents the new chiral magnetic currents that appears from the metric effect and that were not present in previous papers and also compute the magnetic field instability taking into account the potential of Reissner-Nordstrom type of general relativity $\phi = \frac{m_e}{R}$ where this gravitational potential is described by the weak field gravitational Riemannian metric

$$ds^2 = -(1-2\phi)dt^2 + (1+2\phi)[dx^2 + dy^2 + dz^2] \qquad (2)$$

where the here ϕ is the gravitational scalar potential given by $\frac{2GM}{c^2 R}$ which is of course Newtonian gravitational potential. The plan of the paper is as follows: Section 2 addresses the issue of minimal coupling without production of massive photons and its magnetogenesis and charge asymmetry. Section 3 one obtains the magnetogenesis and charge asymmetry in case where massive photons are present, and also discuss in short section 4 the role of minimal torsion-e.m coupling in the case of chiral e.m anomalies. Discussions are left to section 5.

2 Cosmological Magnetic fields in the lab in a metric-torsion background

We shall start here from the Maxwell equation in the flat spacetime endowed by a axial vector and minimal coupling applied to the Maxwell equations

$$\partial_\mu F^{\mu\nu} = J^\nu \qquad (3)$$

Here we shall make use of the minimal coupling principle between flat spacetime and torsion vector and substitute them into Maxwell electrodynamics in order to obtain a new electrodynamics in a torsionful spacetime, where new physical properties appears such as the electric charge fluctuation asymmetry from the torsion fluctuations. Now let us this so-called Maxwell-Cartan expression into $(3+1)-D$ electrodynamics as

$$\nabla \times \mathbf{B} = 4\frac{\pi}{c}\mathbf{J} + 4\frac{\alpha}{3\pi_0}(-g)^{\frac{1}{2}}[Q^0\mathbf{B} + \mathbf{E} \times \mathbf{Q}] \qquad (4)$$

where the axial chiral torsion is given by the expression $Q^\mu = -3\bar{\psi}\gamma^\mu\gamma_5\psi$ in terms of Dirac gamma matrices and Dirac spinors ψ. Here we already note that there is a coupling of the metric field through the term determinant of the metric g to the 0-component of axial torsion , which in turn also represents a natural chiral current in this anomalous equation. The remaining Maxwell s anomalous equations are

$$\nabla.\mathbf{E} + \mathbf{E}.\nabla(-g)^{\frac{1}{2}} = \rho_e + 4\frac{\alpha}{3\pi}(-g)^{\frac{1}{2}}\mathbf{Q}.\mathbf{E} \tag{5}$$

where $\mathbf{E}$ is the electric field and ρ_e is the electric charge density. And the absence of magnetic monopoles is given by

$$\nabla.\mathbf{B} = 0 \tag{6}$$

the Faraday induction law

$$\nabla \times \mathbf{E} = -\partial_t \mathbf{B} \tag{7}$$

where $\mathbf{J}$ is the electric current. Note that from these equations one obtains a simple form of torsion generalization of the charge current conservation $\mathbf{J}_{Ch} = Q_0 \mathbf{B}$ for chiral current in terms of $0-component$ torsion and the traditional chiral current

$$\mathbf{J}_{Ch} = (\mu_5 \mathbf{B} \tag{8}$$

has to be replaced into the Maxwell s equation above to obtain the expression

$$\eta\mu_5^{-1}\nabla \times \mathbf{B} \sim (-g)^{\frac{1}{2}}\partial_t \mathbf{B} \tag{9}$$

This expression is a self-induction magnetic equation sometimes called dynamo equation. where here η is the electric resistivity while μ_5 is the chiral chemical potential. Note that by interaction between metric and torsion one may now to obtain the expression for the magnetic field in terms of the geophysical data as before

$$\alpha\eta\nabla \times \mathbf{B} \sim (-g)^{\frac{1}{2}}Q^0\eta\partial_t \mathbf{B} \tag{10}$$

then the solution of this equation yields the following expression

$$B \sim B_E \alpha \eta Q^0 (-g)^{\frac{1}{2}} \partial_t \mathbf{B} \qquad (11)$$

where the symbol E refers to Earth s data. By substituting the following geophysical data: $M_E \sim 10^{24} kg$ and ratio of the Earth $R_E \sim 10^4 km$ and the Newtonian gravitational constant $G \sim 10^{-8} cgsunits$ one obtains the following expression for the magnetic field at laboratory

$$B_{CMF} \sim 10^{-13} G \qquad (12)$$

Repeating the previous computation but this time without taking into account the small gravitational field of the lab we obtain

$$B \sim B_{seed} \alpha \eta Q^0 \partial_t \qquad (13)$$

where the symbol E refers to Earth s data. By substituting the following geophysical data: $M_E \sim 10^{24} kg$ and ratio of the Earth $R_E \sim 10^4 km$ and the Newtonian gravitational constant $G \sim 10^{-8} cgsunits$ one obtains the following expression for the magnetic field at laboratory

$$B_{CMF} \sim 10^{-16} G \qquad (14)$$

where this time we use a seed field of $B_{seed} \sim 10^{21} G$ and in both cases we made use a cosmic time of the present universe $t_{today} \sim 1 Gyr$.

3 Conclusions and discussions

The conclusions are simple and we have obtained a magnetic field which in the case of pure torsion is amplified 10^5 orders of magnitude characterizing a chiral dynamo. Since torsion is relatively small even at the early universe stages this makes that the amplification growth factor lambda depends upon torsion and it is possible that we obtain slow dynamos or laminar dynamos due the existence of torsion. We also call the attention to the fact that the CMF obtained in laboratory plasmas seems to be similar to the effects discussed here since both cases deal with QED only the one investigated by

Gianluca Grigori and his group [5] uses a dynamo effect driven by lasers shocks in the lab. This is the focus of a current investigation. Several papers [6] where torsion-driven dynamo amplification is used may be found interesting for further laboratory tests for the importance of torsion in the realm of gravitational theory of gravity. Last but not least we must say that different from our paper here spinless particles were used along with the Einstein equivalence principle by Wadud et al [5] to obtain the FRW effective spacetime metric in the lab while here we use the real metric of weak gravitational field limit and spinning particles to generate axial spacetime torsion. Despite of the fact that here we are considering microscopic quantum theory of electrodynamics metric-torsion background the generation of the magnetic fields in plasmas by shock waves described by et al [7] is very important since the problem here can be embedded in a physical hot plasma in the early universe where Chern-Simons QED can be obtained. Last but not least we must mention that the value of CMF obtained here at lab scales has been obtained also by Sigl [8] at the present epoch of the universe at the scales of $100pc$.

4 Acknowledgements

I would like to thank N Mavromatos and J Schoeber for interesting discussions on the subject of this paper.

References

[1] V de Sabbata and C Sivaram, Spin and Torsion in Gravitation, World Sci. (1995).

[2] L C Garcia de Andrade, Eur Phys J C (2018) 78, 254.

[3] B P Dolan, Class Quantum Gravity, 2010.I L Shapiro, Physical Aspects of Spacetime Torsion, Phys Reports, 57,2 (2002).

[4] K Landsteiner, Phys Rev B 89, 075124 (2014).

[5] M Wadud, B King, R Bingham, G Gregori et al, Phys Lett B 777, (2018) 388. P Tzeferacos, A Rigby, A F A Bott et al, Laboratory evidence of dynamo amplification of magnetic fields in a turbulent plasma,Nature communications 9 (2018) 591.

[6] L C Garcia de Andrade, Class and Quantum Grav. (2015) 892. L C Garcia de Andrade, Phys Lett **B** 468, 28 (2011). L Garcia de Anddrade, Lorentz violation bounds from torsion trace and radio galactic dynamos, Phys Rev D (Brief Reports) (2011). L C Garcia de Andrade, JCAP 08 (2014) 23.

[7] M Medvedev, Luis Silva, M Fiori,R A Fonseca, W Mori, J Korean astronomical society 37: 533, (2004).

[8] G Sigl, Phys Rev D 66 (2002) 123002.

Quantum Mechanics, Ontology, and Non-Reflexive Logics*

Décio Krause[†]
Department de Philosophy
Federal University of Santa Catarina
and
Post-Graduate Program in Logic and Metaphysics
Federal University of Rio de Janeiro

88040-900 Florianópolis, SC — Brazil
deciokrause@gmail.com

Abstract

This is a general philosophical paper where I overview some ideas concerning a non-reflexive foundations of quantum mechanics (NRFQM). By NRFQM I roughy speaking mean a formalism and an interpretation that considers an involved ontology of non-individuals as explained in the text. Thus, I do not endorse a purely instrumentalist view of QM, but believe that it speaks of something, and then I try to show that one of the plausible views of this 'something' is as entities devoid of identity conditions.

Keywords: non-reflexive logics, non-reflexive quantum mechanics, non-individuals, quantum ontology, quasi-set theory.

---
*See the warning note at the end.
[†]Partially supported by CNPq.

1 Introduction

A typical way of looking to a scientific theory may be this one:[1] it consists of a triple $\mathcal{T} = \langle \mathcal{F}, \mathcal{M}, \mathcal{R} \rangle$, where $\mathcal{F}$ is a mathematical formalism (the mathematical counterpart of $\mathcal{T}$), $\mathcal{M}$ is the class of its models, that is, the 'realizations' or the 'interpretations' of the theory, and $\mathcal{R}$ represents the set of connection rules which provide the links between the interpretations and the formalism. This is of course just a general scheme that guides us to look to some theories from a very general point of view. Differently from mathematics, physical theories demand the explanation of (at least) one possible realization. Quantum mechanics (QM) would be not different. Below we shall see the details.

If we consider a purely instrumentalist point of view (sometimes associated to Bohr), we can say that QM just provides us with ways of computing probabilities. But this is a radical 'physical' point of view; in general, philosophers aim at to discuss the kind (or kinds) of world(s) QM tells us about. In doing that, we are faced with interesting and counter-intuitive ways of looking to the objects that form our 'quantum reality'. Let me explain this point a little bit, so directing the discussion to that I wish to emphasize. Firstly, I invite you to agree that Sunny Auyang is right when she says that "physical theories are about things" ([Auy95, p. 152]) for, if not, we would be speaking of a purely mathematical theory. But, of which kind of things are we speaking about?

To enlighten the point, let me recall one of Arnold Scharzneger's films, The 6th Day where, coming back to home, he discovers that there is someone in his place, a perfect copy of him.[2] The

---

[1] Several general characterizations of a scientific theories that apply to most of them were proposed in the literature. In a certain sense, all of them coincide in their main aspects to ours. See for instance [DalTor81], [KraAre16].

[2] The story (or history) of Martin Guerre, a Frenchman who lived in the XVI century and which was also substituted by an imposter, is told by Amanda Gefter in "Quantum mechanics is putting human identity on trial: if our particles have no identity, how can we have?", Nautilus (https://goo.gl/vLdrex).

clone act he does, interacts with his wife and sons as he does, puts out the trash as he does, etc. His family does not perceive any difference, and really believe that the person they are interacting with is Adam (the name of the character). But, is he not Adam? To his family, all that imports is that *all happens as if* the person in the house looks as Adam, acts as Adam, etc. That is all. But, you may say, the person in the house *is not* Adam, but just someone quite similar to him.

Let us go to another example: imagine that someone very rich buys a painting (supposedly) by Picasso and then realize that it is not a legitimate Picasso but a copy, a very good one. Okay, you can say: no problem. The copy is not a Picasso, but just a copy. But suppose that Picasso has produced two very similar paintings, one of them acquired by our personage, and that only after the guy has acquired it the second copy appears, and that the specialists are with difficulties to verify the legitimacy of the second painting. What could happen? Probably the the painting (both) would lose their value. Perhaps not, who knows? What imports is that, apparently, identity matters:[3] independently of the similarities, just one of the guys is Adam, just one of the paintings is the original Picasso and in the case of two Picasso's, something else happens: the value of the painting may decrease. Furthermore, the two paintings can be discerned one each other, for some (yet small) difference in tones, in traces or whatever else would be present. With ordinary objects or our scale, apparently Leibniz's celebrated Principle of the Identity of Indiscernibles holds.

Concerning quantum objects, things are different. Although we know the no-cloning theorem [WooZur82], all electrons are electrons-Adam, all protons are protons-Adam, and so on. They don't present any intrinsic difference, but are perfectly alike (supposing of course that it makes sense to speak of them as 'things' of some kind).[4] Yes, you can say that certain quantum objects of the

---

[3]But see below for I will question even this thesis.

[4]An useful characterization of quantum particles, adopted by most physi-

same kind, as two electrons (and for other fermions in general), contrarily to bosons, cannot partake the same quantum state since they must obey Pauli's Principle, *hence* they do present differences for according to this principle, no fermions can partake the same quantum numbers in a same situation. But the problem must be put rightly: think of the two electrons of an Helium atom in its fundamental state. The two electrons are in a superposed state, and yet so they differ by their values of spin in a given direction. But the question is that no one can say which is which.[5] Even if we call one of them 'Paul' and the other 'Peter', before a measurement it is impossible to say which electron is Paul and which one is Peter, contrary to the two supposed Picasso's paintings, for we can write 'Peter' in the back of one of them, something we cannot do with electrons.[6] Things are worst for bosons, for they may share all their quantum numbers (like in a BEC, a Bose-Einstein Condensate [Ket et al.99]), being absolutely indistinguishable by all means provided by the theory and perhaps by any means at all. Identity, here, seems to make no sense.

But, what identity? I mean the intuitive idea of a perfect characterization of an object as a sole object. According to our preferred metaphysics (a Leibnizean one), things having identity are always *different* from any *other* things, and can be recognized as such in different situations. Although sometimes we cannot say in what the difference consists of (more on this below), we tend to suppose that it exists. As we shall see soon, this hypothesis is encoded in our metaphysical pantheon. But let me insist on the nature of quantum objects. Think for instance in the methane combustion:

$$CH_4 + 2O_2 \to CO_2 + 2H_2O.$$

Here, the four oxygen atoms of the two oxygen molecules

---

cists, was given by E. Wigner in 1939 using group theory [Wig59]; see [Cas94].

[5]And this is not a simple epistemological problem. In assuming this, we should agree in admitting hidden variables of a kind.

[6]A long time ago, Shrödinger stressed that "you cannot mark an electron, you cannot paint it red" [Sch53].

will contribute in the reaction so that two of then will form the carbon dioxide molecule and the other two will form the two water molecules. But, which ones? It does not matter. All oxygen (hydrogen, carbon, electrons, protons, etc.) act the same way in the same circumstances. As another exemple, think of an ionization process of a neutral helium atom. One of its electrons may be realized from the atom in order to form a positive ion. Later, an electron can be captured by the ion in order to form a neutral atom again. What is the difference between the first and the second neutral atoms ou between the realized electron and the absorbed one? None. There are no differences at all! If there were differences, chemistry would not work as it does. A long time ago (1803), John Dalton has put things clear when he stressed that

> "[w]hether the ultimate particles of a body, such as water, are all alike, that is, of the same figure, weight, etc. is a question of some importance. From what is known, we have no reason to apprehend a diversity in these particulars: if it does exist in water, it must equally exist in the elements constituting water, namely, Hydrogen and Oxygen. Now it is scarcely possible to conceive how the aggregates of dissimilar particles should be so uniformly the same. If some of the particles of water were heavier than others, if a parcel of the liquid on any occasion were constituted principally of these heavier particles, it must be supposed to affect the specific gravity of the mass, a circumstance not known. Similar observations may be made on other substances. Therefore we may conclude that the ultimate particles of all homogeneous bodies are perfectly alike in weight, figure, etc. In other words, every particle of water is like every other particle of water, every particle of Hydrogen is like every other particle of Hydrogen, etc." [Dal08, pp.142-3]

Quantum objects do not behave as the objects of our scale. They are quite strange.

2 Is identity really so fundamental?

The title of this section is the title of a paper I wrote with my colleague Jonas Arenhart [KraAre15]. We discuss and contest and idea that the notion of identity is fundamental, as advanced by O. Bueno [Bue14]. I will not repeat the arguments of the paper here, but provide a mix of related ideas instead.

Standard objects of our surroundings, we believe, *do have identity*, they are *individuals*. This informally means that they have their *own* characteristics which distinguish them from any other object, although a quite similar one, and they can (in principle) be re-identified several times as being *that* individual. This is due to their *identity*. This belief is one of the most celebrated metaphysical assumptions of Western philosophy already mentioned above, namely, the Principle of the Identity of Indiscernibles (PII), which has its routs in the antiquity, but was celebrated in Leibniz's philosophy.[7] PII says that no two objects can have all the same properties, that there are no *solo numero* distinct objects. Objects having the same properties are *indistinguishable*, or *indiscernible*, while identical objects are the very same object. PII makes these concepts equivalent. As Rodriguez-Pereyra says, "the principle states there cannot be *numerically distinct* but perfectly similar things, or that there cannot be *two* perfectly similar things" [Per.14, p.15 fn.1]. Classical logic and standard mathematics, that is, that one which can be build, say, in a standard set theory like the Zermelo-Fraenkel with the Axiom of Choice system (ZFC), incorporates PII in some way; we can say that it is a theorem of classical logic (and classical mathematics). Every object in ZFC (either if it involves or not the *Urelemente*, entities that are not sets but which can be elements of sets — see [Sup72]) is an *individual* in the sense of obeying PII. More precisely, by obeying the standard theory of identity of ZFC, which if formulated as a first-order theory, formalizes the behavior of a binary primitive relational symbol '=' by means of the following postulates: (1) Reflexivity:

---

[7]Max Jammer suggests that PII has its roots with the Stoics; see [Jam66, p.].

$\forall x(x = x)$; (2) Substitutivity: $\forall x \forall y(x = y \to (\alpha(x) \to \alpha(y)))$, where $\alpha(x)$ is a formula with x free and $\alpha(y)$ results from $\alpha(x)$ by the substitution of y in some free occurrences of x, and (3) the Axiom of Extensionality: $\forall x \forall y (\forall z(z \in x \leftrightarrow z \in y) \to x = y)$. In order to prove that every object in ZFC is an individual, it suffices to acknowledge that given any object a (represented in ZFC either by a set or by an *Urelement*), we can form the unitary set $\{a\}$ and define the following property, which I call 'the identity of a', namely, $I_a(x) \leftrightarrow x \in \{a\}$. It is obvious that the only object obeying I_a is a itself; so, any *other* object will have a difference with a and PII holds.

Good for mathematics. If in discussing arithmetics 'my' number two is different from 'yours', we shall have some troubles, yet we both acknowledge that there are different and non equivalent way of defining 'two' (for instance, Frege's, Russell's, Zermelo's, von Neumann's definitions, and so on). But in a same context, we can agree that the two twos are the same. But for empirical sciences, I guess that we don't need a so strong assumption; worst, in assuming the standard theory of identity as true for empirical objects in general, mainly in regarding quantum objects, we shall face some problems I will touch on below. Firstly let me recall that long time ago David Hume made things clear (in my opinion) when he discussed the identity of objects in his *Treatise* [Hum85, Book I, *passim*]. In short, he said that there is nothing in an object that justifies our belief that it continues to be itself after two successive observations, the second one after an instant in which the object leaves our field of perception. We can say that it is only a postulate of ours that things happen this way. Interestingly enough, Schrödinger had a similar position when he stressed that

> "[w]hen a familiar object reenters our ken, it is usually recognized as a continuation of previous appearances, as being the same thing. The relative permanence of individual pieces of matter is the most momentous feature of both everyday life and scientific experience. If a familiar article, say an earthenware jug, disappears

from your room, you are quite sure that somebody must have taken it away. If after a time it reappears, you may doubt whether it really is the same one — breakable objects in such circumstances are often not. You may not be able to decide the issue, but you will have no doubt that the doubtful sameness has an indisputable meaning — that there is an unambiguous answer to your query. So firm is our belief in the continuity of the unobserved parts of the string!" [Sch98, p.204]

It seems that, for physics at least, identity of objects in time cannot be proven logically; for physics at least (not for art or human relationships) what imports is that once we have an object, an indistinguishable one serves as well. So, we need to assume that indistinguishable objects should exist in some way. The problem with this idea is mathematics (and logic) for, as we have seen, PII entails that indistinguishable objects are *the very same* object; in other words, in ZFC there is not legitimate (*solo numero*) indistinguishable objects. We shall see why next.

3 Indistinguishability in classical logic

Let us consider a structure $\mathcal{E} = \langle S, R_i \rangle$, $i \in I$, built in ZFC. A structure of this kind may admit automorphisms other than the identity function (which is of course an automorphism in every structure). For instance, think of a group structure $\mathcal{G} = \langle G, \star \rangle$. An automorphism of $\mathcal{G}$ is a bijective mapping $h : G \to G$ such that (i) $h(x \star y) = h(x) \star h(y)$, (ii) $h(e) = e$ (where e is the identity element of the group), and (iii) $h(x') = (h(x))'$, where x' is the inverse element of x. For instance, take the group defined by $\mathcal{Z} = \langle \mathbb{Z}, + \rangle$ where $\mathbb{Z}$ is the set of the integers and $+$ the standard addition on this set. Then $h : \mathbb{Z} \to \mathbb{Z}$ defined by $h(x) = -x$ is an automorphism of $\mathcal{Z}$, as is easy to prove. If h is an automorphism of the structure $\mathcal{E}$ and $h(a) = b$ for $a, b \in S$, we say that a and b are

$\mathcal{E}$-indistinguishable. Thus, 2 and -2 are $\mathcal{Z}$-indistinguishable. From the point of view of the structure, that is, *from within* the structure, there are no ways of distinguishing between a and b: they look the same, for they are invariant by the structures' automorphisms. But, are they identical? This of course happens only if there is just one automorphism, namely, the identity function. In this case, we say that the structure is *rigid*. The interesting thing to recall is that in ZFC every structure can be extended, by adding new relations, to a rigid one. This intuitively means that we can always 'go out' the structure and look its elements from the point of view of the outside, and from this point of view, we can realize that the objects we initially thought were indistinguishable, are not indistinguishable at all! For instance, we can 'rigidify' the structure $\mathcal{Z}$ by adding the binary relation $<$, the usual linear order on $\mathbb{Z}$; that is, the extended structure $\mathcal{Z}' = \langle \mathbb{Z}, +, < \rangle$ is rigid, as is easy to see. The idea of 'leaving out' the structure is similar to go to another dimension; in the 1884 book *Flatland* [Abo91], Edwin Abbott created a world in two dimensions, and we can imagine a teenager character of the story who is boring with every one else and decides to keep closed in her room, alone. She supposes nobody can see her. But we, in the third dimension, can. The same happens with our $\mathcal{E}$-indistinguishable objects; looking to the elements of the domain from the extended structure, we can realize that they are not indistinguishable at all. And this can always be done! [CosRod07] Within the ZFC framework, an object is indistinguishable just from itself and remain so independently of what we 'do' with them: they are *individuals*.

The moral of the story is this: within a standard mathematical framework such as ZFC, the only way of considering indiscernible objects is to confine the discussion to a non-rigid structure or something similar which is equivalent to admit as 'identical' those objects belonging to a same equivalence class relative to some equivalence relation or to some congruence (other than identity). This is precisely what the formalism of QM does when postulates that only *certain* states are accessible to quantum objects. For instance, let us consider two indistinguishable bosons and two possible states A and B. The configuration space is the tensor product Hilbert space $\mathcal{H} = \mathcal{H}_1 \otimes \mathcal{H}_2$, where $\mathcal{H}_i$ ($i = 1, 2$) are the state spaces of bosons 1 and 2 respectively. Note that in order to speak of them, we need to name them, say by calling them 'boson 1' and 'boson 2'. But these names cannot make them entities with individuality, so we need to provide a mathematical trick in order these names lose their individuation roles. This is done by assuming that vectors like $\psi_1^A \otimes \psi_2^B$ and $\psi_2^A \otimes \psi_1^B$, meaning respectively that particle 1 is at state A and particle 2 is in B, and that particle 2 is at state A and particle 1 is in B, are not accessible to the particles. Furthermore, these vectors are (in general) different, for the tensor product is not commutative. So, the indiscernibility of the particles cannot be preserved, for it would be a different situation either we consider that particle 1 is at state A and particle 2 is in B or that particle 2 is at state B and particle 1 is in A. These vectors do not represent possible states for the join system; in the terminology introduced by Michael Redhead, these vectors are examples of *surplus structures*, objects resulting from the formalism that have no physical significance (see [RedTel91]). The 'right' states are (1) $\psi_1^A \otimes \psi_2^A$, meaning that both are in A, (2) $\psi_1^B \otimes \psi_2^B$, meaning that both are in B, and (iii) $\frac{1}{\sqrt{2}}(\psi_1^A \otimes \psi_2^B \pm \psi_2^A \otimes \psi_1^B)$, the plus sign holding for bosons (symmetric states) and the minus sign for fermions (anti-symmetric states) — for fermions, this the only available state, due to Pauli's Principle). The situation (iii) says that *one* particle is in A and that the *another one* is in B, but we cannot state which is which. Thus, the formalism preserves Pauli's Principle for fermions; although a permutation of the particles changes the signal of the whole vector, its square remains the same, and that is what imports, for the square of the vector (or wave-function) gives us the probabilities, according to Born's well known rule.

In doing that, the formalism can be written having ZFC as its mathematical (and logical) basis. Of course alternative frameworks could be invoked instead, but ZFC suffices and is sufficiently general for the considerations we have in mind. So, we are performing a trick in assuming that only symmetric and antisymmetric vectors are available for quanta. This is similar to the confinement of the discussion to a non-rigid structure.[8]

4 Ontology

The word 'ontology' has acquired a number of different meanings in philosophy. Traditional philosophy has qualified it as that part of metaphysics that studies the general structures of what there is. In this sense, there cannot be distinct ontologies, for what there is is what there is and things, in the sense of *being*, cannot have two distinct natures. But today we have relativized (or 'naturalized') the word to a certain theory or conception; thus we can say that, given a scientific theory, its ontology is described by specifying the kind of entities the theory is compromised with. In this sense, we can say that the standard formalism of QM is compatible with many non-equivalent ontologies (see [FreKra06] for a wide discussion) which can be aggregated to the formalism as a kind of interpretation of it (an intended semantics). The standard formalism (Hilbert spaces) does not speak of quantum objects strictly speaking, but just of *states* and *observables*, and we need to provide a parallel discussion (sometimes disliked by physicists) in order to answer simple questions such as 'states of what'? But, 'logically speaking', how can we provide a semantics for the formalism? According to standard semantic procedures (formal semantics), the first step is to define a *domain of discourse*. For instance, we may suppose that we are speaking of a collection of bosons or of another quantum system. Let us fix a collection of indistinguishable bosons (all in the same quantum state). Would this domain be a set? Remember what Georg Cantor, the founder father of set theory, said about sets: "by an 'aggregate' (*Menge*) we

---

[8]In [Dom et al.08] — see also [Dom et al.10] — we have started the development of a formalism that dispenses labels to the particles. This is something to be further explored.

are to understand any collection into a whole (*Zusammenfassung zu einein Ganzen*) M of *definite and separate objects m* of our intuition or our thought", my emphasis [Can55, p.85]. In ZFC, due to the Axiom of Extensionality, the set $\{1,1,2,3,3,3\}$ is identical with the set $\{1,2,3\}$ and has cardinal 3.[9] So, how to provide an adequate semantics for, say, indistinguishable bosons? Hermann Weyl has found a way: he has taken a set S, say with n elements, and an equivalence relation $\sim$ defined on S. Then the equivalence classes $C_1, \ldots, C_k$ have cardinals $n_1, \ldots, n_k$ so that we have an 'ordered decomposition' $n = \sum_{i=1}^{k} n_i$ (see [Wey49, p.240]) and, as he says, this would be what imports to QM, that is, *the quantity* of elements in each state (equivalence class), and not their individual description. The equivalent classes play the role of characterizing indistinguishable elements of the collection, that is, elements of S that belong to a same equivalence class are taken as indistinguishable. For a discussion of this case, see [Kra91], [FreKra06]. But this is a trick, for *we know* that the elements of S are all distinct from one each other by fiat! Leibniz's principle holds, and if we have a finite number of objects, we may even ask for the differences. But, as we know from QM, in certain situations there are none!

We may say that the differences are in logic. But, is logic measurable? Things became difficult if we try to push deeper this philosophical question. Thus, let me suggest an alternative. Today most logicians and philosophers have no more fears in admitting different kinds of heterodox logics, that is, systems that depart from classical logic in some way. Intuitionistic logic do not accept the general validity of the excluded middle principle; paraconsistent logics do not accept the general validity of the principle of contradiction. But the laws of identity are still taken as a taboo. No one I know accepts to question them, and more, to question

---

[9]Let me recall that there is a *multiset theory* where an element can occur more than once in a multiset, so that $\{1,1,2,3,3,3\}$ nas cardinal 6 [Bli88]. But this does not fit QM (see [Kra91]), for the repeated objects are *the very same* object, while in QM no one will agree that in a collection of indistinguishable bosons, they are all the same boson. As we shall see below, words like 'the same', 'identical', 'different' causes troubles here.

the ancient metaphysical rule that there cannot be truly indistinguishable objects. Why not to admit them, at least logically? In the same vein as there is no logical proof that there is no another perfectly similar Guernica hidden somewhere, an ontology composed by truly indistinguishable objects, not made ad hoc by some trick as shown above, looks reasonable. Perhaps such a metaphysics would fit well the claims of QM. So, we are free to try to found a semantics for QM whose domain comprises collections of indistinguishable objects. By the way, as said David Hilbert, the mathematician (and the philosopher) should investigate all possible theories [Hil02].

5 The inadequacy of the standard theory of identity

Standard theory of identity says that two *distinct* objects present, at least in principle, a difference in their qualities (that is, there is a property obeyed by just one of them). If we have two of them, they are different in this sense, yet we would not be able to specify the distinctive property. For instance, the bosons in a collection of bosons in the same state, if represented in ZFC, do present differences, yet QM cannot tell us what are they. Remember: in ZFC, if we have a set with more than one element, its elements are *different*.

Let us fix this typical case, namely, a collection of indistinguishable bosons, say in a Bose-Einstein Condensate (a 'BEC').[10] A BEC can be obtained by freezing molecules or atoms near to the absolute zero; in such a situation, the objects start acting as if they were just one thing, a 'big molecule' [Ket et al.99]. They are in the same quantum state and do not present any differences, *but they are not the same object!*. But, let me insist, if we suppose they obey the standard theory of identity, the differences exist, yet

---

[10]A very interesting and didactic page on BECs is http://www.colorado.edu/physics/2000/bec/what_is_it.html.

not perceptible to us. In this case, we need to assume that there is something (some kind of 'variable') hidden in the quantum mechanical description (in the case, a quantum field theoretical description). But no physicist (I suppose) is comfortable in assuming this. The bosons in a BEC do not present any difference, even a hidden one. So, it seems that they should not belong to the class of objects that obey the standard theory of identity. Of course Bohmian quantum mechanics (BQM) says differently, for it agrees with the 'classical' metaphysics of individuals. In BQM, all particles have positions that distinguish them one another. The problem is that as Carlo Rovelli says, the particles "do not revel" their positions; they are hidden to us (and, we could add, also to the gods) [Rov18, p.269, fn.55].

5.1 My proposal

What, them? In my opinion, the better way to deal with entities of this kind is to separate the two notions that are merged in the standard theory of identity: indistinguishability and identity. Indistinguishable objects share properties; you and me apparently are indistinguishable regarding the 'property' *to have interest in quantum physics*. But for sure we have several other differences. 'Truly' indistinguishable objects do not present any difference by definition; they share *all* their properties. Identical objects are not distinct objects, but the same one. In other words, there is no more than one object. In my opinion, the first concept is useful in quantum physics, while the second one causes troubles, and is useful (perhaps) only in mathematics, art and human relationships. In fact, if we assume the standard theory of identity for bosons, we need to assume also the corresponding 'theory of difference' for them, which says that if we have two of them, a difference exists. But, which one? So, out of a purely metaphysical hypothesis, in order to speak of a suitable semantics for quantum languages, we need a mathematical theory were these two concepts are not taken as equivalent. This theory is called *quasi-set theory* we shall see below. But, first, let me say something about the underlying logic of such a theory, a *non-reflexive logic*.

6 Non-reflexive logics, and Schrödinger

Generally speaking, non-reflexive logics are non-classical logics that deviate from classical logic with respect to the notion of identity. Since we can have classical logic *without* identity [Men79], in order to characterize them it is necessary to provide a way to modify in some sense the way classical logic deals with identity. And this may go as follows. If the logic does not contain a primitive binary predicate to be interpreted as identity, we chose a binary predicate of the language and associate to it the diagonal of the domain of the interpretation D, namely, the set $\Delta_D = \{\langle x,x \rangle : x \in D\}$. If the logic comprises a primitive binary predicate of identity, we associate the same set to it. The problem is that, in usual parlance, identity (the informal notion of 'being the very same') cannot be axiomatized. It means that we can never know if the associated set is in fact the diagonal of the domain or another set characterized as the quotient set of the domain by some congruence relation; the structures are elementary equivalent, what means that the same sentences are true in both of them (for details, see [Hod83], [Men79, p.83]). So, from the point of view of the first order language, we cannot distinguish between the two structures and, then, we never know if the predicate of identity really stands for the diagonal of the domain, that is, either we are speaking of the individuals of the domain or of equivalence classes of them.

A typical way of departing from the standard notion of identity is to try to violate the Principle of Identity in some way. But there is no *the* Principle of Identity, for it can be formulated in several non equivalent ways. For instance, at the propositional level, we can write $p \to p$ where p is a propositional variable. In this case, we can interpret the implication as *cause*, prescribed by suitable axioms (see [SylCos88]). That is, $p \to q$ is read p *causes* q, and it is assumed that nothing can case itself, so, $p \to p$ does not hold. Other systems can be obtained by considering the Principle of Identity as formulated in a first order language, namely, $\forall x(x = x)$, where x is an individual variable, also called the reflexive law of identity. The negation of this rule reads $\exists x(x \neq x)$. But we are not claiming that there is something which is not identical to itself; the principle can be violated simply by assuming that the predicate of identity does not hold in general, that is, that there may exist objects in the domain to which it does not make sense to say that they are either equal or different. In this case, $x = y$ simply does not have sense. A typical case is Schrödinger's idea that the notion of identity does not have sense for elementary particles in quantum mechanics. This is the case we have in mind. Inspired in this idea, we have developed Schrödinger Logics in which the Principle of Identity in this first-order form does not hold in general (for details and historical references, see [FreKra06, chap.8]). The theory to be presented below incorporates this idea.

Other discussions on non-reflexive logics can be seen in some papers mentioned in our references and in the indications therein; look at [CosBue09], [KraAre18], [Kra94].

7 Quasi-set theory

In this section I shall sketch a minimal nucleus of quasi-set theory just to give to the reader a general idea of how it works. Later, I shall say something about 'quantum semantics'. In what follows, ZFU stands for the Zermelo-Fraenkel set theory with *Urelemente* [Sup72]. Thus, our intended domain comprises different kinds of entities; standard ZFU is compatible with the existence of *sets* and the *Urelemente*, or simply *atoms*. In the theory sketched below, $\mathfrak{Q}$, there are two kinds of atoms; the M-atoms play the role of the *Urelemente* of ZFU (and in the intended semantics stand for the usual objects of our surroundings — at our scale, the *individuals*), while the m-atoms have a different behavior, and will be thought of as representing elementary particles (either in non-relativistic quantum mechanics or in quantum field theories; both situations can be covered by the formalism, although we shall be speaking more of quantum mechanics). I will modify a little some already presented versions of the theory, mainly in postulate (qc_1), where I have admitted the possibility that a collection may do not have a cardinal (as in the case of quantum field theories, where creation and anhilation operators are introduced). Another modification is concerning the definition of extensional identity given below (definition 7.1v).

Let us call $\mathfrak{Q}$ a first order theory whose primitive vocabulary contains, beyond the vocabulary of standard first order logic without identity (propositional connectives, quantifiers, etc. −see [Men79]), we have the following specific symbols: (1) three unary predicates m, M, Z, (2) two binary predicates $\in$ and $\equiv$, (3) one unary functional symbol qc. Notice once again that identity is not part of the primitive vocabulary, and that the only terms in the language are variables and expressions of the form $qc(x)$, where x is an individual variable, and not a general term.[11] The intuitive meaning of the primitive symbols is given as follows:

(i) $x \equiv y$ (x is indiscernible from y)

(ii) $m(x)$ (x is a 'micro-object', or an m-atom)

(iii) $M(x)$ (x is a 'macro-object' or an M-atom)

(iv) $Z(x)$ (x is a 'set' − a copy of a ZFU set)

(v) $qc(x)$ (the quasi-cardinal of x)

The underlying logic of $\mathfrak{Q}$ is a kind of *non-reflexive logic*, where the standard theory of identity does not hold (see [FreKra06, chap. 8], [Are14], [CosBue09]). Now, we introduce some definitions, with the intuitive interpretation attributed to them.

Definition 7.1

---

[11]This restriction avoids that, for instance, $qc(qc(x))$ turns to be a term.

(i) $Q(x) := \neg(m(x) \lor M(x))$ (x is a qset)

(ii) $P(x) := Q(x) \land \forall y(y \in x \to m(y)) \land \forall y \forall z(y \in x \land z \in x \to y \equiv z)$
(x is a pure qset, having only indiscernible m-atoms as elements.)

(iii) $D(x) := M(x) \lor Z(x)$
(x is a *Ding*, a 'classical object' in the sense of Zermelo's set theory, namely, either a set or a 'macro *Urelement*'.)

(iv) $E(x) := Q(x) \land \forall y(y \in x \to Q(y))$
(x is a qset whose elements are qsets.)

(v) $x =_E y := (Z(x) \land Z(y) \land \forall z(z \in x \leftrightarrow z \in y)) \lor (M(x) \land M(y) \land \forall_Z z(x \in z \leftrightarrow y \in z))$ (Extensional identity)— we shall write simply $x = y$ instead of $x =_E y$ from now on. Notice that the expression $x = y$, when either x or y is an m-atom, yet it can be written, it does not have any meaning in the theory.[12] The notion of identity applies just to sets and to M-atoms. Furthermore, just to explain the terminology, sometimes I use relativized quantifiers: for instance, $\forall_Q x \gamma$ means $\forall x(Q(x) \to \gamma)$, while $\exists_Q x \gamma$ means $\exists x(Q(x) \land \gamma)$; these same for predicates other than Q.

(vi) $x \subseteq y := \forall z(z \in x \to z \in y)$ (subqset)
Important to realize here the conditional in this definition. Having no identity, we may be in trouble in trying to prove that a certain m-atom belongs to a quasi-set, for it should be *identical* to some element of it. This fact does not matter for our purposes. The definition says that *if* z belongs to x then z belongs to y. In $\mathfrak{Q}$, it suffices to prove (or to assume) that there is an indiscernible from z in x. For instance, in a Litium atom $1s^2 2s^1$,

---

[12]This is similar to name $\mathcal{R}$ the collection $\mathcal{R} = \{x : x \notin x\}$ (Russell's set), which can be expressed in the language of ZFC but is not a *set* of this theory, supposed consistent.

it suffices to say that there is one electron in the outer shell; it does not matter which one.

As I have said, $\mathfrak{Q}$ is a theory compatible with the existence two kinds of ur-elements, the *m*-atoms and the *M*-atoms, and also collections formed by either atoms or other collections, the qsets, or by both, atoms and qsets. The theory does not postulate the existence of atoms, as in the standard presentations of ZFU. Some qsets are specially important: when their transitive closure does not contain *m*-atoms, they contain only what we call 'classical objects' of the theory (objects satisfying *D*); items fulfilling this condition satisfy the predicate *Z* and coincide with the sets in ZFU. So, classical mathematics can be built inside $\mathfrak{Q}$, in its *classical part*.

The main idea motivating the development of the theory is that some items are non-individuals (roughly speaking, entities for which the standard notion of identity does not apply), and does not obey the notion encapsulated in the definition of extensional identity. As explained above this concept is not defined for *m*-atoms, the items which intuitively represent quantum indistinguishable objects. Thus, on one side, these things 'do not have identity', that is, it does not make sense to say they are identical or different and, on the other side, the indistinguishability relation holds for every item of the theory, so *m*-atoms may be indistinguishable without being identical. Important to notice that in saying that some entities are non-individuals, we are not supposing that we cannot speak of them; really, we *can* speak of them, that is, we can *write* $x = y$ even for *m*-atoms, but this expression does not have a sense according to the theory: it says nothing. For instance, a qset of indiscernible *m*-atoms may have a quasi-cardinal greater than one, say 5, and so we can think of five entities in some situation, although they cannot be discerned then in any way. Furthermore, quantified expressions must be interpreted adequately; again in considering a qset of indistinguishable objects (say, bosons in a BEC), the universal quantifier says 'all elements of the BEC', while the existential quantifier says 'some element of the BEC'. Thus, universal quantification does not mean 'each' element of the qset (which would presuppose identity) as the standard interpretation suggest (for more on this point, see [KraAre15]).

7.1 The postulates of $\mathfrak{Q}$

Besides postulates for classical first-order logic without identity (which we shall not list here), we introduce the specific postulates for $\mathfrak{Q}$.

($\equiv_1$) $\forall x(x \equiv x)$

($\equiv_2$) $\forall x \forall y(x \equiv y \to y \equiv x)$

($\equiv_3$) $\forall x \forall y \forall z(x \equiv y \wedge y \equiv z \to x \equiv z)$

($=_4$) $\forall x \forall y(x = y \to (\alpha(x) \to \alpha(y)))$, with the usual restrictions.

The first three postulates say that indistinguishability is an equivalence relation. Now, this relation is not necessarily compatible with other primitive predicates; so we can keep identity and indistinguishability as distinct concepts. In fact, if x and y are indistinguishable m-atoms and being z a qset, $x \in z$ does not entail that $y \in z$, and conversely. The fourth postulate says that substitutivity holds only for identical things, that is, for 'classical' things.

Remark: Someone may say that we are presupposing identity in the metalanguage when we say that variables x and y are different. This is true, but does not collapse the theory. A similar situation occurs for instance with some paraconsistent logics [CosKraBue06]. These are logics apt to deal with contradictory sentences. That is, the Principle of Contradiction in the form $\neg(\alpha \wedge \neg \alpha)$ does not hold in general. But, in elaborating such systems, that is, in the metalevel, we do use the principle as being true, for no one would suggest that some expression (finite sequence of symbols of the language) is a formula *and* is not a formula, say. In other words, nothing is a formula and not a formula at once. This 'constructive' character of scientific theories (and of logics) is discussed in the Chapter 3 of [KraAre16].

Other postulates are:

($\in_1$) $\forall x \forall y (x \in y \to Q(y))$
If something has an element, then it is a qset; in other words, the atoms have no elements (in terms of the membership relation).

($\in_2$) $\forall_D x \forall_D y (x \equiv y \to x = y)$
Indistinguishable *Dinge* are extensionally identical. This makes $=$ and $\equiv$ coincide for this kind of entities.

($\in_3$) $\forall x \forall y [(m(x) \land x \equiv y \to m(y)) \land (M(x) \land x = y \to M(y)) \land (Z(x) \land x = y \to Z(y))]$

($\in_4$) $\exists x \forall y (\neg y \in x)$
This qset can be proved to be a set (in the sense of obeying the predicate Z), and it is unique, as it follows from the axiom of weak extensionality we shall see below. Thus, from now own we shall denote it, as usual, by '$\emptyset$'.

($\in_5$) $\forall_Q x (\forall y (y \in x \to D(y)) \leftrightarrow Z(x))$
This postulate grants that something is a set (obeys Z) iff its transitive closure does not contain *m*-atoms. That is, *sets* in $\mathfrak{Q}$ are those entities obtained in the 'classical' part of the theory.

($\in_6$) $\forall x \forall y \exists_Q z (x \in z \land y \in z)$

($\in_7$) If $\alpha(x)$ is a formula in which x appears free, then

$$\forall_Q z \exists_Q y \forall x (x \in y \leftrightarrow x \in z \land \alpha(x)).$$

This is the Separation Schema. We represent the qset y as follows:
$$[x \in z : \alpha(x)].$$
When this qset is a set, we write, as usual, $\{x \in z : \alpha(x)\}$.

($\in_8$) $\forall_Q x(E(x) \to \exists_Q y(\forall z(z \in y \leftrightarrow \exists w(z \in w \land w \in x))))$.
The union of x, written $\bigcup x$. Usual notation is used in particular cases.

7.2 Some basic concepts

From ($\in_6$), by the Separation Scheme using $\alpha(w) \leftrightarrow w \equiv x \lor w \equiv y$, we get a subqset of z which we denote
$$[x,y]_z$$
which is the qset of the indiscernibles of either x or y that belong to z. When $x \equiv y$, this qset reduces to
$$[x]_z$$
called the qset of the indiscernibles from x that belong to z. The qset $[x,y]_z$ does not have necessarily only *two* elements (that is, we may have $qc([x,y]_z) > 2$), for there may be more than just one indistinguishable from x or y in z. Given the qset z and one of its elements, x, the collections $[x]$ and $[x]_z$ stand for *all* indiscernible from x and the qset of the indiscernible from x that belong to z respectively. (Usually, $[x]$ is too big to be a qset — as in general are collections of *all* objects so and so, as in standard set theory.)

Later, with the postulates of quasi-cardinals, we will be able to prove $[x]_z$ has a subqset whose quasi-cardinal equals to 1, written
$$[\![x]\!]_z.$$

We call it the *strong singleton* of x (really, *a* strong singleton of x, for we cannot grant that it is unique). It has just one element, and we can think of this element *as if* it were x, but it follows from the definition that all we can know is that $[\![x]\!]_z$ contains *one object of the 'species'* x. That is, $qc([\![x]\!]_z) = 1$, so there is one item indistinguishable from x in this qset. To prove that this element is x, we need identity.

7.3 Other postulates and definitions

($\in_9$) $\forall_Q x \exists_Q y \forall z (z \in y \leftrightarrow w \subseteq x)$,
The power qset of x, denoted $\mathcal{P}(x)$. Interesting here is that we would be in trouble to teach quasi-set theory to children. For instance, take a qset x with cardinal 2 so that its elements (call them y and z) are indistinguishable. Now try to write the qset $\mathcal{P}(x)$. You can't do it significantly. Really, it results that the two subsets with quasi-cardinal 1 are indistinguishable (by the Weak Extensionality Axiom), so, something like $\mathcal{P}(x) = [\emptyset, [y], [z], x]$ has no clear sense. Even so, as we shall see from axiom (qc_7), the quasi-cardinal of $\mathcal{P}(x)$ is 4.

($\in_{10}$) $\forall_Q x (\emptyset \in x \land \forall y (y \in x \rightarrow y \cup [y]_x \in x))$,
The infinity axiom.

($\in_{11}$) $\forall_Q x (E(x) \land x \neq \emptyset \rightarrow \exists_Q y (y \in x \land y \cap x = \emptyset))$,
The axiom of foundation, where $x \cap y$ is defined as usual.

Definition 7.2 (Weak ordered pair)

$$\langle x, y \rangle_z := [[x]_z, [x, y]_z]_z \qquad (1)$$

Then, $\langle x, y \rangle_z$ takes all indiscernible from either x or y that belong to z, and it is called the 'weak' ordered pair, for it may have more than two elements. Sometimes the sub-indice z will be left implicit.

Definition 7.3 (Cartesian Product) *Let z and w be two qsets. We define the cartesian product $z \times w$ as follows:*

$$z \times w := [\langle x, y \rangle_{z \cup w} : x \in z \land y \in w] \qquad (2)$$

Functions and relations cannot also be defined as usual, for when there are m-atoms involved, a mapping may not distinguish between arguments and values. Thus we provide a wider definition for both concepts, which reduce to the standard ones when restricted to classical entities. Thus,

Definition 7.4 (Quasi-relation) *A qset R is a binary quasi-relation between to qsets z and w if its elements are weak ordered pairs of the form $\langle x, y \rangle_{z \cup w}$, with $x \in z$ and $y \in w$.*

Definition 7.5 (Quasi-function) *f is a quasi-function among q-sets A and B if and only if f is quasi-relation between A and B such that for every $u \in A$ there is a $v \in B$ such that if $\langle u, v \rangle \in f$ and $\langle w, z \rangle \in f$ and $u \equiv w$ then $v \equiv z$.*

In words, a quasi-function maps indistinguishable elements to indistinguishable elements. An interesting question concerns the more specific kinds of functions, that is, injections, surjections and bijections. One can, with some restrictions, define the corresponding concepts, but we shall not present them here (see [FreKra06, chap. 7]).

7.4 Postulates for quasi-cardinals

Notice that in $\mathfrak{Q}$ the standard notion of identity is not defined for some entities (definition 7.1v). Now, the identity concept is essential to define many of the usual set theoretic concepts of standard mathematics, such as well order, the ordinal attributed to a well ordered set, and the cardinal of a collection. Since identity is to be senseless for some items in $\mathfrak{Q}$, how can we employ these notions? One alternative would be to look for different formulations employing methods that do not rely on identity. Another possibility would be to introduce these concepts as primitive and give adequate postulates for them. Concerning the notion of cardinal, there are interesting issues we should acknowledge. First of all, in $\mathfrak{Q}$, there cannot be well-orders on quasi-sets of indistinguishable

m-atoms. Really, a well-order would imply, for example, that there is a least element relative to this well order, a notion which could only be formulated if identity was defined for m-atoms, for this element would be different from any *other* element in the quasi-set. Second, the usual claim that aggregates of quantum entities can have a cardinal but not an ordinal demands a distinction between the notions of ordinal and of cardinal of a quasi-set; this distinction is made in $\mathfrak{Q}$ by the introduction of cardinals as a primitive notion, called quasi-cardinals.[13]

Let us see the postulates for quasi-cardinals; for details and motivations, see [FreKra06, Chap.7], [FreKra10]. Here α, β, ... stand for cardinals (defined as usual in the classical part of the theory, that is, in the theory $\mathfrak{Q}$ when we rule out the m-atoms):

(qc_1) $\forall_Q x (\exists_Z y (y = qc(x)) \to \exists! y (Cd(y) \wedge y = qc(x) \wedge (Z(x) \to y = card(x))))$

In words, if the qset x has a quasi-cardinal, then its (unique) quasi-cardinal is a cardinal (defined in the 'classical' part of the theory) and coincides with the cardinal of x stricto sensu if x is a set. As recalled above, this axiom does not grant that every qset has a well defined quasi-cardinal.

(qc_2) $\forall_Q x (\exists y (y = qc(x)) \to x \neq \emptyset \to qc(x) \neq 0))$.
Every non-empty qset that has a quasi-cardinal has a non-null quasi-cardinal.

(qc_3) $\forall_Q x (\exists_Z \alpha (\alpha = qc(x)) \to \forall \beta (\beta \leq \alpha \to \exists_Q z (z \subseteq x \wedge qc(z) = \beta)))$
If x has quasi-cardinal α, then for any cardinal $\beta \leq \alpha$, there is a subqset of x with that quasi-cardinal.

In the remaining axioms, for simplicity, we shall write $\forall_{Q_{qc}} x$ (or $\exists_{Q_{qc}} x$) for quantifications over qsets x having a quasi-cardinal.

---

[13] As shown by Domenech and Holik, we can define quasi-cardinals for finite qsets in $\mathfrak{Q}$, without resulting that the qset will have an associated ordinal in the usual sense; see [DomHol07].

(qc_4) $\forall_{Q_{qc}} x \forall_{Q_{qc}} y (y \subseteq x \to qc(y) \leq qc(x))$

(qc_5) $\forall_{Q_{qc}} x \forall_{Q_{qc}} y (Fin(x) \land x \subset y \to qc(x) < qc(y))$

It can be proven that if both x and y have a quasi-cardinal, then $x \cup y$ has a quasi-cardinal. Then,

(qc_6) $\forall_{Q_{qc}} x \forall_{Q_{qc}} y (\forall w (w \notin x \lor w \notin y) \to qc(x \cup y) = qc(x) + qc(y))$

In the next axiom, $2^{qc(x)}$ denotes (intuitively) the quantity of subquasi-sets of x. Then,

(qc_7) $\forall_{Q_{qc}} x (qc(\mathcal{P}(x)) = 2^{qc(x)})$

This last axiom enables us to think of subqsets of a given qset in the usual sense; for instance, if $qc(x) = 3$, the axiom says that there exists $2^3 = 8$ subqsets, and axiom (qc_3) enables us to think that there are subqsets with 0, 1, 2 and 3 elements. Furthermore, as we have seen above, in $\mathfrak{Q}$ we can prove that given any object $a \in z$ (either an m-atom, M-atom or quasi-set) we may obtain the strong singleton of a, $[\![a]\!]_z$ whose quasi-cardinal is 1. Important to insist that there is no sense of saying, within $\mathfrak{Q}$, that a is the only element of $[\![a]\!]$, for in order to prove it we need identity. Anyway, $\mathfrak{Q}$ is consistent with this idea and we may reason *as if* this is really so. So, we can think that within $\mathfrak{Q}$ that we may have a certain m-atom, without identifying it, except that it has some characteristics or properties, and not others (for instance, it may be discernible from another m-atom b). That m-atoms may have different properties can be seen from the fact of $\mathfrak{Q}$ that $\mathfrak{Q}$ *doesn't prove* the Substitutivity of Indiscernibles, that is,

$$\mathfrak{Q} \nvdash a \equiv b \to \forall_Q z (a \in z \leftrightarrow b \in z).$$

To prove this result, it suffices to take $[\![a]\!]_z$. Since $qc([\![a]\!]_z) = 1$, a and b cannot belong both to this qset, except if a is identical to b, which cannot be assumed in the case of m-atoms. So, in an extensional context (and $\mathfrak{Q}$ is also a kind of an extensional theory, although this should be qualified), we can read $a \in z$ as a having a certain 'property' (whose 'extension' would be z). So, even indistinguishable m-atoms may have distinct properties, as the two electrons in an Helium atom in its fundamental state have different values of spin in a given direction. As for bosons, let a and b name two bosons in a BEC. The first think to acknowledge is that these names do not make sense, for they cannot individualize the named bosons; furthermore, in $\mathfrak{Q}$, the strong singletons $[\![a]\!]_z$ and $[\![b]\!]_z$ are indistinguishable (by the Weak Extensionality Axiom below), hence there are no differences among them (yet they are not *the same* quasi-set).

7.5 The Weak Extensionality Axiom

The weak extensionality axiom generalizes the usual extensionality axiom. Intuitively, it grants us that two q-sets with the same quantity of the same kinds of elements are indistinguishable. For that, we need two extra definitions, the notion of similarity between q-sets, denoted by Sim, and the notion of Q-similarity, denoted Qsim. Intuitively speaking, similar q-sets have elements of the same kind, and q-similar q-sets have elements of the same kind, and in the same quantity:

Definition 7.6

(i) $\text{Sim}(x,y) := \forall z \forall w (z \in x \wedge w \in y \to z \equiv y)$;

(ii) $\text{Qsim}(x,y) := \text{Sim}(x,y) \wedge qc(x) = qc(y)$.

The weak extensionality axiom reads as follows:

$(\equiv_{12})$ $\forall_Q x \forall_Q y ((\forall z (z \in x/\equiv \to \exists t (t \in y/\equiv \wedge \text{Qsim}(z,t)))) \wedge \forall t (t \in y/\equiv \to \exists z (z \in x/\equiv \wedge \wedge \text{Qsim}(t,z))) \to x \equiv y)$

Intuitively speaking, qsets that have 'the same quantity' (given by their q-cardinals) of elements of the same kind are indiscernible.

The following theorem express the invariance by permutations in $\mathfrak{Q}$, and with this result we finish our revision. To prove it, we shall assume another result, namely, that $y \subseteq t$ entails $qc(x - y) = qc(x) - qc(y)$; let us call this result Theorem $(\star)$ (the proof can be found in [FreKra06, chap.7]). The theorem goes as follows:

Theorem 7.1 (Invariance by Permutations) *Let x be a finite qset such that $\neg(x = [z]_t)$ for some t and let z be an m-atom such that $z \in x$. If $w \in t$, $w \equiv z$ and $w \notin x$, then there exists $[\![w]\!]_t$ such that*

$$(x - [\![z]\!]_t) \cup [\![w]\!]_t \equiv x$$

Proof: Case 1: the only element of $[\![z]\!]_t$ does not belong to x. Then $x - [\![z]\!]_t = x$. Let w be so that its only element belongs to x (for instance, it may be z). Then $(x - [\![z]\!]_t) \cup [\![w]\!]_t = x$, hence the theorem. Case 2: the only element of $[\![z]\!]_t$ belongs to x. Then $qc(x - [\![z]\!]_t) = qc(x) - 1$ by the mentioned theorem $(\star)$. Let $[\![w]\!]_t$ be such that its only element is w itself, so $(x - [\![z]\!]_t) \cup [\![w]\!]_t = \emptyset$. Hence, by Postulate (qc_7), $qc(x - [\![z]\!]_t) = qc(x)$. Thus, by the Weak Extensionality Axiom, the theorem follows. For more details, see [FreKra06, chap.7]. ∎

In words, two indiscernible elements z and w, with $z \in x$ and $w \notin x$, expressed by their strong-singletons $[\![z]\!]_t$ and $[\![w]\!]_t$, are 'permuted' and the resulting qset x remains indiscernible from the original one. The hypothesis that $\neg(x = [z]_t)$ grants that there are indiscernible from z in t which do not belong to x. This theorem has a 'physical' interpretation: the qset x must be a neutral atom which is to be ionized by realizing an electron in order to become a negative ion. Thus the m-object z would represent an electron in the outer shell, while w is 'another' electron not in the atom (these words are to be understood metaphorically). Thus, the electron z is realized and, in another experiment, an electron is captured again so that the atom becomes neutral again. The question is: is this last neutral atom *the same* (identical) to the first one? Of course, this would be so if and only if the captured electron is, *ceteris paribus*, exactly the same as the realized one. But, is there any sense in saying that the realized electron is *identical* with the captured one? Quasi-set theory escapes from this dilemma by assuming that the basic notion is that of indiscernibility; the electrons are indiscernible, so as the neutral atoms. And this is enough for physics. Philosophically, we have again our main thesis: the notion of identity is just a useful notion, not an essential one (in [KraAre15], we discuss this thesis with some care).

8 Quasi-set semantics for quantum languages

The word 'semantics' has also acquired a lot of meanings (as 'ontology' did, as we have seen). In the context that interests us here, it refers to the possible links between a certain language, in general a formal language, and certain 'realities' that lie outside the language. In other words, the task is to attribute *meaning* to certain terms of the language in order to say that by using its resources we can *speak* of certain entities that form the *domains of application* of the language. Of course a formal language, and the logic of axiomatized theories in general, can make reference to infinitely many different domains. Usually we chose one of them to be our *intended interpretation*. For first order languages we have a rather well developed theory involving semantics in this sense, Model Theory. But for more sophisticated languages (higher-order languages) there is no a general theory of its *models*, that is, interpretations where the postulates of the theory are *true*. By 'more sophisticated languages' I mean mainly the languages of physical theories, whose models are in general not first-order structures or, as I prefer to call them, *order-1* structures, composed by a domain (or several domains) and relations and operations relating and operating with the elements of these domains only. In physics (and even in mathematics), in general the relations and operations we have relate (and operate) also with sets of such elements, with functions and matrices formed with them, and so on. These *order-n* structures ($n > 1$) need to be dealt with case by case. This is particularly so with QM. So, let us analyse minimally how we can deal with this question in the scope of quasi-set theory.[14]

In this section I shall sketch a minimal semantics for QM build in $\mathfrak{Q}$. I will postpone to another opportunity the details and the formal proofs. I will simply justify the general idea of using quasi-sets to make sense the existence of indistinguishable but not identical objects. In speaking of 'quantum languages', we need to take some care. Yuri Manin has recalled that quantum mechanics has no its *own* language, making use of a fragment of the language of standard functional analysis (the Hilbert space formalism) [Man10, p.80]. But, inspired by Dalla Chiara and Toraldo di Francia [DalTor93] and by G. Cattaneo [Catt93], we can suppose a suitable language incorporating the standard logical vocabulary, plus the following nonlogical symbols:

(i) A collection of monadic predicates P_i ($1 = 1, \ldots, n$) to represent 'meaningful properties' of the quantum systems, the *observables*. A typical case may be 'the value of the spin in the z-direction'. Let me call P this collection.

(ii) The quantum systems are referred to in different instants of time (according to the intended interpretation) by individual parameters (generic names) $a_1, a_2, \ldots, a_m$. A parameter acts here as when we write the equation of a straight line in Analytic Geometry as $ax + by + c = 0$ and say that a, b, c are parameters ranging on real numbers; we are not specifying particular numbers, but just making reference to them. So, the parameters of our language simply refer to quantum systems without naming them; these parameters are other kind of variables. It seems clear that the more interesting situations are when the quantum

---

[14]Updated: in 2016, Jonas Arenhart and I published a book where more details on this discussion are given; see [KraAre16].

systems for a collection of indiscernible entities, like a BEC. In cases like this one is that $\mathfrak{Q}$ is useful.

(iii) A ternary functional symbol $\mathcal{P}$ to be interpreted as *probability* in a way to be described below.

The semantics goes as follows. In the classical part of $\mathfrak{Q}$, we can consider all sets referred to below, as for instance the set $\mathcal{B}(\mathbb{R})$ of the Borelians of the real number line. Thus we consider the following structure as our *quantum structure*:

$$\mathcal{QM} = \langle S, \{H_i\}, \{A_j\}, \{U_k\}, \mathcal{B}(\mathbb{R}) \rangle,$$

where S is a quasi-set suitable for representing the quantum systems, the Hs are Hilbert spaces, the As are Hermitian operators defined on suitable Hs, the Us are unitary operators which provide the dynamic of the system (Schrödinger's equation), and $\mathcal{B}(\mathbb{R})$ is the set of all Borel sets of the real number line. The rules of interpretation are defined as follows:

(i) The parameters a_1, a_2 etc. are interpreted either as elements of S or as subsets of S. Let me observe that in assuming the structure above, we are introducing explicitly the quantum systems in the semantic considerations, something that is omitted in the usual approaches (see [DalTor93], [Catt93], [vanF75], [Ish95, pp.203ff]).

(ii) Each element $s \in S$ (or $s \subseteq S$) is associated to a unitary Hilbert space $\mathcal{H} \in \{H_i\}$, represented by a unitary vector $|\psi\rangle \in \mathcal{H}$. Composed quantum systems (when $s \subseteq S$) are associated to tensor product of Hilbert spaces, as usual.

(iii) The predicates P_j are associated to Hermitian operators $\hat{A} \in \{A_j\}$ of the attributed Hilbert space. The eingenvalues of $\hat{A}$ are the possible outcomes of the measurements made on P_j.

(iv) For any triple $\langle s, P, \Delta \rangle \in S \times \mathsf{P} \times \mathcal{B}(\mathbb{R})$, we have that $\mathcal{P}(s, P, \Delta) \in [0, 1]$, and this number is interpreted as the probability that for the physical system in state $|\psi\rangle$ (associated to s), a measurement made in P gives a value in Δ. As in the standard formulations of QM, we can assume the notion of probability as given in some suitable way (yet the topic is controversial, as is well known). The postulates describing the behavior of $\mathcal{P}$) are those of [Mac63, pp.62ff].

Then we can proceed as usual with the quantum formalism and intended interpretation, but now with the quantum systems playing a formal role in the developments. Of course more should be said, but I think that the general idea of using quasi-set theory is done. Anyway, some simple question can be envisaged, as the following ones.

8.1 Questions

Some questions are in order, and then we shall see why I have proposed the use of quasi-sets in the semantics presented above. As Dalla Chiara and Toraldo di Francia have emphasized, there are two basic questions to be answers by any semantics for such a language, namely (here adapted):

(i) Does S or some of its sub collections determine a set of m elements in the standard set-theoretical sense?

(ii) Can a parameter a_i determine a well defined element of S?

As they conclude, "...both these questions have a negative answer" [DalTor93, p.276]. The reasons are easy to find. If indistinguishable, a collection of quantum systems should not be taken as a set of a standard set theory (recall that any attempt in this sense will be in need of some mathematical trick); secondly, the denotation function (the rules of interpretation) cannot be defined as a standard function, for it will not distinguish the elements of S to which attribute the a_i. More could be said, but this is enough to sustain our argumentation that quasi-set theory provides a more suitable framework to developed a semantics for quantum languages.

As I have said, all of this of course deserve further explanations and details. But it is clear that QM works as usual, but now with a more suitable form of semantics which serves to make sense of the informal claims about quantum systems. Note that S may have no a well-defined cardinal number, so this semantics may fit also the case of relativistic quantum mechanics, at least for free fields, that is, the Hilbert spaces can be taken as Fock spaces (see [FreKra06, chap.9]). In [Dom et al.08] and in [Dom et al.10], more is said about the development of QM in $\mathfrak{Q}$.

Warning note

This is a revised version of a paper with the same name that was written by invitation to be published in a book titled *The Mammoth Book on Quantum Mechanics Interpretations*, edited by Open Academic Press, Berlin, and having as editor a certain Ulf Edvinsson, who has invited me. The book was announced in the page of OAP and should appear by 2016. This never happened. Later I discovered that OAP is in a list of predatory editorial houses and that "Ulf Edvinsson" is (apparently) a fake name. Furthermore, I couldn't contact anyone responding by OAP to remove my name from the announcement of the book and for impeding them to publish the paper. I strongly apologize for such a fault, which is completely mine. Since the subject presented here has been among my preoccupations ever since I met Chico Doria for the first time (in 1987), it is a pleasure to dedicate the stuff to him.

References

[Abo91] ABBOT, E. A. [1991], *Flatland: a Romance of Many Dimensions*. With a new introduction by Thomas Banchoff. Princeton: Princeton Un. Press.

[Are14] ARENHART, J. R. B. [2014], Semantic analysis of non-reflexive logics. *Logic JnL IGPL* 22 (4): 565-84.

[Auy95] AUYANG, S. [1995], *How is Quantum Field Theory Possible*. Princeton: Princeton Un. Press.

[Bli88] BLIZARD, W. D. [1988], Multiset theory. *Notre Dame Journal of Formal Logic* 30 (1): 36-66.

[Bue00] BUENO, O. [2000], Quasi-truth in quasi-set theory. *Synthese* 125: 33-53..

[Bue14] BUENO, O. [2014], Why identity is fundamental. *American Philosophical Quarterly* 51, 325-332.

[Can55] CANTOR, G. [1955], *Contributions to the Founding of the Theory of Transfinite Numbers*. New York: Dover Pu.

[Cas94] CASTELLANI, E. [1994], Sulla nozione di oggetto nella fisica classica e quantistica. In Cellucci, C., Di Maio, M. C. e Roncaglia, G., (eds.), *Logica e Filosofia della Scienza: problemi e prospettive*, (Atti del Congresso Triennale della Società Italiana di Logica e Filosofia delle Scienze, Lucca, Gennaio 1993), Ed. ETS, Pisa: 385-392.

[Catt93] CATTANEO, G. [1993], The 'logical' approach to axiomatic quantum theory. In In Corsi, G. et al. (eds.), *Bridging the gap: philosophy, mathematics, and physics*.Kluwer Ac. Pu. (Boston Studies in the Philosophy of Science, 140), pp. 225-260.

[CosBue09] DA COSTA, N. C. A AND BUENO, O. [2009], Non-reflexive logics. *Revista Brasileira de Filosofia* 232: 181-196.

[CosRod07] DA COSTA, N. C. A. AND RODRIGUES, A. M. N. [2003], Definability and invariance, *Studia Logica* 82, 1-30.

[CosKraBue06] DA COSTA, N. C. A., KRAUSE, D. AND BUENO, O. [2006], "Paraconsistent logics and paraconsistency", in in D. Jacquette, D.M.Gabbay, P.Thagard and J.Woods (eds.), *Philosophy of Logic*, Elsevier, 2006, in the series Handbook of the Philosophy of Science, v. 5, 655-781.

[DalTor81] DALLA CHIARA, M. L. AND TORALDO DI FRANCIA, G. [1981], *Le Teorie Fisiche: Un' Analisi Formale*, Torino: Boringhieri.

[DalTor93] DALLA CHIARA, M. L. AND TORALDO DI FRANCIA, G. [1993], Individuals, kinds and names in physics. In Corsi, G. et al. (eds.), *Bridging the gap: philosophy, mathematics, and physics*.Kluwer Ac. Pu. (Boston Studies in the Philosophy of Science, 140), pp.261-84.

[Dal08] DALTON J. [1808], *A new system of chemical philosophy.* London: S. Russell, 1808.

[DomHol07] DOMENECH, G., HOLIK, F. [2007], A discussion on particle number and quantum indistinguishability, *Foundations of Physics* **37** (6): 855-78.

[Dom et al.08] DOMENECH, G ET AL. [2010], $\mathfrak{Q}$-spaces and the foundations of quantum mechanics. *Found. Phys* 38 (11): 969-94.

[Dom et al.10] DOMENECH, G ET AL. [2010], No labelling quantum mechanics of indiscernible particles. *Int. J. Theor. Phys.* 49: 3085-91.

[FreKra06] FRENCH, S. AND KRAUSE, D. [2006], *Identity in Physics: A Historical, Philosophycal, and Formal Analysis.* Oxford: Oxford Un. Press.

[FreKra10] FRENCH, S. AND KRAUSE, D. [2010], Remarks on the theory of quasi-sets. *Studia Logica* 95 (1-2): 101-124.

[Hil02] HILBERT, D., [1902], Mathematical problems, *Bull. American Mathematical Society* **8**, 437-479, translated by M. W.

Nelson from 'Mathematische probleme', *Archiv der Math. u. Phys.* **1**, 1901, 44-63 and 213-237. Reprinted in Browder, F. E. (ed.), [1976], *Mathematical Problems Arising from Hilbert Problems*, Proceedings of Symposia in Pure Mathematics, Vol. XXVIII, Providence, American Mathematical Society, pp. 1-34.

[Hod83] HODGES, W. [1983], Elementary Predicate Logic. In: D. M. Gabbay and F. Guenthner, (eds.) *Handbook of Philosophical Logic - Vol. I: Elements of Classical Logic*, pp. 1-131. D. Reidel: Dordrecht.

[Hum85] HUME, D. [1985], *Treatise of human nature*. 2nd. Ed. L. A. Selby-Bigge. Oxford: Oxford Un. Press.

[Ish95] ISHAN, C. J. [1995], *Lectures on quantum theory : mathematical and structural foundations*. London: World Scientific Pu.

[Jam66] JAMMER, M. [1966], *The conceptual development of quantum mechanics*, New York, McGraw Hill.

[Ket et al.99] KETTERLE, W., DURFEE, D. S., AND STAMPER-KURN, D. M. [1999], aking, probing and understanding Bose-Einstein condensates. arXiv.1999.

[Kra91] KRAUSE, D. [1991], Multisets, quasi-sets and Weyl's aggregates. *Journal of Non-Classical Logic* 8 (2): 9-39.

[Kra94] KRAUSE, D., Non-reflexive logics and the foundations of physics. In Cellucci, C., Di Maio, M. C. e Roncaglia, G., (eds.), *Logica e Filosofia della Scienza: problemi e prospettive*, (Atti del Congresso Triennale della Società Italiana di Logica e Filosofia delle Scienze, Lucca, Gennaio 1993), Ed. ETS, Pisa: 393-405.

[KraAre15] KRAUSE, D. AND ARENHART, J. R. B. [2015], Is identity really so fundamental? Preprint, http:

//philsci-archive.pitt.edu/11295 (forthcoming in *Foundations of Science*).

[KraAre16] KRAUSE, D. AND ARENHART, J. R. B. [2016], *The Logical Foundations of Scientific Theories: Languages, Structures, and Models*. London: Routledge.

[KraAre18] KRAUSE, D. AND ARENHART, J. R. B. [2018], Presenting Nonreflexive Quantum Mechanics: Formalism and Metaphysics. *Cadernos de História e Filosofia da Ciência* 4 (2) Jan-Jun. 2016, pp. 59-91.

[Mac63] MACKEY, G. [1963], *Mathematical Foundations of Quantum Mechanics*. New Yourk and Amsterdam: W. A. Benjamim.

[Man10] MANIN, YU. I. [2010], *A course in mathematical logic for mathematicians*. 2nd ed., Springer.

[Men79] MENDELSON, E. [1979], *Introduction to mathematical logic*. 2a. ed. Monterey: Wadsworth Advanced Books & Software.

[Piag86] PIAGET, J. [1986], *The Construction of Reality in the Child*. Ballantine Books.

[Per.14] RODRIGUEZ-PEREYRA, G. [2014], *Leibniz's Principle of Identity of Indiscernibles*. Oxford: Oxford Un. Press.

[RedTel91] REDHEAD, M. AND TELLER, P. [1991], Particles, particle labels, and quanta: the toll of unacknowledged metaphysics, *Foundations of Physics* 21, 43-62.

[Rov18] ROVELLI, C. [2018], *The Order of Time*. New York: Riverhead Books.

[Sch53] SCHRÖDINGER, E. [1953], What is matter?, *Scientific American* Sept., pp.52-57.

[Sch98] SCHRÖDINGER, E. [1998], What is an elementary particle? In Castellani, E. (ed.) *Interpreting Bodies: Classical and Quantum Objects in Modern Physics*. Princeton: Princeton Un. Press, pp. 197-210.

[Sup72] SUPPES, P. [1972], *Axiomatic Set Theory*. New York: Dover Pu.

[SylCos88] SYLVAN, R. AND DA COSTA, N. C. A. [1988], Cause as an implication. *Studia Logica* 47 (4): 413-28.

[vanF75] VAN FRAASSEN, B. [1975], The labyrinth of quantum logics. In C. A. Hooker (ed.), *The Logico-Algebraic Approach to Quantum Mechanics, Vol. I: Historical Evolution*. Dordrecht: D. Reidel, pp. 575-607.

[Wey49] WEYL, H. [1949], *Philosophy of Mathematics and Natural Science*. Princeton: Princeton Un. Press.

[Wig59] WIGNER, E. [1959], *Group Theory and its Applications to the Quantum Mechanics of Atomic Spectra*. New York and London: Academic Press.

[WooZur82] WOOTERS, W. K. AND ZUREK, W. H. [1982], A single quantum cannot be cloned. *Nature* 299: 802-3 (28 Oct.1982)

On the existence of hidden machines in computational time hierarchies

Felipe S. Abrahão, Klaus Wehmuth, and Artur Ziviani

ABSTRACT. Challenging the standard notion of totality in computable functions, one has that, given any sufficiently expressive formal axiomatic system, there are total functions that, although computable and "intuitively" understood as being total, cannot be proved to be total. In this article we show that this implies the existence of an infinite hierarchy of time complexity classes whose representative members are hidden from (or unknown by) the respective formal axiomatic systems. Although these classes contain total computable functions, there are some of these functions for which the formal axiomatic system cannot recognize as belonging to a time complexity class. This leads to incompleteness results regarding formalizations of computational complexity.

1. INTRODUCTION

The standard notion of a function being total is one of the defying counterintuitive phenomena in axiomatizations of computer science (within Zermelo-Fraenkel with the Axiom of Choice (ZFC) or any other standard axiomatics for set theory). As shown in [6], one of the damaging difficulties for axiomatizations based on first-order theories, encompassing Peano Arithmetic (PA) or ZFC, is the fact that, for every such a theory, there are total computable functions that are not recognizable as total recursive/computable functions. This occurs in the context of fast-growing functions in such a way that, beyond a certain growth rate, some total computable functions cannot be proved to be total in sufficiently expressive formal axiomatic systems. These functions can be computed for each input, but, although computable and total, the respective expression "function f is total" cannot be a theorem. Thus, assuming the consistency of the chosen formal axiomatic systems, one can show there are incompleteness results regarding the totality of computable functions [6].

(Felipe S. Abrahão, Klaus Wehmuth, Artur Ziviani) NATIONAL LABORATORY FOR SCIENTIFIC COMPUTING (LNCC), 25651-075 – PETROPOLIS, RJ – BRAZIL
E-mail addresses: fsa@lncc.br, klaus@lncc.br, ziviani@lncc.br.
Authors acknowledge the partial support from CNPq through their individual grants: F. S. Abrahão (313.043/2016-7), K. Wehmuth (312599/2016-1), and A. Ziviani (308.729/2015-3). Authors acknowledge the INCT in Data Science – INCT-CiD (CNPq 465.560/2014-8). Authors also acknowledge the partial support from FAPESP (2015/24493-1), and FAPERJ (E-26/203.046/2017).

In this article, we show that the same kind of phenomenon also strikes the very foundation of computational complexity. In particular, instead of formalizing the notion of "function f is total", we investigate the notion of "function f belongs to a time complexity class" in formal axiomatic systems. Not only we demonstrate incompleteness of certain formulas about time complexity classes for axiomatizations based on first-order theories, but we also show the existence of a whole denumerable hierarchy of time complexity classes for which there are members that are not recognizable as belonging to a time complexity class.

2. Preliminaries

2.1. Computable functions and formal axiomatic systems. Regarding some basic notation, let $\{0,1\}^*$ be the set of all binary strings. Let $|x|$ denote the length of a string $x \in \{0,1\}^*$. Let $(x)_2$ denote the binary representation of the number $x \in \mathbb{N}$. In addition, let $(x)_L$ denote the representation of the number $x \in \mathbb{N}$ in language L.

Definition 2.1. Let $\mathbf{M}(x)$ denote the output of a Turing machine (TM) $\mathbf{M}$ when x is given as input in its tape. Thus, $\mathbf{M}(x)$ denotes a *partial recursive* function

$$\varphi_\mathbf{M}: \mathbf{L} \to \mathbf{L},$$
$$x \mapsto y = \varphi_\mathbf{M}(x)$$

where $\mathbf{L}$ is a language.

In particular, $\varphi_\mathbf{U}(x)$ is the *universal* partial recursive function [10] and $\mathbf{L_U}$ denotes a universal programming language for a universal Turing machine $\mathbf{U}$. Note that, if x is a non-halting program on $\mathbf{M}$, then this function $\mathbf{M}(x)$ is undefined for x. Wherever $n \in \mathbb{N}$ or $n \in \{0,1\}^*$ appears in the domain or in the codomain of a partial (or total) recursive function

$$\varphi_\mathbf{M}: \mathbf{L} \to \mathbf{L},$$
$$x \mapsto y = \varphi_\mathbf{M}(x)$$

where $\mathbf{M}$ is a Turing machine, running on language $\mathbf{L}$, it actually denotes $(n)_\mathbf{L}$.

Let $\{e\}$ denote the partial computable function $\varphi_\mathbf{M}$ for which e is its index (e.g., its Gödel number) such that the Kleene's predicate $T(e, x, z)$ has a well defined value z whenever there is a y such that $\varphi_\mathbf{M}(x) \equiv \mathbf{M}(x) \equiv \{e\}(x) = y$. Note that we are employing e for symbolizing the encoding of a *deterministic* TM, for example by employing Gödel numbers.

With respect to weak asymptotic dominance of function f by a function g, we employ the usual $f(x) = \mathbf{O}(g(x))$ for the big $\mathbf{O}$ notation when f is asymptotically upper bounded by g; and with respect to strong asymptotic dominance by a function g, we employ the usual $f(x) = \mathbf{o}(g(x))$ when g dominates f.

Let S denote any sufficiently expressive formal axiomatic system (FAS) in the language L such that there is an interpretation of Zermelo-Fraenkel with the Axiom of Choice (ZFC) into S. Note that there is then an interpretation of Peano Arithmetic (PA) into S, and therefore Kleene's predicate is definable in the language of S. As in [6], we denote the well-formed formula (wff) expressing "$\{e\}$ is a total function" in the language of S (i.e., $\forall x \exists z T(e,x,z)$) by $[\{e\} \text{ is total}]$. From the Kleene's predicate $T(e,x,z)$, one can also construct a wff (e.g., denoted by $[\{e\}(x) = y]$) in the language of S that defines the function relation $\{e\}(x) = y$ wherever there is a y given a x. And, since S encompasses PA, then $[\text{Prov}_S(h,x)] \in L$, where $\text{Prov}_S(h,x)$ is the wff in PA that represents the existence of a proper deductive proof of the wff encoded by x in which h is the Gödel number of the sequence proof steps from the axioms of S that end in the wff encoded by x [1, 11].

In addition, we have that the partial computable function $\textbf{\textit{time}}(e,x)$ that returns the computation running time (or *running time* for short) of the partial computable function $\{e\}$, if it defined for x, can also be defined in the language of S, which we denote by $[\textbf{\textit{time}}(e,x) = y]$ [3, 2]. Note that $\textbf{\textit{time}}(e,x)$ is the running time of deterministic TMs with input x and that, since $\{e\}$ is computable,

$$\textbf{\textit{time}}(e,x) \text{ is defined iff } \{e\}(x) \text{ is also defined.}$$

For example, a TM of index t_e that computes $\textbf{\textit{time}}(e,x)$ can be the one that receives x as input, emulates $\mathbf{M}$ of index e with input x, with an additional tape for counting the number of computation steps of e, and then returns the value from this additional counting tape whenever $\mathbf{M}$ reaches a halting state. In particular, we know that the running time for calculating this emulation (wherever the value of $\{e\}(x)$ is defined) is in the worst case cubic, that is, $\textbf{\textit{time}}(t_e, x) = \mathbf{O}\left((\textbf{\textit{time}}(e,x))^3\right)$ [8].

2.2. **Computational time complexity.** We base our notion of time complexity classes in a traditional manner as in the literature [8, 9]. The time complexity class $\textbf{TIME}\,(f(|x|))$ is the class of decision problems that can be solved by deterministic TMs in $\mathbf{O}(f(|x|))$ computation steps. Analogously, we also have the time complexity class $\textbf{NTIME}\,(f(|x|))$ for non-deterministic TMs. This way, the polynomial time complexity $\textbf{P-TIME}$ of all decision problems that can be computed by deterministic TMs in polynomial computation steps as a function of the input length is given by the parametrized time complexity $\textbf{P-TIME} = \bigcup_{j>0} \textbf{TIME}\,(|x|^j)$ and the same applies analogously to the non-deterministic case $\textbf{NP-TIME}$.

In addition to decision problems (i.e., TMs that always decide correctly whether an input belongs or not to a language L), one also has *function problems*. Thus, the class $\textbf{FP-TIME}$ is the class of partial functions $\{e\}$ such that, for every x for which there is $y = \{e\}(x)$, one has that $\textbf{\textit{time}}(e,x) = \mathbf{O}\left(|x|^k\right)$ is dominated by a polynomial, where $k \in \mathbb{N}$. More specifically, the time complexity class $\textbf{FTIME}\,(f(|x|))$ is the class of function problems that can be solved by deterministic TMs in $\mathbf{O}(f(|x|))$ computation steps. Analogously, the class $\textbf{FNP-TIME}$ is the class of partial relations R such that, for every x for which there is y with $(x,y) \in R$, one has that there is a non-deterministic polynomially time-bounded TM that can find at least one y such that $(x,y) \in R$.

The class of all function problems in **FNP-TIME** that are total (i.e., for every x there is at least one y with $(x,y) \in R \in$ **FNP-TIME**) is denoted by **TFNP-TIME**. The same way, the class of all function problems in **FP-TIME** that are total (i.e., for every x there is *only one* y with $(x,y) \in f \in$ **FP-TIME**) is denoted by **TFP-TIME**. And, as usual, the time complexity class **TFTIME**$(f(|x|))$ is the class of total function problems that can be solved by deterministic TMs in $O(f(|x|))$ computation steps. An interesting theorem that already is known to hold is that [8]:

FP-TIME = **FNP-TIME** $\iff$ **P-TIME** = **NP-TIME** .

It is straightforward to show that, since both the time complexity of deciding whether a TM halts or not in polynomial time and the time complexity of emulating a TM's running time are cubic, one has that

TFP-TIME $\subseteq$ **FP-TIME** $\subseteq$ **TFNP-TIME** $\subseteq$ **FNP-TIME**

hols.

All of these classes can also be analogously extended to the exponential time complexity classes **EXP-TIME**, **FEXP-TIME**, **TFEXP-TIME**, and so on. In this article, we will deal only with total function problems. In particular, our proofs hold for deterministic Turing machines.

3. A LOGIC DEPENDENCE OF COMPUTER SCIENCE FROM THE TOTALITY OF FUNCTIONS

In [6], it is shown the existence of function that defies the axiomatization of computer science, so that even "fairly intuitive" notions cannot be grasped by sufficiently expressive formal axiomatic systems. A computable function can be constructively defined, i.e., programmed on an universal TM, such that there is a diagonalization procedure that eventually lands on every total computable function f that S can prove it is total (i.e., $S \vdash [f$ is total$]$) and then maximizes its value. In other words, it is a computable function that eventually grows faster than any other total computable function that S proves is total. A function F with these properties was constructed as follows:

Definition 3.1. Let F be a function whose values are given by the TM $\mathbf{M}_F$ that receives $n \in \mathbb{N}$ as input and:

(1) enumerates all e such that $S \vdash [\text{Prov}_S(y, [\{e\} \text{ is total}])]$ and $y \leq n$;
(2) constructs a finite list $(e_1, \ldots, e_z)$ of these functions $\{e\}$, where $z \in \mathbb{N}$;
(3) returns $\max \{y \mid y = \{e\}(x) + 1 \wedge x \leq n\}$.

Note that we have that F is intuitively total in the sense that, if S is indeed consistent, then every computable function indexed by e that S proves is total will always return a value. This way, a maximization from the constructed list of these e's will be always possible.

Moreover, F is clearly a partial computable function. There is a TM $\mathbf{M}_F$ such that, for every given particular n as input, it will effectively calculate the value of $F(n)$. The question is whether or not the fact that F is total can be proved in S: and the answer is negative. To this end, just note that, if S eventually proves [$\{e_F\}$ is total], where e_F is the index of the TM $\mathbf{M}_F$, there will be a n_0 from which, for every $n \geq n_0$, $F(n) > F(n)$. Thus, one can individually check that, for each natural number n, $F(n)$ exists and can be computed. However, S cannot "join" all those results together to show that F is total.

This kind of diagonalization on growth rate can be also found in the Busy Beaver function. Specially in the form $BB: \mathbb{N} \to \mathbb{N}$ as a function that returns the largest integer that a program $p \in \mathbf{L_U}$ with length $\leq N \in \mathbb{N}$ can output running on machine $\mathbf{U}$ [4]. Such a Busy Beaver function has several interesting properties. For example, although function BB eventually grows faster than any other computable function f_c (that is, for every computable function $f_c: \mathbb{N} \to \mathbb{N}$, there is $N_0 \in \mathbb{N}$ such that, for every $N \geq N_0$, $BB(N) > f_c(N)$), it is a *scalable* uncomputable function, i.e., for every $N \in \mathbb{N}$, there is a program $p \in \mathbf{L_U}$ such that $\mathbf{U}(p) = BB(N)$ (in particular, $|p| \leq N$). Moreover, function BB is an incompressible function, i.e., there is constant $c \in \mathbb{N}$ such that, for every $N \in \mathbb{N}$, $\mathbf{I}_A(BB(N)) \geq N - c$, where $\mathbf{I}_A(\cdot)$ is the algorithmic information or algorithmic complexity of an object [7, 5].

Now, note that BB eventually grows faster than function F, since F is computable. Indeed, function F defined in [6] also has that "ungraspable but eventually reachable" property that BB has. In other words, although function F eventually grows faster than any other S-provenly total computable function f_c, it is a *scalable* total computable function. However, what one may deem to be even more counter-intuitive in the case of F, is that it can be indeed computed by a effectively constructible TM, wheres BB is an uncomputable function. Thus, as BB works like a "ceiling" function for every computable function—function F included—, function F also works like a "ceiling" function, but for S-provenly total computable functions.

This way, as pointed out in [6], any attempt to find a FAS for computer science in which the naive intuition of e.g. totality of functions can be grasped, faces difficulties of the same order of incompleteness in mathematical logic. Other incompleteness phenomena related to axiomatization of computer science were also presented in [6], such as the relation between the totality of function F and Σ_1-soundness and recognition of sets of polynomially time-bounded TMs. With respect to the latter, we shall show in this article other incompleteness phenomena in computer science that some may deem to be even more dramatic. In other words, going even further into computer science, we shall show later on that this "odd" function F is also related to incompleteness phenomena in one of the central subjects in theoretical computer science: computational complexity (or algorithm analysis).

4. General time complexity classes

We aim to show that there are time complexity classes of function problems that contain functions associated with so fast-increasing running time that the expression in S that defines "this function belongs to a time complexity class" cannot be proved, although it would be intuitively true in a standard model of arithmetic for example. For this purpose, the main idea is that a recognizable time complexity class in S is the one for which every TM M in this class can be proved to belong to by S. In this article, we are only dealing with deterministic TMs and with total function problems. Thus, more formally:

Definition 4.1. Let L be the language of the FAS S. Let $[\{e\} \in \mathbf{TF}X\text{-}\mathbf{TIME}]$ denote

$$\exists z (z \in X \wedge \forall x \exists h, k ([\mathbf{time}(e,x) = k] \wedge k \leq h \wedge [\{z\}(|x|) = h]))$$

in the language L, where X is a free variable. We say that S *recognizes* $\{e\}$ as belonging to a deterministic time complexity class X iff $S \vdash [\{e\} \in \mathbf{TF}X\text{-}\mathbf{TIME}]$.

In general, X can be any set definable in the language of S. This is the case of for example $X = \{f_p\}$, $X = \{f_{exp}\}$, or $X = \{f\}$, where f_p is a polynomial, f_{exp} is an exponentiation, and f is any time-constructible [9] (or proper complexity function [8]). Thus, we immediately obtain from Definition 4.1 that

$$\mathbf{TF}\{f\}\text{-}\mathbf{TIME} = \mathbf{TFTIME}(f) \ .$$

One can also define X as a union of functions in order to cover what is called *parametrized time complexity classes*. For example: if $X = \{f | f(x) = \mathbf{O}\left(x^k\right) \wedge x, k \in \mathbb{N}\}$ is the set of all polynomials, then $\mathbf{TF}X\text{-}\mathbf{TIME} = \mathbf{TFP}\text{-}\mathbf{TIME}$; if $X = \{f | f(x) = \mathbf{O}\left(k^{(x^m)}\right) \wedge x, k, m \in \mathbb{N}\}$ is the set of all exponentiation, then $\mathbf{TF}X\text{-}\mathbf{TIME} = \mathbf{TFEXP}\text{-}\mathbf{TIME}$; and so on.

Now, instead of a specific set X of running time functions, one could also investigate whether or not S can recognize a TM belonging to an arbitrary deterministic time complexity class. Indeed, this is easily defined by:

Definition 4.2. Let L be the language of the FAS S. We say that S *recognizes* $\{e\}$ as belonging to *at least one* deterministic time complexity class iff $S \vdash \exists X [\{e\} \in \mathbf{TF}X\text{-}\mathbf{TIME}]$.

5. Non-recognizable Turing machines in a time complexity hierarchy

In this section, we tackle the main objective in this article. We aim to investigate total time-bounded TMs for which S cannot recognize as belonging to *at least one* deterministic time complexity class. Indeed, TM $\mathbf{M}_F$ is one of these:

Theorem 5.1. *There is a total computable function $\{e\}$ such that, if S is consistent, then*

(1) $$S \nvdash \exists X [\{e\} \in \mathbf{TF}X\text{-}\mathbf{TIME}]$$

and

(2) $\quad S \nvdash \neg \exists X [\{e\} \in \mathbf{TF}X\text{-}\mathbf{TIME}]$.

Proof. We employ in this proof the total computable function F. Let $\{e\}$ denote F. First, Equation (2) trivially follows from the fact that, if $S \vdash \neg \exists X [\{e\} \in \mathbf{TF}X\text{-}\mathbf{TIME}]$ and S is consistent, then we would have that F is not a total function in a standard model of S for example. Then, it remains to prove that $S \nvdash \exists X [\{e\} \in \mathbf{TF}X\text{-}\mathbf{TIME}]$. To this end, suppose $S \vdash \exists X [\{e\} \in \mathbf{TF}X\text{-}\mathbf{TIME}]$ holds. Now, note from Definitions 4.2 and 4.1 that

(3) $\quad S \vdash (\exists X [\{e\} \in \mathbf{TF}X\text{-}\mathbf{TIME}] \to [\mathbf{time}(e,\cdot) \text{ is total}])$

and

(4) $\quad S \vdash ([\mathbf{time}(e,\cdot) \text{ is total}] \to [\{e\} \text{ is total}])$

hold. Therefore, we would have that

(5) $\quad S \vdash [\{e\} \text{ is total}]$,

which we already know it is false. $\square$

In addition, we can extend function F in order to include $\exists X [F \in \mathbf{TF}X\text{-}\mathbf{TIME}]$:

Definition 5.1. Let $F^{(1)}$ be a function whose values are given by the TM $\mathbf{M}_{F^{(1)}}$ that receives $n \in \mathbb{N}$ as input and:
(1) enumerates all e such that $S \vdash [\mathrm{Prov}_{S^{(1)}}(y, [\{e\} \text{ is total}])]$ and $y \leq n$, where $S^{(1)} = S + \exists X [F \in \mathbf{TF}X\text{-}\mathbf{TIME}]$;
(2) constructs a finite list $(e_1, \ldots, e_z)$ of these functions $\{e\}$, for some $z \in \mathbb{N}$;
(3) returns $\max \{y \,|\, y = \{e\}(x) + 1 \wedge x \leq n\}$.

And, by continuing this process, we will obtain a sequence $F, F^{(1)}, F^{(2)}, \ldots$ of functions defined respectively for $S, S^{(1)}, S^{(2)}, \ldots$, where $S = S^{(0)}$, $F = F^{(0)}$ and

$$S^{(k+1)} = S^{(k)} + \exists X \left[F^{(k)} \in \mathbf{TF}X\text{-}\mathbf{TIME} \right] .$$

This way, for each iteration one obtains another $\exists X \left[F^{(k+1)} \in \mathbf{TF}X\text{-}\mathbf{TIME} \right]$ such that

(6) $\quad S^{(k+1)} \nvdash \exists X \left[F^{(k+1)} \in \mathbf{TF}X\text{-}\mathbf{TIME} \right]$

and

(7) $\quad S^{(k+1)} \nvdash \neg \exists X \left[F^{(k+1)} \in \mathbf{TF}X\text{-}\mathbf{TIME} \right]$.

Furthermore, since the time complexity overhead in simulating a TM is cubic, we will have that

(8) $\quad S^{(k+1)} \vdash \exists X \left[\mathbf{time}(e_{F^{(k)}}, \cdot) \in \mathbf{TF}X\text{-}\mathbf{TIME} \right]$

In fact, any time-constructible total function f composed with $\mathbf{time}(e_{F^{(k)}}, \cdot)$ can be proved by $S^{(k+1)}$ to be in a time complexity class, that is

(9) $\quad S^{(k+1)} \vdash \exists X \left[f\left(\mathbf{time}(e_{F^{(k)}}) \right), \cdot) \in \mathbf{TF}X\text{-}\mathbf{TIME} \right]$.

For example, one has that

(10) $\quad S^{(k+1)} \vdash \exists X \left[\text{hyperexp}\left(\textit{\textbf{time}}(e_{F^{(k)}}), \cdot\right) \in \textbf{T}\textbf{F}X\textbf{-TIME} \right]$,

where hyperexp($\cdot$) is the hyperexponentiation function.

Thus, the following theorem establishes an infinite denumerable time hierarchy built from $F, F^{(1)}, F^{(2)}, \ldots$:

Theorem 5.2. *Let X be an arbitrary set of functions such that $\textit{\textbf{time}}(e_{F^{(k)}}, \cdot)$ is the faster growing function in X. Then, there is a set Y of functions such that $X \subseteq Y$ and $F^{(k+1)} \notin \textbf{T}\textbf{F}X\textbf{-TIME}$ and $F^{(k+1)} \in \textbf{T}\textbf{F}Y\textbf{-TIME}$.*

Proof. The main idea of the proof is to show that, if $F^{(k+1)} \in \textbf{T}\textbf{F}X\textbf{-TIME}$, then it would eventually grow faster than it could. Suppose that $F^{(k+1)} \in \textbf{T}\textbf{F}X\textbf{-TIME}$. Then, by our construction of X, we would have $\textit{\textbf{time}}(e_{F^{(k+1)}}, x) = \mathbf{O}\left(\textit{\textbf{time}}(e_{F^{(k)}}, x)\right)$. But, from Equation (10), we have that

$$S^{(k+1)} \vdash \exists X \left[\text{hyperexp}\left(\textit{\textbf{time}}(e_{F^{(k)}}), \cdot\right) \in \textbf{T}\textbf{F}X\textbf{-TIME} \right] .$$

Now, remember that $F^{(k+1)}$ eventually grows faster than any $S^{(k+1)}$-provenly total computable function. Therefore, the value returned by $F^{(k+1)}$ would eventually become so large that even the length of its binary representation (which is in a logarithmic order of the value) would strongly dominate the number of computation steps (which by construction is in $\mathbf{O}\left(\textit{\textbf{time}}(e_{F^{(k)}}, x)\right)$) that are generating it, which is a contradiction. Finally, in order to prove there is a Y such that $F^{(k+1)} \in \textbf{T}\textbf{F}Y\textbf{-TIME}$ and $X \subseteq Y$, we define $Y \coloneqq X \cup \{\mathbf{O}\left(\textit{\textbf{time}}(e_{F^{(k+1)}}, \cdot)\right)\}$. $\square$

6. Conclusion

By formalizing the general notion of a computable function belonging to a time complexity class in first-order language, we showed in Theorem 5.1 that, for any sufficiently expressive formal axiomatic system (e.g., ZFC) believed to be consistent, there is a function that belongs to a time complexity class but that this formal axiomatic system cannot prove it belongs to a time complexity class. Such a phenomenon is a new type of incompleteness, but that occurs in computational complexity expressed by formal axiomatic systems. In addition, we showed in Theorem 5.2 that there is an infinite sequence of total computable functions, each within respectively higher time complexity classes, such that none of these functions can be recognized as belonging to a time complexity class by sufficiently expressive formal axiomatic systems. In other words, for every sufficiently expressive formal axiomatic system believed to be consistent, there are total computable functions that are hidden from belonging to time complexity classes, which in turn are ordered in an infinite denumerable time complexity hierarchy. Thus, together with the logic dependence of computer science with respect to the concept of totality as in [6], the present article highlights the presence of such a logic dependence also in computational complexity.

References

[1] Felipe S. Abrahão. *Demonstrando a consistência da aritmética*. Universidade Federal do Rio de Janeiro (UFRJ), Brazil, 2011. doi: 10.5281/zenodo.1213459. URL https://doi.org/10.5281/zenodo.1213459. Master dissertation.

[2] Felipe S. Abrahão. *Metabiologia, Subcomputação e Hipercomputação: em direção a uma teoria geral de evolução de sistemas*. Ph.d. thesis, Universidade Federal do Rio de Janeiro (UFRJ), Brazil, 2015. Federal University of Rio de Janeiro (UFRJ), Rio de Janeiro. Available at http://objdig.ufrj.br/10/teses/832593.pdf.

[3] Felipe S. Abrahão. The "paradox" of computability and a recursive relative version of the Busy Beaver function. In Cristian Calude and Mark Burgin, editors, *Information and Complexity*, chapter 1, pages 3–15. World Scientific Publishing, Singapure, 1 edition, 2016. ISBN 978-9813109025. doi: 10.1142/9789813109032_0001.

[4] Felipe S. Abrahão, Klaus Wehmuth, and Artur Ziviani. Algorithmic networks: Central time to trigger expected emergent open-endedness. *Theoretical Computer Science*, 785:83–116, sep 2019. ISSN 03043975. doi: 10.1016/j.tcs.2019.03.008.

[5] Felipe S. Abrahão, Klaus Wehmuth, Hector Zenil, and Artur Ziviani. Algorithmic information and incompressibility of families of multidimensional networks. arXiv Preprints. Research report no. 8/2018, National Laboratory for Scientific Computing (LNCC), Petrópolis, Brazil, 2020. URL https://arxiv.org/abs/1810.11719v9.

[6] Walter Carnielli and Francisco Antônio Dória. Are the foundations of computer science logic-dependent? In Cedric Degremont, Laurent Keiff, and Helge Ruckert, editors, *Dialogues, Logics and Other Strange Things–Essays in Honour of Shahid Rahman*, pages 87–107. College Publications, London, 2008. ISBN 978-1904987130.

[7] Gregory Chaitin. *Algorithmic Information Theory*. Cambridge University Press, 3 edition, 2004. ISBN 0521616042.

[8] Christos H. Papadimitriou. *Computational Complexity*. Addison-Wesley Publishing, 1994. ISBN 0-201-53082-1.

[9] Elaine A Rich. *Automata Theory and Applications*. Prentice-Hall, Inc., USA, 2007. ISBN 0132288060.

[10] Hartley Rogers Jr. *Theory of Recursive Functions and Effective Computability*. MIT Press, Cambridge, MA, USA, 1987. ISBN 0-262-68052-1.

[11] C. Smorynski. The Incompleteness Theorems. In Jon Barwise, editor, *Handbook of mathematical logic*, volume 90 of *Studies in Logic and the Foundations of Mathematics*, chapter D.1, pages 821–865. Elsevier Science Publishers, 1977. doi: 10.1016/S0049-237X(08)71123-6. URL https://linkinghub.elsevier.com/retrieve/pii/S0049237X08711236.

What is an Axiom?

Jean-Yves Beziau

Dedicated to Francisco Antônio Dória for his 75th birthday

Abstract

After some methodological considerations, we start by a general overview of the meaning of the notion of axiom. We then examine two kinds of axioms, on the one hand the Zurich axioms, on the other hand the axiom of choice. After that we discuss two trinities: definition/axiom/proof, axiom/proof/truth. We conclude by some remarks on the rise and fall of the axiomatic method. We end by a few recollections about Francisco Dória.

Keywords Axiom, Definition, Proof, Truth, Axiomatic Method

Hymne à l'axiome par le Baron de Chambourcy
Ce qui est évident n'est pas forcément vrai,
Ne nous emmène pas souvent très loin.
Ce n'est pas non plus nécessaire de faire des hypothèses très compliquées,
Suppositions douteuses qui ne feront qu'embrouiller notre esprit.
L'idée est plutôt de s'élever au-dessus du tumulte,
Avoir une perspective plus claire et plus générale,
Qui nous permettra de guider notre pensée afin d'atteindre la compréhension.
Resdecendre dans la vallée et d'en apprécier la beauté,
Sans tomber dans les pièges du détail.
Axiome, aile qui nous fait survoler les platitudes.

1 A Four-Dimensional Perspective

"Axiom" is first of all a word. This in fact can be said of many words, and this can be seen as rather tautologous or/and nonsensical. But "axiom" is emblematic in the sense that it is a word, by contrast for example to "God", "Cat" or "Banana", that has no real translation in other languages, only alphabetic adaptations.

Originally we have "$ἀξίωμα$" in Greek, a word with the direct transliteration "axioma" in Spanish, Portuguese or Interlingua; and with small variations: "assioma" in Italian, "Аксиóма" in Russian, "axiome" in French, "Axiom" in German, English, Swedish, "Axiomo" in Ido, "Aksiyom" in Turkish.

It therefore works rather like a name of a country, a city, a river, a person, i.e. a proper name. Also similar to "axiom" are "logic" and "mathematics", which by chance have some direct connections with it.

However, despite the primacy of the word, we will not stay at this superficial level, but we will go deeper and deeper, higher and higher, trying to eradicate the root, going up to the fresh spring, opening some new perspectives.

In a previous paper we have promoted the pyramid of meaning [8]. This pyramid has a triangular basis: word, idea, thing (language / thought / reality). The top of the pyramid, supervising and synthetizing this triangle, we decided to call it "notion". The notional viewpoint reduces neither to language, nor to thought, nor to reality. It encompasses the three.

In the present paper we are dealing not only with the word "axiom", but also with the idea and reality attached to it, i.e. with the notion of axiom (no quotes!). To investigate a notion, we can go in four complementary directions:
• searching for a definition,
• understanding the relation of this notion with other notions,
• investigating the use of it,
• making a theory that will fix and clarify its meaning.

We will develop here this fourfold approach in a general perspective where these four aspects are intertwined.[1]

2 General Meaning

Before developing a critical discussion about a notion or elaborating a theory of it, it is always good to open our eyes to see how this notion has been conceived. This is important to avoid to be too idiosyncratic, on the other hand it is also important to avoid to sink in a descriptive ocean, an enumeration of all the ways a notion has been conceived, or can be thought in all possible worlds. To do that we will highlight here some definitions, quotes and associated words.

2.1 Definition

A definition can be found in a dictionary or in an encyclopedia. We have made a selection of 6 of them, 3 dictionaries and 3 encyclopedias.

Dictionary.com
1. a self-evident truth that requires no proof.
2. a universally accepted principle or rule.
3. *Logic, Mathematics.* a proposition that is assumed without proof for the sake of studying the consequences that follow from it.

Cambridge
1. formal
a statement or principle that is generally accepted to be true, but need not be so: *It is a widely held axiom that governments should not negotiate with terrorists.*
2. science, specialized
a formal statement or principle in mathematics, science, etc., from which other statements can be obtained: *Euclid's axioms form the foundation of his system of geometry.*

Merriam-Webster
1. a statement accepted as true as the basis for argument or inference: *one of the axioms of the theory of evolution*
2. an established rule or principle or a self-evident truth: *no one gives what he does not have*
3. a maxim widely accepted on its intrinsic merit: *the axioms of wisdom*

Encyclopedia Britannica
Axiom, in logic, an indemonstrable first principle, rule, or maxim, that has found general acceptance or is thought worthy of common acceptance whether by

---
[1] This paper, as two other recent papers of mine ([9], [10]), has a strong methodological and pedagogical aspect: at the same time that I am dealing with a topic, I am trying to investigate how this can be done.

virtue of a claim to intrinsic merit or on the basis of an appeal to self-evidence. An example would be: "Nothing can both be and not be at the same time and in the same respect."

In Euclid's Elements the first principles were listed in two categories, as postulates and as common notions. The former are principles of geometry and seem to have been thought of as required assumptions because their statement opened with "let there be demanded" (etestho). The common notions are evidently the same as what were termed "axioms" by Aristotle, who deemed axioms the first principles from which all demonstrative sciences must start; indeed Proclus, the last important Greek philosopher ("On the First Book of Euclid"), stated explicitly that the notion and axiom are synonymous. The principle distinguishing postulates from axioms, however, does not seem certain. Proclus debated various accounts of it, among them that postulates are peculiar to geometry whereas axioms are common either to all sciences that are concerned with quantity or to all sciences whatever.

In modern times, mathematicians have often used the words postulate and axiom as synonyms. Some recommend that the term axiom be reserved for the axioms of logic and postulate for those assumptions or first principles beyond the principles of logic by which a particular mathematical discipline is defined. Compare theorem.

Wikipedia

An axiom or postulate is a statement that is taken to be true, to serve as a premise or starting point for further reasoning and arguments. The word comes from the Greek axioma ($\alpha\xi\iota\omega\mu\alpha$) "that which is thought worthy or fit" or "that which commends itself as evident."

The term has subtle differences in definition when used in the context of different fields of study. As defined in classic philosophy, an axiom is a statement that is so evident or well-established, that it is accepted without controversy or question. As used in modern logic, an axiom is a premise or starting point for reasoning.

As used in mathematics, the term axiom is used in two related but distinguishable senses: "logical axioms" and "non-logical axioms". Logical axioms are usually statements that are taken to be true within the system of logic they define and are often shown in symbolic form (e.g., (A and B) implies A), while non-logical axioms (e.g., $a + b = b + a$) are actually substantive assertions about the elements of the domain of a specific mathematical theory (such as arithmetic).

When used in the latter sense, "axiom", "postulate", and "assumption" may be used interchangeably. In most cases, a non-logical axiom is simply a formal logical expression used in deduction to build a mathematical theory, and might or might not be self-evident in nature (e.g., parallel postulate in Euclidean geometry). To axiomatize a system of knowledge is to show that its claims can be derived from a small, well-understood set of sentences (the axioms), and there may be multiple ways to axiomatize a given mathematical domain.

Any axiom is a statement that serves as a starting point from which other statements are logically derived. Whether it is meaningful (and, if so, what it means) for an axiom to be "true" is a subject of debate in the philosophy of mathematics

The Columbia Electronic Encyclopedia
Axiom, in mathematics and logic, general statement accepted without proof as the basis for logically deducing other statements (theorems). Examples of axioms used widely in mathematics are those related to equality (e.g., "Two things equal to the same thing are equal to each other"; "If equals are added to equals, the sums are equal") and those related to operations (e.g., the associative law and the commutative law). A postulate, like an axiom, is a statement that is accepted without proof; however, it deals with specific subject matter (e.g., properties of geometrical figures) and thus is not so general as an axiom. It is sometimes said that an axiom or postulate is a self-evident statement, but the truth of the statement need not be evident and may in some cases even seem to contradict common sense. Moreover, a statement may be an axiom or postulate in one deductive system and may instead be derived from other statements in another system. A set of axioms on which a system is based is often wished to be independent; i.e., no one of its members can be deduced from any combination of the others. (Historically, the development of non-Euclidean geometry grew out of attempts to prove or disprove the independence of the parallel postulate of Euclid.) The axioms should also be consistent; i.e., it should not be possible to deduce contradictory statements from them. Completeness is another property sometimes mentioned in connection with a set of axioms; if the set is complete, then any true statement within the system described by the axioms may be deduced from them.

2.2 Quotes

Quotes is also an interesting tool which is generally not used systematically. This may be fine or/and funny to put one or two quotes at a beginning of a paper. But here what we are doing is different: we present a selection of quotes. The idea is to choose a set of quotes which are complementary (in a classical or Bohrian way) and exhaustive, i.e. encompass the main meanings of the notion, avoiding repetition. The authors of the quotes are important in the sense that they are considered as great and/or famous minds in a given field, and here also it is nice to have complementarity and exhaustivity. In this spirit we have selected the following nine quotes.

Henry Adams
I tell you the solemn truth, that the doctrine of the Trinity is not so difficult to

accept for a working proposition as any one of the axioms of physics.

Alonzo Church
The only thing that might have annoyed some mathematicians was the presumption of assuming that maybe the axiom of choice could fail, and that we should look into contrary assumptions.

Tom Stoppard
I am not a mathematician, but I was awere that for centuries, mathematics was considered the queen of sciences because it claimed certianty. It was grounded on some fundamental certainties - axioms - that led to others.

Gustave Flaubert
There are neither good nor bad subjects. From the point of view of pure Art, you could almost establish it as an axiom that the subject is irrelevant, style itself being an absolute manner of seeing things.

L.Ron Hubbard
A science is something which is constructed from truth on workable axioms. There are 55 axioms in scientology which are very demonstrably true, and on these can be constructed a great deal.

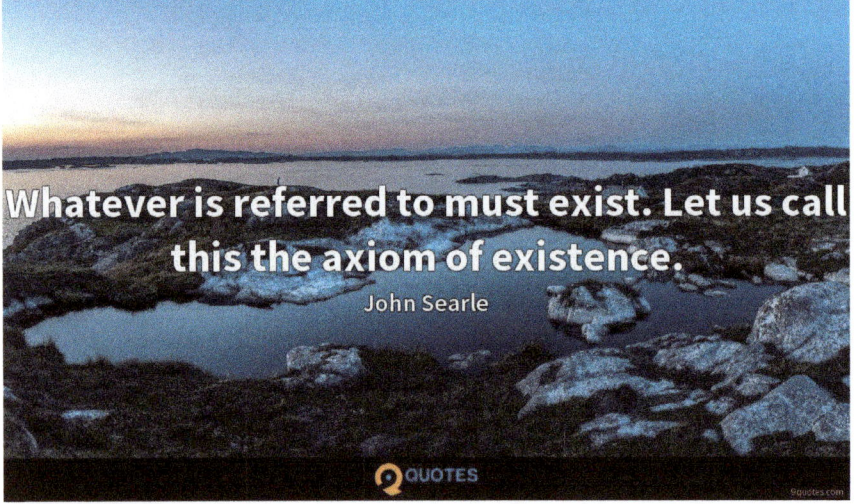

Albert Einstein
The grand aim of all science it to cover the greatest number of empirical facts by logical deduction from the smallest number of hypotheses of axioms.

Gottfried Leibniz
Finally there are simpl ideas of which no definition can be given, there are also

axioms or postulates, or in word primary principles, which cannot be proved and have no need of proof.

Vincent McNabb
We shall not hold the dangerous axiom that 'truth is the best policy', because policy is but a means to an end; and truth is an end, not a means.

2.3 An Axiomatic Cloud

We are now ready to present a cloud of notions linked to the notion of axiom. This idea of cloud is related with the work I have done with Patrick Suppes at Stanford centered on the notion of *association* [26]. A cloud can be seen as a first step to establish a *network*. The notions in a cloud of a notion X are those which are most commonly associated to X, but the cloud is not articulated in the form of a hierarchy.

2.4 What an Axiom is Not

When examining a notion, it is also good to investigate opposite notions. At a linguistic level, this can be done by designing a cloud of antonyms. But we will not use this methodology here. We will give typical examples of statements

which are not axioms. Here is a short list:
- *Tomatoes are not potatoes*
- $2 + 2 = 4$
- $\exists x \exists y \exists z (x < y) \wedge (y < z) \wedge \neg(x < z)$

Why are those statements not axioms? Here three reasons explaining the exclusion from the universe of axioms of these statements:
- A axiom is not a simple truth.
- An axiom is something that leads you somewhere: it should be useful as a principle of action, or deduction.
- An axiom should be pretty general.

The problem here is not the question of the distinction between *axiom* and *postulate*, or between an *absolute truth* and a *hypothesis*, or a *degree of generality*. About these questions and the evolution of the meaning of the notion of axiom, the reader may consult for example [20].

3 Two Axiomatic Examples

3.1 Zurich Axioms

Before going directly to hardcore problems of logic and mathematics, let us deal with a lighter topic, banking. On the one hand for having a general perspective, on the other hand because it may keep some readers on the way. Our paper is not directed only for professional logicians and mathematicians but for a larger audience.

There is a book entitled *The Zurich Axioms*. It has been written by Max Gunther (1927-1998), the son of a famous Swiss banker, trying to explain how his father and other bankers succeeded to earn lots of money. Here is how he presents the book and the origin of the use of the word "axiom" in the title:

> The Swiss, observing all this, conclude that the sensible way to conduct one's life is not to shun risk, but to expose oneself to it deliberately. To join the game; to bet. But not in the caterpillars mindless way. To bet, instead, with care and thought. To bet in such a way that large gains are more likely than large losses. *To bet and win.*
>
> Can this be done? Indeed. There is a formula for doing it. Or perhaps formula is the wrong word, since it suggests mechanical actions and a lack of choice. A better word might be 'philosophy'. This formula or philosophy consists of twelve profound and mysterious rules of risk-taking called the Zurich Axioms.
>
> The list of rules evolved gradually. It grew shorter, sharper, tidier, and more useful as time went on. Nobody remembers who coined the term 'Zurich Axioms?', but that is the name by which the rules came to be known and are still known.

Here are some examples of these axioms:

- Always play for meaningful stakes.
- When the ship starts to sink, don't pray. Jump.
- Chaos is not dangerous until it begins to look orderly.

We have here *rules* or *principles* that are used to develop a certain activity. It can be an explicit command or a statement that will guide your action. We are therefore in a sphere similar to the one of a game or a religion. You may believe that if you obey the ten commandments you will go to heaven, or at least avoid to go to hell. At the level of gaming, these Zurich axioms are not like the rules you must obey to play e.g. chess, but some principles you may follow to win the game.

We can use the same generic word, "rule", for both cases to manifest the interplay between the two, but it also important to emphasize the distinction, or even the opposition: deontic rules vs. strategic rules (see [21]). Strategic rules are neither obvious nor mechanical. If it were the case everybody will be millionaire and chess champion. It is not like a cooking recipe: you follow the rules and the result is a nice cake!

An important feature of the Zurich axiom is that they are a synthesis of different ideas that gives a *direction*, an *orientation*. Axioms in logic and mathematics can also be seen in this way.

3.2 The Axiom of Choice

Let us now, before going to the general abstract notion of axiom in mathematics and logic, study a specific case of "real" axiom. Generality is nice, but it is also good to focus on a particular example, not a sample case, but a typical case.

The word "axiom" is widely known, its precise logico-mathematical meaning, if any, not so much. And if we ask for an example of axiom, it is not clear if something will promptly emerge, will be distinguished. There is of course the *axiom of parallel*, but it is rather known as a *postulate*. We will not choose it here as a guinea pig due to its long history. We will choose an axiom of modern times, *the axiom of choice*.

The name is famous, but few people know exactly what this axiom is, in particular for two reasons: on the one hand the name of this axiom is rather ambiguous and does not properly reflect its meaning, on the other hand there are many different equivalent versions of this axiom (see [23]), each having a different meaning. Only a hard extensionalist can claim there is only one meaning behind all the versions of this axiom.

The qualifier "choice" is highly ambiguous, because it gives the impression that it has to do with que difficult choice question we may face in every day life: how I will choose to go on the left rather on the right, decide to live in Paris rather than in New York, to become a dentist rather than a singer, etc.

The ambiguity of the meaning of this axiom turns the question of its truth or/and obviousness problematic. There can be two ways out: equivalence or derivation.

The equivalence way is to find a formulation of it which is more obvious than the other ones. One of the most famous formulations of the axiom of choice, Zorn's lemma (a partially ordered set containing upper bounds for every chain contains at least one maximal element), gives the impression that this axiom is rather artificial, far-fetched. On the other hand Edison Farah proved that general distributivity of conjunction relatively to disjunction is equivalent to the Axiom of Choice [15].[2] This seems rather "natural".

Another way out is to derive the axiom of choice from other more elementary axioms. This is what people have tried to do. However the fact that this axiom is independent of the axioms of the most famous set theory, ZF, shows that there is no easy way to derive this axiom from some basic axioms whose meaning and truth is rather simple and obvious.

Independently of its own meaning, one of the reasons to support this axion can be pragmatic: with it we can do lots of interesting things, we can prove many interesting results that cannot be proven without it. It can therefore be welcome from the viewpoint of a winning strategy. But among mathematicians there are some people who don't like this pragmatic view and this can be for quite different reasons. A "Platonist" may say that this axiom does not correspond to abstract reality of the mathematical universe. A "Constructivist" may argue that this axiom does not correspond to any constructive procedure.

4 Two Trinities

4.1 Definition, Axiom, Proof

There is a famous DAP trilogy: definition, axiom, proof. It was promoted in particular by Blaise Pascal (1623-1662) as the basic methodology for mathematics. The ideas of Pascal [19] can be summarized in the following table:

---

[2]Edison Farah was the host of André Weil in São Paulo. Weil had the idea that this conjecture proven by Farah was false.

	PASCAL 8 RULES
Rules for Definitions	Not to undertake to define any of the things so well-known of themselves that clearer terms cannot be had to explain them.
	Not to leave any terms that are at all obscure or ambiguous without definition.
	Not to employ in the definition of terms any words but such as are perfectly known or already explained.
Rules for Axioms	Not to omit any necessary principle without asking whether it is admitted, however clear and evident it may be.
	Not to demand, in axioms, any but things that are perfectly evident of themselves.
Rules for Proofs	Not to undertake to demonstrate <u>any thing</u> that is so evident of itself that nothing can be given that is clearer to prove it.
	To prove all propositions at all obscure, and to employ in their proof only very evident maxims or propositions already admitted or demonstrated.
	To always mentally substitute definitions in the place of things defined, in order not to be misled by the ambiguity of terms which have been restricted by definitions.

The *DAP* trilogy is the basis of the *axiomatic method*, despite the fact that this terminology just keeps explicit the notion of axiom. It is a bit weird, but we see a similar situation in the *holy trinity*, where the holy spirit predominates over the father and the son.

And, as in the case of Christianity, the relations between the three notions of the *DAP* trinity is quite complex. If we have, let us say, a group of axioms for beer and sausage, can't we say that these axioms are defining what beer and sausage are? Later on, on the basis of these definitional axioms, it is possible to introduce by definition, without further axioms, lots of different new entities, all kinds of goulash, by mixing these two ingredients. Related to this question is the question of primitivity. Is there a primordial soup of terms and axioms?

This point was seriously challenged in modern mathematics and logic. Alfred Tarski (1901-1983) was very much inspired by Blaise Pascal to develop the axiomatic method and he insisted that there were not absolute primitives (see [28], [27]). And in fact this relativity applies also to the distinction between proof and axiom, because a proof, yes, can be considered as an axiom and vice versa!

Confusions regarding the *DAP* trinity show up when talking about set theory as a basis for all mathematics. It is true that axiomatic set theory, let us say the most famous one, *ZF*, permits to define many concepts of mathematics, like for example the concept of group. But this does not mean that the axioms of group are a consequence of *ZF* axioms (see [2]). We cannot say that set theory allows to define everything in the universe and that through it we can axiomatize the whole universe. This would be exaggerated in many senses.

Also it is exaggerated to say that we can apply the axiomatic method to everything. In philosophy, Spinoza (1632-1677) was the only one, who really tried to do so. But Pascal, his contemporary, explicitly said that he did not believe that this method applies to everything, and that in particular it makes no sense to try to define the notion of human being.

4.2 Axiom, Proof, Truth

The axiomatic method was improved in many ways in modern logic and modern mathematics. Lots of new distinctions and concepts appeared. There is of course the question of *formalization*. The way to express propositions was systematically developed in more details, leading in particular to the language of first-order logic.

An important distinction is between proof theory and model theory. These two theories give two completely different visions of what an axiom is. This is why the completeness theorem which links the two is so spectacular.

The proof-theoretical vision is close to the traditional one: using some rules, we deduce step by step some theorems from axioms or/and definitions. The main difference between the "tradition" and modern times is that was developed a systematic theory, called *proof theory*. In particular in this theory the "steps" were made clearer and more explicit.

The model-theoretical vision is fairly new because it establishes the relation between axioms and "reality". What kind of reality? That's the question! This reality is structural: we have some structures, that can be conceived from a set-theoretical point of view and that can correspond to the real world in its various aspects. If a structure STR obeys a group of axioms AXI, we say that STR is a model of AXI, and that AXI are true in STR. The heart of model theory is to articulate truth in a structure, and it took several decades before this was made clear by Tarski (cf. [18]).

A group of axioms AXI may have a variety of models quite different from each other. For example if we consider the basic axioms ORD for the notion of order, antisymmetry and transitivity, we have a good variety of heterogeneous models: partial order vs. linear order, discrete vs. dense order, with end points vs. without end points, and so on. We can nevertheless say that we have *axiomatized the notion of order*. But that is different from *axiomatizing the reality of a given structure*, let's say the structure of natural numbers, STN. To axiomatize STN means to find a group of axioms AXN such that the only model of AXN is STN.

Thoralf Skolem (1887-1963) proved an important metatheorem [25] about non-standard models of arithmetic, using the compactness theorem. He showed that we cannot axiomatize STN model-theoretically: given a set of axioms for STN, it is always possible to find a model which is radically different, in the sense that there are some non-standard numbers, coming after all the standard numbers.

On the other hand Kurt Gödel (1906-1978) with his incompleteness metatheorem [16] showed that it is not possible to proof-theoretically axiomatize the natural numbers. Considering any axioms of a proof-system for STM, there will be a proposition such that neither this proposition, nor its negation can be deduced from them, the famous Gödel's proposition inspired by the liar paradox.

In both cases these results are related to first-order logic and recursion. "Recursive" is a technical term to talk about mechanization or computability. Modern logicians were able to precisely characterize through recursion theory these informal notions. We will not here enter in details. But what is important is the set of axioms to be sizable. If we take all the propositions true in STN, this is obviously an axiomatization of STN. The other aspect of recursion is also about "sizability": if we consider that a proof is any sequence of propositions, then everything is proof-theoretically axiomatizable.

As we have pointed out, the word "axiom" in modern logic is used in two very different ways. The completeness theorem establishes a bridge between the two but not equate them. It is very important to keep in mind this distinction. To use two different words would be a bit artificial, but when necessary, in case of ambiguity, it is important to specify what we are talking about.

How to export the axiomatic method outside of mathematics? One may want to find some axioms describing the structure of the universe, in the sense that the universe is the only model of these axioms. And then we can deduce all the truths about the universe from these axioms. We see here the intertwining between the proof-theoretical and the model-theoretical notions of axiom.

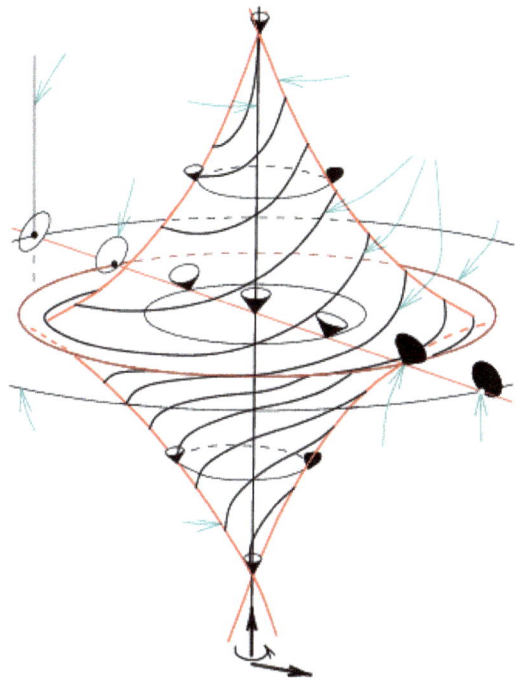

Dória together with Newton da Costa proved the incompleteness of classical mechanics and other physical theories (see [14] and [13]). And Gödel has a famous result showing that Einstein's theory of relativity admits non rotating universe as models [17]. It is important to clearly distinguish two aspects of this result. The first is than Einstein's theory does not axiomatize the universe, rotating or not, because there are radically different models of this theory. The second point is to know if it is natural or not to have rotating universes and how to eliminate them if considered as parasitic.

5 The Rise and Fall of the Axiomatic Method

The axiomatic method is not anymore at the center of the stage and probably will never be again. The golden age of the axiomatic method was the first half of the 20th century. The axiomatic method was strongly promoted in modern logic and modern mathematics, through in particular the work of Hilbert and Bourbaki. And there was also the will to apply it to all sciences: biology, physics, economics, etc.

We have now a clear idea that it is not possible to find a single axiom, or a group of axioms, from which everything can be deduced, computationally or not, or which characterizes, describes the universe of sets or monkeys, or the universe in its entirety. This impossibility does not prevent to use this method in an interesting way, and not only in logic and mathematics. We can also fruitfully use it in any field, or to develop the understanding of fundamental notions, such as for example causality.

It is important to still promote and use the axiomatic method, we cannot go back to prehistoric time where this method was not known. The axiomatic method, born in Ancient Greece, is certainly part of the seven wonders of thought.

But there are other wonders. One of them is conceptualization. They can go hand to hand, they are not opposed, rivals, enemy sisters, like Kyana and Keyla. They rather are in a subcontrary opposition, than a contrary or contradictory opposition (about these notions, see e.g. [7]). Moreover conceptualization can go alone by herself and need not be subordinated to axiomatization.

Garrett Birkhoff
Universal Algebra

$$\mathcal{A} = \langle A; f_{i(i \in I)} \rangle$$

This is was clearly shown by the work by Garret Birkhoff on *universal algebra* [12]. He was able to conceptualize notions like those of subalgebra and morphism, without any axioms. It is indeed easier to conceptualize these notions with no axioms, considering an algebra just as a family of operators on a set, obeying, yes, no axioms! Farewell to universal algebra in the sense of Sylvester and Whitehead.[3]

Inspired by Birkhoff, I decided to develop *universal logic* along the same conceptual line of thought, considering a logical structure, as a structure obeying

---

[3]Category theory is also going in this direction, see [24] and [22].

no axioms [1], by contrast to Tarski's general approach to the metalogic theory of consequence operator [3], based on three axioms [4].

Tarski's approach is already breaking the paradigm of *laws of thought*, in particular because it is an abstract framework with no logical operators. At best Tarski's axioms can be considered as *laws of consequence*. But although Tarski had a the very general perspective of a general theory of science, he was focusing on deductive sciences, cf. the expression "Methodology of deductive sciences", which is part of the title of his introductory book of logic [28] and of the title of the series of events he launched, CLMPS: *Congress of Logic, Methodology and Philosophy of Sciences*.

If we go to empirical sciences there are good reasons to reject the second Tarskian axiom, the basis of a class of logic has been nicely negatively baptized by John McCarthy (1927-2011): *non-monotonic logics*. There are also good reasons to reject the two other axioms, reflexivity and transitivity, both for theoretical and practical reasons. We can promote what I have called *axiomatic emptiness* [6].

If we want to develop a very general theory of reasoning, in an abstract and mathematical way, it is good to just start with concepts. Then we can fruitfully use and apply these concepts in some specific cases where axioms will show up. This methodology is neither in favor of monism or pluralism. The idea is to encompass and systematize the various systems of logic in one unified general approach which is itself not one system, based on some axioms, describing one way of reasoning.

6 Dedication and Personal Recollections

I have known Francisco Antônio Dória since many years. I don't remember exactly when and where I met him for the first time, probably in 1991 or 1992 in São Paulo. I came for the first time to Brazil in August 1992 to work with Newton da Costa at the University of São Paulo and Dória at this time was one of his main collaborators working on physical matters. Dória was living in Rio de Janeiro but used to come to São Paulo to work with da Costa.

Dória was member of my PhD jury at the University of São Paulo and we edited together a special issue of *Logique et Analyse* [11]. I moved to Rio de Janeiro in 1995 and I regularly met him in Rio or in Petrópolis, where he had a country house, discussing about all kinds of topics.

In 2000 I went for two years at Stanford University to work with Patrick Suppes. Dória had been there before and one of his PhD students, the editor of the present volume, Acacio de Barros, was at the time a main collaborator of Suppes and helped me to settle in the Farm. Going back to Brazil Acacio sold me his horse, called *Voyage*.

Acacio and I organized an event in honor of Dória December 8 and 9, 2018 at the *Brazilian College of Advanced Studies* of the *Federal University of Rio de Janeiro*, sponsored by the *Brazilian Academy of Philosophy* of which he is a member (Chair 21). As illustrated by the above poster, Dória is a true polymath.

References

[1] J.-Y. Beziau, "Universal Logic", in *Logica '94 - Proceedings of the 8th International Symposium*, T.Childers and O.Majers (eds), Czech Academy of Science, Prague, 1994, pp.73–93.

[2] J.-Y. Beziau, "La théorie des ensembles et la théorie des catégories: présentation de deux soeurs ennemies du point de vue de leurs relations avec les fondements des mathématiques", *Boletín de la Asociación Matemática Venezolana*, 9 (2002), pp.45-53.

[3] J.-Y.Beziau, "From consequence operator to universal logic: a survey of general abstract logic", in *Logica Universalis: Towards a general theory of logic*, Birkhäuser, Basel, 2005, pp.3-17.

[4] J.-Y. Beziau, "Les axiomes de Tarski", in R.Pouivet and M.Rebuschi (eds), *La philosophie en Pologne 1918-1939*, Vrin, Paris, 2006, pp.135-149.

[5] J.-Y. Beziau, "13 Questions about universal logic", *Bulletin of the Section of Logic*, 35 (2006), pp.133-150.

[6] J.-Y.Beziau, "What is a logic ? - Towards axiomatic emptiness", *Logical Investigations*, vol.16 (2010), pp.272-279.

[7] J.-Y.Beziau, "Round squares are no contradictions", in *New Directions in Paraconsistent Logic*, Springer, New Delhi, 2015, pp.39-55.

[8] J.-Y.Beziau, "Possibility, imagination and conception", *Principios*, vol.23 n.40 (2016), pp.59-95.

[9] J.-Y.Beziau, "The Pyramid of Meaning", in J.Ceuppens, H.Smessaert, J. van Craenenbroeck and G.Vanden Wyngaerd (eds), *A Coat of Many Colours - D60*, Brussels, 2018.

[10] J.-Y.Beziau, "Dice: a hazardous symbol for chance?", in *Logic, Intelligence and Artifices: Tributes to Tarcsio H. C. Pequeno*, College Publication, London, 2018, pp.365-385.

[11] J.-Y.Beziau and F.A.Dora (eds), Special issue of *Logique et Analyse*, Contemporary Brazilian Research in Logic, Part. 1, 153-154, 1996.

[12] G.Birkhoff, "Universal algebra", in *Comptes Rendus du Premier Congres Canadien de Mathématiques*, Presses de l'Université de Toronto, Toronto. 1946. P.310-326

[13] G.Chaitin, F.A.Dória and N.C.A. da Costa, *Goedel's Way: Exploits Into an Undecidable World*, CRC Press, New York, 2011.

[14] N.C.A. da Costa and F.A.Dória, "Undecidability and incompleteness in classical mechanics", *International Journal of Theoretical Physics*, 30 (1991), pp.1041–1073.

[15] E.Farah, *Algumas proposiões equivalentes ao axioma da escolha*, University of São Paulo, São Paulo, 1954.

[16] K.Gödel, "Über formal unentscheidbare Sätze der Principia Mathematica und verwandter Systeme, I.", *Monatshefte für Mathematik und Physik*, 38 (1931), pp.173–198.

[17] K.Gödel, "An example of a new type of cosmological solutions of Einsteins field equations of gravitation", *Reviews of Modern Physics*, 21 (1949), pp.447450.

[18] W.Hodges, "Truth in a Structure", *Proceedings of the Aristotelian Society*, 86 (1986), pp.135-152.

[19] B.Pascal, *De l'esprit géométrique et de l'art de persuader*, 1657.

[20] J.Borges de Paula, *O termo 'axioma', considerando a realação entre a filosofia e a matemática alicerçada no pensamento sobre complementaridade 'otteano'*, PhD, Federal University of Mato Grosso, Cuiabá, 2014.

[21] T.Pequeno and J.-Y.Beziau, "Rules of the game", in J.-Y.Beziau and M.E.Coniglio (eds), *Logic without frontiers*, College Publication, London, 2011, pp.131-144.

[22] A.Rodin, *Axiomatic method and category theory*, Springer, Cham, 2014.

[23] H.Rubin and J.E.Rubin, *Equivalents of the axiom of choice I and II*, North Holland, Amsterdam, 1963 and 1985.

[24] S.Schanuel and W.Lawvere, *Conceptual Mathematics: A First Introduction to Categories*, Cambridge University Presse, Cambridge 2009.

[25] T.Skolem, *Über die Nicht-charakterisierbarkeit der Zahlenreihe mittels endlich oder abzählbar unendlich vieler Aussagen mit ausschliesslich Zahlenvariablen*, Fundamenta Mathematicae, 23 (1934), pp.150–161.

[26] P.Suppes and J.-Y.Beziau, "Semantic computation of truth based on associations already learned", *Journal of Applied Logic*, 2 (2004), pp.457–467.

[27] A.Tarski, "Sur la méthode déductive", in *Travaux du IXe Congrès International de Philosophie, VI*, Hermann, Paris, 1937, pp.95–103.

[28] A.Tarski, *O logice matematycznej i metodzie dedukcyjnej*, Atlas, Lvóv and Warsaw (4th English edition by Jan Tarski: *Introduction to Logic and to the Methodology of the Deductive Sciences*, Oxford University Press, Oxford, 1994).

Jean-Yves Beziau
UFRJ - Federal University of Rio de Janeiro, RJ, Brazil
CNPq - Brazilian Research Council
ABF - Brazilian Academy of Philosophy
jyb@ufrj.br

Galois pairs with the modal operators of paraconsistent logic J_3

Itala M. Loffredo D'Ottaviano
Centre for Logic, Epistemology and the History of Science
Department of Philosophy
University of Campinas
itala@cle.unicamp.br

Hércules de Araujo Feitosa
Department of Mathematics
College of Sciences
São Paulo State University
haf@fc.unesp.br

Abstract

The paraconsistent logic J_3 was introduced, by D'Ottaviano and da Costa, as a solution to a problem proposed by Jaśkowski. This system, more than paraconsistent, is also many-valued and modal; and has been studied in the literature under distinct motivations and denominations. In this paper, we emphasize the modal aspect of J_3 and, since Galois pairs are abundant in mathematical contexts, we exhibit a Galois adjunction with the modal operators of the original version of J_3.

Introduction

The objective of this paper consists in showing that the modal operators of the paraconsistent logic J_3 characterize as a Galois pair.

The system J_3 was introduced by D'Ottaviano and da Costa [17], in 1970, from a three-valued matrix semantics. It was conceived as a solution to a problem proposed by Jaśkowski ([24] and [25]), involving aspects of the recently created paraconsistent logics.

The system J_3, more than paraconsistent and trivalent, in its original version characterizes also as a modal system.

D'Ottaviano, in her doctoral thesis, deepened its study, obtained new results on the system J_3, and introduced the class of J_3-theories, which provided the first works in the literature with results for a modal theory for paraconsistent logics ([13], [14], [15] and [16]).

The system J_3 has been studied by other authors (see for example [1], [2], [22] and [33]). In various papers, the logic J_3 was introduced and studied under distinct motivations, denominations and approaches (see for example [5] and [7]).

In this paper, we emphasize the modal aspects of J_3. Instead of the operators ∇ and Δ used in the original version [17], we use the operators with the alethic understanding for J_3, that is, the operators $\Box$ (necessary) and $\Diamond$ (possible).

In the first section, we present elements of the theory on pairs of functions that maintain relations with Galois connections.

In the Section 2, we introduce the logic J_3 from its matricial semantics, according to the original version of [17], as well as some motivations and examples.

In the next sections, we present the tableaux system for J_3, denoted by TJ_3, introduced by Silva, Feitosa and Cruz [33], with the addition of new rules for formulas whose main operators are modal ones, which emphasize aspects of the modal character of J_3. This tableaux system allows us the characterization of the modal operators of J_3 as a Galois pair.

1 Galois pairs

In this section we present some basic algebraic concepts and the pairs of functions motivated in the Galois connections, which will be used for further developments. For references on algebraic logic we indicate the books [19], [29] and [30]; for elements of Galois pairs we suggest [26], [31], [32] and [34]. The book [19] also develops on Galois connections.

Definition 1.1. *A binary relation $\leq$ on a set A is a partial order if it is reflexive, antisymmetric and transitive, that is, respectively:*
  (i) *for all $a \in A$, $a \leq a$;*
  (ii) *for all $a, b \in A$, if $a \leq b$ and $b \leq a$, then $a = b$;*
  (iii) *for all $a, b, c \in A$, if $a \leq b$ and $b \leq c$, then $a \leq c$.*

Definition 1.2. *A partially ordered set (poset) is a pair $\langle A, \leq \rangle$, in which A is a non-empty set and $\leq$ is a partial order over A.*

Definition 1.3. *Let $\langle A, \leq \rangle$ be a poset and $a, b \in A$. The supremum of the pair $\{a, b\}$, in case there is one, is the element $c \in A$ such that:*
  (i) $a \leq c$ *e* $b \leq c$;
  (ii) *if $a \leq d$ and $b \leq d$, then $c \leq d$.*

Definition 1.4. *Let $\langle A, \leq \rangle$ be a poset and $a, b \in A$. The infimum of the pair $\{a, b\}$, in case there is one, is the element $e \in A$ such that:*
  (i) $e \leq a$ *and* $e \leq b$;
  (ii) *if $f \leq a$ and $f \leq b$, then $f \leq e$.*

Usually, we denote the supremum of $\{a,b\}$ by $\sup\{a,b\}$ or $a \curlyvee b$, and the infimum of $\{a,b\}$ by $\inf\{a,b\}$ or $a \curlywedge b$. The supremum of $\{a,b\}$ is the least upper bound of $\{a,b\}$, the infimum of $\{a,b\}$ is the greatest lower bound of $\{a,b\}$.

Definition 1.5. *If $\langle R, \leq \rangle$ is a poset, for which given any $a, b \in R$ there are the $\inf\{a,b\}$ and the $\sup\{a,b\}$, we name lattice the algebraic structure $\langle R, \curlywedge, \curlyvee \rangle$, in which:*

$$a \curlywedge b = \inf\{a,b\} \quad \text{and} \quad a \curlyvee b = \sup\{a,b\}.$$

Proposition 1.6. *If $\langle R, \leq \rangle$ is a lattice, then, for all $a, b, c \in R$, hold the laws:*
R_1 $(a \wedge b) \wedge c = a \wedge (b \wedge c)$ *and* $(a \vee b) \vee c = a \vee (b \vee c)$ *[associativity];*
R_2 $a \wedge b = b \wedge a$ *and* $a \vee b = b \vee a$ *[commutativity];*
R_3 $(a \wedge b) \vee b = b$ *and* $(a \vee b) \wedge b = b$ *[absorption].* ∎

Definition 1.7. *If $f : (A, \leq_A) \to (P, \leq_P)$ is a function between two partially ordered sets, then:*
   (i) *the function f preserves the orders, if $a \leq_A b$ implies $f(a) \leq_P f(b)$;*
   (ii) *the function f inverts the orders, if $a \leq_A b$ implies $f(b) \leq_P f(a)$.*

In the context of mathematical analysis, usually these functions are called increasing and decreasing, respectively; but this denomination is less usual in the context of Galois connections. Other times they are called isotone and antitone, being that the first ones also occur with the name of monotone.

Definition 1.8. *Let $f : (A, \leq_A) \to (A, \leq_A)$ be a function. Then:*
   (i) *f is idempotent, if $f \circ f = f$;*
   (ii) *f is extensive or inflationary if, for all $a \in A$, $a \leq f(a)$;*
   (iii) *f is deflationary if, for all $a \in A$, $f(a) \leq a$.*

Definition 1.9. *Let $f : (A, \leq_A) \to (A, \leq_A)$ be a function. Then:*
   (i) *f is a Tarski operator (deductive closure operator) if is extensive (or inflationary), preserves orders and is idempotent;*
   (ii) *f is an interior operator if is deflationary, preserves orders and is idempotent.*

When we analyse the definition of Galois connection, we will be able to make four simple permutations, what will generate us other pairs of functions, which maintain some similarity with the definition of connection.

As usual, the symbol ⇔ must be understood as 'if, and only if'.

Definition 1.10. If $(A, \leq_A)$ and $(P, \leq_P)$ are partially ordered sets, $a \in A$ and $p \in P$ are any elements and $f : A \to P$ and $g : P \to A$ are functions, then:

(i) the pair (f, g) is a Galois connection if $a \leq_A g(p) \Leftrightarrow p \leq_P f(a)$;

(ii) the pair $(f, g)^d$ is a dual Galois connection if $g(p) \leq_A a \Leftrightarrow f(a) \leq_P p$;

(iii) the pair $[f, g]$ is an adjunction if $a \leq_A g(p) \Leftrightarrow f(a) \leq_P p$;

(iv) the pair $[f, g]^d$ is a dual adjunction if $g(p) \leq_A a \Leftrightarrow p \leq_P f(a)$.

The name adjunction comes from the theory of categories. In many texts on the subject, the pair $[f, g]$ is also called *residuated*.

If $(A, \leq_A)$ is a partially ordered set, then we denote the inverse order of $\leq_A$ by $\leq_A^{op}$ and, thereby, $(A, \leq_A^{op}) = (A, (\leq_A)^{-1})$. Thus:

$$a \leq_A b \Leftrightarrow b \leq_A^{op} a.$$

From this definition stem the following results.

Proposition 1.11. Let $(A, \leq_A)$ and $(P, \leq_P)$ be partially ordered sets and $f : A \to P$ and $g : P \to A$ functions:

(i) if $[f, g]$ is an adjunction, then $[g, f]^d$ is a dual adjunction;

(ii) if $[f, g]^d$ is a dual adjunction, then $[g, f]$ is an adjunction;

(iii) if (f, g) is a Galois connection, then (g, f) is also a Galois connection;

(iv) if $(f, g)^d$ is a dual Galois connection, then $(g, f)^d$ is also a dual Galois connection. ∎

Proposition 1.12. If (f, g) is a Galois connection for $(A, \leq_A)$ and $(P, \leq_P)$, then:

(i) $(f, g)^d$ is a dual Galois connection for $(A, \leq_A^{op})$ and $(P, \leq_P^{op})$;

(ii) $[f, g]$ is an adjunction for $(A, \leq_A)$ and $(P, \leq_P^{op})$;

(iii) $[f, g]^d$ is a dual adjunction for $(A, \leq_A^{op})$ and $(P, \leq_P)$. ∎

Following, we emphasize the adjunctions, as particular cases of Galois pairs. We enunciate various results, which, with the proper particularities, can be applied to the remaining Galois pairs.

In general, we do not indicate the orders $\leq_A$ and $\leq_P$, because the context allows for the identification of over which set we address of the order in question.

The following proposition gives us conditions to have an adjunction.

Proposition 1.13. *Let $(A, \leq_A)$ and $(P, \leq_P)$ be two partial orders, $f : A \to P$ and $g : P \to A$ functions, with $a, b \in A$ and $p, q \in P$. Then, the pair $[f, g]$ is an adjunction if, and only if, the following conditions hold:*
  (i) $a \leq g(f(a))$;
  (ii) $f(g(p)) \leq p$;
  (iii) $a \leq b \Rightarrow f(a) \leq f(b)$;
  (iv) $p \leq q \Rightarrow g(p) \leq g(q)$. ∎

Thus, we have another way of defining adjunction: the pair $[f, g]$ is an adjunction if the functions f and g preserve the orders and the composites $g \circ f$ and $f \circ g$ are, respectively, inflationary and deflationary.

Proposition 1.14. *If the pair $[f, g]$ is an adjunction for the partial orders $(A, \leq_A)$ and $(P, \leq_P)$, then $f(a) = f(g(f(a)))$ e $g(b) = g(f(g(b)))$.* ∎

Proposition 1.15. *If $[f, g]$ is an adjunction for $(A, \leq_A)$ and $(P, \leq_P)$, then the two compositions $g \circ f$ and $f \circ g$ are Tarski and interior operators, respectively, on A and P.* ∎

Proposition 1.16. *If the pair $[f, g]$ is an adjunction for the lattices $(A, \curlywedge, \curlyvee)$ e $(P, \curlywedge, \curlyvee)$, then:*
  (i) $f(x \curlyvee y) = f(x) \curlyvee f(y)$;
  (ii) $g(x \curlywedge y) = g(x) \curlywedge g(y)$. ∎

Proposition 1.17. *If $[f, g_1]$ and $[f, g_2]$ are adjunctions for $(A, \leq_A)$ and $(P, \leq_P)$, then $g_1 = g_2$. If $[f_1, g]$ and $[f_2, g]$ are adjunctions for $(A, \leq_A)$ e $(P, \leq_P)$, then $f_1 = f_2$.* ∎

Proposition 1.18. *If $[f,g]$ is an adjunction for $(A, \leq_A)$ and $(P, \leq_P)$, then:*
  (i) $a \in g(P) \Leftrightarrow g(f(a)) = a$;
  (ii) $p \in f(A) \Leftrightarrow f(g(p)) = p$;
  (iii) $f(A) = f(g(P))$;
  (iv) $g(P) = g(f(A))$. ∎

Thus, each point $a \in g(P)$ is a fixed point of the function $g \circ f$, end each point $p \in f(A)$ is a fixed point of the function $f \circ g$.

Proposition 1.19. *If $[f,g]$ is an adjunction for $(A, \leq_A)$ and $(P, \leq_P)$, then:*
  (i) $f(a) = min\{p \in P : a \leq g(p)\}$;
  (ii) $g(p) = max\{a \in A : f(a) \leq p\}$. ∎

Proposition 1.20. *If $[f_1, g_1]$ is an adjunction for $(A, \leq_A)$ and $(B, \leq_B)$, and $[f_2, g_2]$ is an adjunction for $(B, \leq_B)$ and $(C, \leq_C)$, then $[f_2 \circ f_1, g_1 \circ g_2]$ is an adjunction for $(A, \leq_A)$ and $(C, \leq_C)$.* ∎

Each Galois pair has similar results to the ones here enunciated on the adjunctions.

Following, we will see Galois pairs in J_3.

2 The logic J_3

The three-valued paraconsistent logic J_3 was introduced by D'Ottaviano and da Costa, in [17], in the propositional language $L = \{\neg, \Diamond, \vee\}$, in which the two first operators are unary and the last one is binary, from a matrix semantics.

The matrix semantics of J_3 is presented below.

We indicate by $Var(J_3) = \{p_1, p_2, p_3, ...\}$ the set of the propositional variables of J_3 and by $For(J_3)$ the set of the formulas of J_3, defined the usual way, as in [20] and [23].

Notions on matrix models can be found in [8], [9] and [28].

Definition 2.1. *A restrict valuation for J_3 is a function*

$$v : Var(J_3) \to \{0, \tfrac{1}{2}, 1\}.$$

Definition 2.2. *A valuation for J_3 is a function that extends, in an unique way, the restrict valuation for the whole set $For(J_3)$, according to the matrixes below:*

	$\neg$
0	1
$\tfrac{1}{2}$	$\tfrac{1}{2}$
1	0

$\vee$	0	$\tfrac{1}{2}$	1
0	0	$\tfrac{1}{2}$	1
$\tfrac{1}{2}$	$\tfrac{1}{2}$	$\tfrac{1}{2}$	1
1	1	1	1

	$\Diamond$
0	0
$\tfrac{1}{2}$	1
1	1

Definition 2.3. *The matrix semantics of J_3 is characterized by the matrix:*

$$\mathcal{M}_{J_3} = (\{0, \tfrac{1}{2}, 1\}, \{\tfrac{1}{2}, 1\}, \neg, \vee, \Diamond),$$

in which the set of designated values is $D = \{\tfrac{1}{2}, 1\}$.

Thus, each formula of J_3 has to assume one of the three values of the set $\{0, \tfrac{1}{2}, 1\}$, but the two elements of the set D are the distinguished or designated values to represent truth in J_3.

Definition 2.4. *A formula $\varphi \in For(J_3)$ is valid, according to $\mathcal{M}_{J_3}$, if, for every J_3-valuation v, $v(\varphi) \in D$.*

The propositional operators $\neg$, $\vee$ and $\Diamond$ formalize, respectively, the notions of negation, disjunction and possible truth (possibility), in distinction to necessary truth (necessity), which we will see ahead.

If a proposition φ assumes, according to a given valuation, the value $\tfrac{1}{2}$, then its negation $\neg\varphi$ also has the value $\tfrac{1}{2}$, and this emphasizes the paraconsistent aspect of the negation of J_3, that is, a proposition and its negation can be both true.

The operator $\Diamond$ separates the designated elements from the undesignated ones, but also points out the modal aspect of J_3. If a formula $\Diamond\varphi$ assumes a value in D, then it is possible, true and possible.

If $\Gamma \subseteq For(J_3)$, then $v(\Gamma) = \{v(\gamma) : \gamma \in \Gamma\}$.

The relation of logical implication or semantic consequence for J_3 is given as follows.

Definition 2.5. *If $\Gamma \cup \{\varphi\} \subseteq For(J_3)$, then Γ implies logically φ, or φ is a semantic consequence of Γ, when, for every J_3-valuation v, if $v(\Gamma) \subseteq D$, then $v(\varphi) \in D$.*

Thus, we have that for every valuation v:

$$\Gamma \vDash \varphi \iff (v(\Gamma) \subseteq D \Rightarrow v(\varphi) \in D).$$

Beyond these basic operators, the following operators are defined in J_3.

Conjunction: $\varphi \wedge \psi =_{def} \neg(\neg\varphi \vee \neg\psi)$
Strong negation: $\sim \varphi =_{def} \neg\Diamond\varphi$
Necessary: $\Box\varphi =_{def} \neg\Diamond\neg\varphi$
Conditional: $\varphi \to \psi =_{def} \neg\Diamond\varphi \vee \psi$
Bi-conditional: $\varphi \leftrightarrow \psi =_{def} (\varphi \to \psi) \wedge (\psi \to \varphi)$
Consistency: $\circ\varphi =_{def} \sim (\varphi \wedge \neg\varphi)$.

The meaning of these new operators is given by the following tables:

$\wedge$	0	$\frac{1}{2}$	1
0	0	0	0
$\frac{1}{2}$	0	$\frac{1}{2}$	$\frac{1}{2}$
1	0	$\frac{1}{2}$	1

	$\sim$
0	1
$\frac{1}{2}$	0
1	0

	$\Box$
0	0
$\frac{1}{2}$	0
1	1

$\to$	0	$\frac{1}{2}$	1
0	1	1	1
$\frac{1}{2}$	0	$\frac{1}{2}$	1
1	0	$\frac{1}{2}$	1

$\leftrightarrow$	0	$\frac{1}{2}$	1
0	1	0	0
$\frac{1}{2}$	0	$\frac{1}{2}$	$\frac{1}{2}$
1	0	$\frac{1}{2}$	1

	$\circ$
0	1
$\frac{1}{2}$	0
1	1

By the definition of conjunction, from disjunction, we observe that in J_3 the De Morgan laws work:
$$\neg(\varphi \land \psi) \Leftrightarrow (\neg\varphi \lor \neg\psi) \text{ and } \neg(\varphi \lor \psi) \Leftrightarrow (\neg\varphi \land \neg\psi).$$

The strong negation, a defined operator, behaves like classic negation. The strong negation of a true proposition corresponds to a falsehood, and vice-versa. With this negation we rescue the bivalent character of the logic, inside a part of J_3. Furthermore, under this aspect, J_3 may be understood as an extension of the Aristotelian classical propositional logic [21].

The operator of necessity indicates that a necessarily true proposition assumes only the logical value 1, while the possibly true assumes also the value $\frac{1}{2}$.

The conditional and bi-conditional operators of J_3 coincide with the respective ones of the three-valued Łukasiewcz' logic $Ł_3$ (see [27], [3] and [8]).

The operator $\circ$ was originally introduced in the language of the logics of the hierarchy of paraconsistent propositional calculi C_n, $1 \le n \le \omega$, by da Costa [10], in 1963 (see also [11] and [12]), as the defined operator of 'good behavior': if the proposition assumes only the values 0 and 1, then it is 'well behaved' or acts in accordance to classical logic.

D'Ottaviano and Epstein [18], in 1988, and Epstein [21], in 1990, present the system J_3 through two distinct axiomatics: in one of them, considering as primitive the operators $\neg, \land$ and $\to$; in another one, considering as primitive, more than the operators $\neg, \land, \lor$ and $\to$, also the operator $\circ$, denominated consistency operator.

In 1999, Carnielli and Amo [4] present, motivated by some problems of databases in informatics, a trivalent system which corresponds exactly to the logic J_3. Carnielli and Marcos [6] and Carnielli, Coniglio and Marcos [5] introduce a family of paraconsistent logics, the logics of formal inconsistency (LFI's), in which the consistency operator is internalized in the language as a primitive operator; the first logic of the family of LFI's, the LFI1 logic, is exactly the logic J_3.

Finally, the consistency operator, according to the most recent literature, or the operator of 'good behaviour' of past original versions, indicates that if the proposition assumes only the values 0 or 1, then it is well behaved or acts accordingly to the classical logic.

Other interesting operators, which can be defined in J_3, are the operators of contingency and the Ł-implication of Lukasiewcz, with its respective tables:

Contingency: $\bullet \varphi =_{def} \Diamond \varphi \wedge \neg \Box \varphi$
2-implication: $\varphi \rightarrowtail \psi =_{def} \Diamond \neg \varphi \vee \psi$
Ł-implication: $\varphi \rightarrow_3 \psi =_{def} (\varphi \rightarrowtail \psi) \wedge (\neg \psi \rightarrowtail \neg \varphi)$.

$\rightarrowtail$	0	$\frac{1}{2}$	1
0	1	1	1
$\frac{1}{2}$	1	1	1
1	0	$\frac{1}{2}$	1

$\rightarrow_3$	0	$\frac{1}{2}$	1
0	1	1	1
$\frac{1}{2}$	$\frac{1}{2}$	1	1
1	0	$\frac{1}{2}$	1

	$\bullet$
0	0
$\frac{1}{2}$	1
1	0

The operator of contingency is the negation (classical or of J_3) of the operator $\circ$, and will be seen as an inconsistency operator.

The operator $\rightarrow_3$ will allow us to define a strict implication, as in the modal logics.

The validity of the propositions written with the operators of J_3 can be tested through the three-valued matrix, as in the following example.

(a) $\varphi \rightarrow (\psi \rightarrow \varphi)$

φ	$\rightarrow$	(ψ	$\rightarrow$	φ)
0	1	0	1	0
0	1	$\frac{1}{2}$	0	0
0	1	1	0	0
$\frac{1}{2}$	1	0	1	$\frac{1}{2}$
$\frac{1}{2}$	$\frac{1}{2}$	$\frac{1}{2}$	$\frac{1}{2}$	$\frac{1}{2}$
$\frac{1}{2}$	$\frac{1}{2}$	1	$\frac{1}{2}$	$\frac{1}{2}$
1	1	0	1	1
1	1	$\frac{1}{2}$	1	1
1	1	1	1	1

Since the concluding column of the previous table entails only the values 1 and $\frac{1}{2}$, then the formula is valid according to the matrix model $\mathcal{M}_{J_3}$.

It stems from the definition of valuation that every formula of J_3 that is valid according to a trivalent valuation $v : Var(J_3) \to \{0, \frac{1}{2}, 1\}$, is also valid according to the Boolean restriction of v. That is, according to $v : Var(J_3) \to \{0, 1\}$ with the Boolean meaning of the operators $\neg, \vee, \wedge$ and $\to$, in which the value $\frac{1}{2}$ is 'erased'. Thus, every formula J_3-valid is a tautology.

As we can verify in the texts of the bibliography on J_3, beyond the formula of the example (a) above, the following formula is also J_3-valid .

(b) $(\varphi \to (\psi \to \sigma)) \to ((\varphi \to \psi) \to (\varphi \to \sigma))$.

We also have that in J_3 the MP Rule is the only basic rule. More than that, the MP preserves the J_3-validity of the deduction.

(c) (MP) $\varphi, \varphi \to \psi/\psi$.

Therefore, in J_3 it holds the Theorem of Deduction relative to the operator $\to$.

3 Tableaux for J_3

A system of tableaux for the logic J_3 was introduced by Silva, Feitosa and Cruz in [33]. We present, below, this system.

In D'Ottaviano and Epstein [18] and Epstein [21], the logic J_3 is characterized with the following set of basic operators: $\neg, \circ, \wedge, \vee, \to$. We would not need the modal operators $\square$ and $\Diamond$ as basic, but, on the other hand, the operator $\circ$ is primitive, which may also be understood in a certain way as a modal operator.

The system of tableaux $T(J_3)$, introduced in [33], has as reference this deductive system for J_3, whose basic operators are: $\neg, \circ, \wedge, \vee$ and $\rightarrow$.

The rules of expansion for the $T(J_3)$ system are classified in three kinds:

Type α - the consequences of the formula are direct and the expansion is not ramified;

Type β - the expansion is ramified in two distinct branches;

Type γ - the expansion is ramified in three distinct branches.

The rules of type γ are of a type not contemplated in classical tableaus, and stem from the trivalent aspects of the valuations of J_3.

Definition 3.1. *A branch of a tableau of $T(J_3)$ is closed when one of the following cases happen:*
  (i) *a formula occurs with distinct values in the branch;*
  (ii) *the formula marked $\frac{1}{2}$ $\circ \varphi$ occurs in the branch.*

As the tables of J_3, indicate, there are no cases in which any formula with the consistency operator $\circ$ assumes the value $\frac{1}{2}$.

However, as in the development of a tableau this may happen in an expansion, so the condition (ii) needs to be included in the definition of the closure of a branch in $T(J_3)$.

Definition 3.2. *A tableau of $T(J_3)$ is closed if all of its branches are closed.*

We introduce, following, the rules of expansion for the system $T(J_3)$.

Negation:

$$[\,0\,\neg\,]\;\dfrac{0\;\;\neg\varphi}{1\;\;\varphi} \qquad [\,\tfrac{1}{2}\,\neg\,]\;\dfrac{\tfrac{1}{2}\;\;\neg\varphi}{\tfrac{1}{2}\;\;\varphi} \qquad [\,1\,\neg\,]\;\dfrac{1\;\;\neg\varphi}{0\;\;\varphi}$$

Consistency:

$$[0 \circ] \frac{0 \quad \circ\varphi}{\frac{1}{2} \quad \varphi} \qquad\qquad [1 \circ] \frac{1 \quad \circ\varphi}{0 \; \varphi \mid 1 \; \varphi}$$

Conjunction:

$$[0 \wedge] \frac{0 \; \varphi \wedge \psi}{0 \; \varphi \mid 0 \; \psi} \quad [\tfrac{1}{2} \wedge] \frac{\tfrac{1}{2} \; \varphi \wedge \psi}{\tfrac{1}{2} \varphi \mid \tfrac{1}{2} \varphi \mid 1 \varphi \\ 1 \psi \mid \tfrac{1}{2} \psi \mid \tfrac{1}{2} \psi} \quad [1 \wedge] \frac{1 \; \varphi \wedge \psi}{\begin{array}{c} 1 \quad \varphi \\ 1 \quad \psi \end{array}}$$

Disjunction:

$$[0 \vee] \frac{0 \; \varphi \vee \psi}{\begin{array}{c} 0 \quad \varphi \\ 0 \quad \psi \end{array}} \quad [\tfrac{1}{2} \vee] \frac{\tfrac{1}{2} \; \varphi \vee \psi}{0 \varphi \mid \tfrac{1}{2} \varphi \mid \tfrac{1}{2} \varphi \\ \tfrac{1}{2} \psi \mid \tfrac{1}{2} \psi \mid 0 \psi} \quad [1 \vee] \frac{1 \; \varphi \vee \psi}{1 \; \varphi \mid 1 \; \psi}$$

Conditional:

$$[0 \to] \frac{0 \; \varphi \to \psi}{\begin{array}{c} 0 \quad \psi \\ \tfrac{1}{2} \varphi \mid 1 \varphi \end{array}} \quad [\tfrac{1}{2} \to] \frac{\tfrac{1}{2} \; \varphi \to \psi}{\begin{array}{c} \tfrac{1}{2} \quad \psi \\ \tfrac{1}{2} \varphi \mid 1 \varphi \end{array}} \quad [1 \to] \frac{1 \; \varphi \to \psi}{0 \; \varphi \mid 1 \; \psi}$$

Certainly, these rules of expansion of the system $T(J_3)$ were obtained by means of the analysis of the trivalent matrix of the logic J_3.

In the paper [33] is shown that this tableaux system is adequate to the logic J_3.

Following, we will use the constructions in tableaux to address the modal operators of J_3, which according to some presentations of this logic have been disregarded by the operator of consistency ∘.

4 On the modal operators of J_3

We begin with the introduction of specific rules of expansion, in $T(J_3)$, for the modal operators.

Possible:

$$[\,0\,\Diamond\,] \; \frac{0 \;\; \Diamond\varphi}{0 \;\; \varphi} \qquad\qquad [\,1\,\Diamond\,] \; \frac{1 \;\; \Diamond\varphi}{\frac{1}{2}\;\varphi \,|\, 1\;\varphi}$$

Necessary:

$$[\,0\,\Box\,] \; \frac{0 \;\; \Box\varphi}{0 \;\; \varphi \,|\, \frac{1}{2}\;\varphi} \qquad\qquad [\,1\,\Box\,] \; \frac{1 \;\; \Box\varphi}{1 \;\; \varphi}$$

We need as well to include a specific condition to the closure of the new tableaux.

Definition 4.1. *A branch of a tableau of $T(J_3)$ is closed when one of the following cases happen:*
(i) *a formula occurs with distinct values in the branch;*
(ii) *it occurs in the branch a formula marked $\frac{1}{2}$ $\circ\,\varphi$, $\frac{1}{2}$ $\Diamond\varphi$, or $\frac{1}{2}$ $\Box\varphi$.*

Certainly, no formula with the operators $\circ$, $\Diamond$ or $\Box$ can receive the value $\frac{1}{2}$.

Now, we will use these rules to show some results of J_3.

We begin with a surprising result, from the point of view of the modal logics - in J_3, any possible proposition must value:

(T_1) ⊩ $\Diamond\varphi \to \varphi$

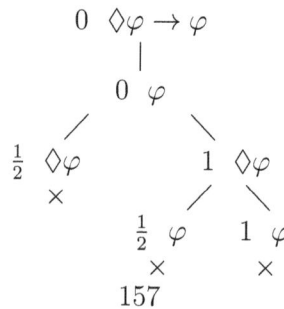

Hence follows a second deduction rule, presented in [14]:

(T_2) $\Diamond\varphi \Vdash \varphi$

A version of the modal axiom D also holds in J_3:

(T_3) $\Vdash \Box\varphi \to \Diamond\varphi$

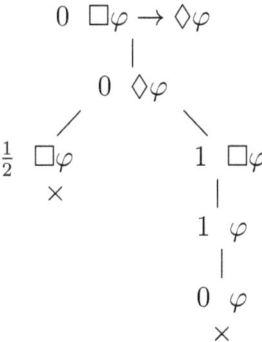

Therefore, we also have:

(T_4) $\Vdash \Box\varphi \to \varphi$

(T_5) $\Box\varphi \Vdash \Diamond\varphi \Vdash \varphi$

(T_6) $\Vdash \varphi \to \Diamond\varphi$

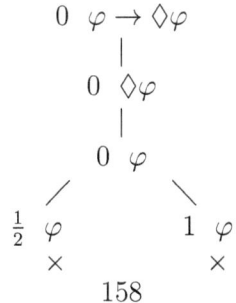

(T_7) ⊩ $\varphi \Leftrightarrow$ ⊩ $\Diamond\varphi$

To verify the following non-validity, it suffices to consider $v(\varphi) = \frac{1}{2}$:

(T_8) ⊮ $\varphi \to \Box\varphi$

Now, the search for conditions of adjunction with the operators $\Diamond$ and $\Box$.

(T_9) ⊩ $\Diamond\Box\varphi \to \varphi$

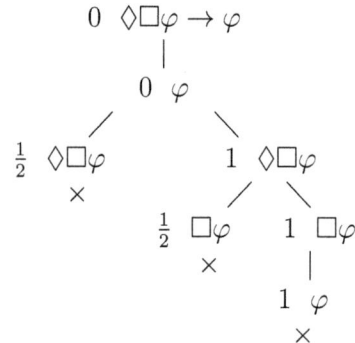

(T_{10}) ⊩ $\varphi \to \Box\Diamond\varphi$

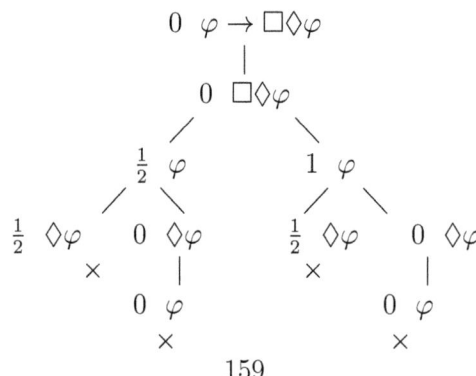

However, let us observe that the following proposition does not hold, when $v(\varphi) = \frac{1}{2}$:

(T_{11}) $\nVdash \varphi \to \Diamond\Box\varphi$

Nevertheless:

(T_{12}) If $\Vdash \varphi \to \psi$, then $\Vdash \Diamond\varphi \to \Diamond\psi$

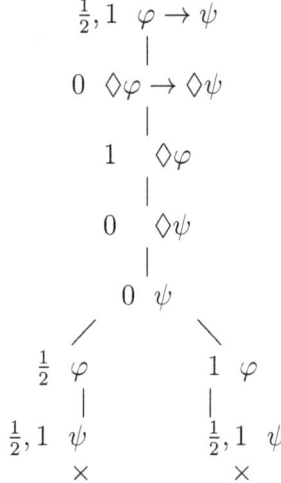

If $v(\varphi) = 1$ and $v(\psi) = \frac{1}{2}$, it is verified that the following implication does not hold:

(T_{13}) $\Vdash \varphi \to \psi \nRightarrow \Vdash \Box\varphi \to \Box\psi$.

For now, we do not succeeded in the obtainment of a Galois pair represented by the operators $\Diamond$ and $\Box$.
But we can still refine this question.

We define now the strict modal implication.

Definition 4.2. *Strict implication:* $\varphi \supset \psi \Leftrightarrow \Box(\varphi \to_3 \psi)$.

The table of the strict implication is the following:

⊃	0	½	1
0	1	1	1
½	0	1	1
1	0	0	1

This implication sends us immediately to a partial order, because we have that, if $v(\varphi) \leq v(\psi)$, then $v(\varphi \supset \psi) = 1$ and, thence, $\varphi \supset \psi$; if $v(\varphi) > v(\psi)$, then $v(\varphi \supset \psi) = 0$ and, thence, $\varphi \not\supset \psi$.

Thus, hold:

(i) $\varphi \supset \varphi$;
(ii) $\varphi \supset \psi$ and $\psi \supset \sigma \Rightarrow \varphi \supset \sigma$;
(iii) $\varphi \supset \psi$ and $\psi \supset \varphi \Leftrightarrow v(\varphi) = v(\psi)$.

Here we introduce derived rules specific for strict implication:

$$[\,0\,\supset\,]\,\dfrac{0\ \varphi \supset \psi}{\tfrac{1}{2}\ \varphi, 0\ \psi \mid 1\ \varphi, \tfrac{1}{2}\ \psi \mid 1\ \varphi, 0\ \psi}$$

$$[\,1\,\supset\,]\,\dfrac{1\ \varphi \supset \psi}{0\ \varphi \mid \tfrac{1}{2}\ \varphi, \tfrac{1}{2}\ \psi \mid 1\ \psi}$$

Following, an interaction between the two implications $\supset$ and $\rightarrow$:

(T_{14}) If $\Vdash \varphi \supset \psi$, then $\Vdash \varphi \rightarrow \psi$

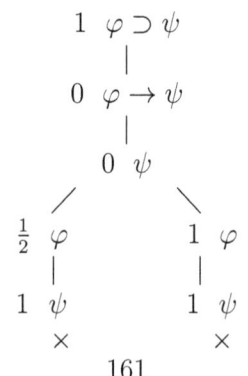

We have yet:

(T_{15}) If $\Vdash \varphi \supset \psi$, then $\Vdash \Diamond\varphi \supset \Diamond\psi$

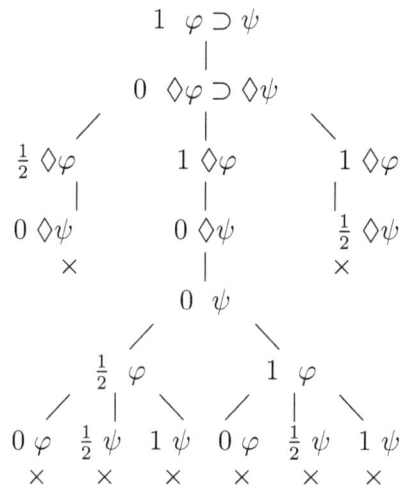

(T_{16}) If $\Vdash \varphi \supset \psi$, then $\Vdash \Box\varphi \supset \Box\psi$

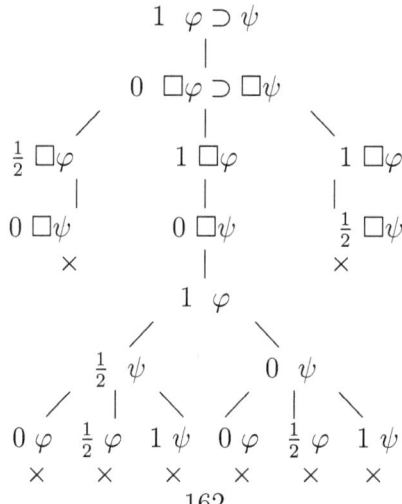

(T_{17}) ⊩ $\Diamond\Box\varphi \supset \varphi$

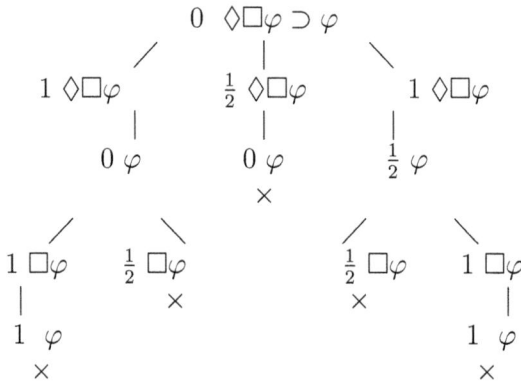

(T_{18}) ⊩ $\varphi \supset \Box\Diamond\varphi$

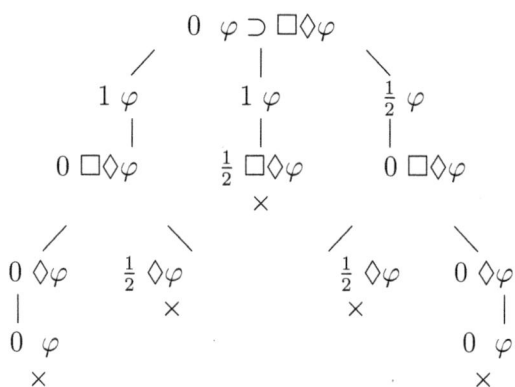

Therefore, we now have a Galois adjunction, if we consider $f = \Diamond$ and $g = \Box$.

The results on adjunction may be replicated here.

(T_{19}) ⊩ $\varphi \supset \Box\psi \Leftrightarrow \Diamond\varphi \supset \psi$

(T_{20}) ⊩ $\Diamond(\varphi \vee \psi) \Leftrightarrow (\Diamond\varphi \vee \Diamond\psi)$

(T_{21}) ⊩ $\Box(\varphi \wedge \psi) \Leftrightarrow (\Box\varphi \wedge \Box\psi)$

(T_{22}) ⊩ $\Diamond\varphi \Leftrightarrow \Diamond\Box\Diamond\varphi$

(T_{22}) ⊩ $\Box\varphi \Leftrightarrow \Box\Diamond\Box\varphi$.

Final considerations

We have applied the J_3-tableaux of Silva, Feitosa and Cruz [33], expanded by specific rules to the modal operators of J_3.

With this, we obtained several results about the modal part of J_3, including another example of a Galois pair, as in the many and varied mathematical theories.

We believe that the analysis of the modal operators of J_3 still deserves to be deepened.

This paper is a sincere and heartfelt tribute to our dear friend Francisco Antonio Dória.

References

[1] AVRON A. Natural 3-valued logics - Characterization and proof theory. *The Journal of Symbolic Logic*, v. 56, n. 1, p. 276-294, 1991.

[2] BATENS D. Paraconsistent extensional propositional logics. *Logique et Analyse*, v. 90-91, p. 195-234, 1980.

[3] BORKOWSKI, L. (Ed.) *Selected works of J. Łukasiewicz*. Amsterdam: North-Holland, 1970. (English version of [27]).

[4] CARNIELLI, W. A.; AMO, S. A logic-based system for controlling inconsistencies in evolutionary databases. *Proc. of the VI Workshop on Logic, Language, Information and Computation* (WoLLIC'99), p. 89-101, 1999.

[5] CARNIELLI, W.; CONIGLIO, M. E; MARCOS, J. Logics of formal inconsistency. In GABBAY, D.; GUENTHNER, F. (Eds.) *Handbook of Philosophical Logic*, 2nd. ed., v. 14, p. 1-93, 2007.

[6] CARNIELLI, W. A.; MARCOS, J. A taxonomy of C-systems. Paraconsistency: the logical way to the inconsistent. *Proc. of the II World Congress on Paraconsistency* (WCP'2000), p. 1-94, Marcel Dekker, 2001.

[7] CARNIELLI, W. A.; MARCOS J.; AMO S. Formal inconsistency and evolutionary databases. *Logic and Logical Philosophy*, v. 8, p. 115-152, 2000.

[8] CIGNOLI, R. L. O.; D'OTTAVIANO, I. M. L.; MUNDICI, D. *Álgebras das lógicas de Lukasiewicz*. Campinas: UNICAMP, Centro de Lógica, Epistemologia e História da Ciência, 1994. (Coleção CLE, v. 12).

[9] CIGNOLI, R. L. O.; D'OTTAVIANO, I. M. L.; MUNDICI, D. *Algebraic foundations of many-valued reasoning*. Dordrecht: Kluwer Academic Publishers, 2000. (Trends in Logic, v. 7).

[10] da COSTA, N. C. A. Calculs propositionnels pour les systèmes formels inconsistants. *Comptes Rendus de l'Académie des Sciences de Paris* (A-B), v. 257, p. 3790-3793, 1963.

[11] da COSTA, N. C. A. *Sistemas formais inconsistentes* (Inconsistent formal systems). Ph. Thesis in Portuguese. Curitiba: Universidade Federal do Paraná, 1963.

[12] da COSTA, N. C. A. On the theory of inconsistent systems. *Notre Dame Journal of Formal Logic*, v. 11, p. 497-510, 1974.

[13] D'OTTAVIANO, I. M. L. *Sobre uma teoria de modelos trivalente* (On a three-valued model theory). Ph. Thesis in Portuguese. Campinas: Universidade Estadual de Campinas, 1982.

[14] D'OTTAVIANO, I. M. L. The completeness and compactness of a three-valued first-order logic. *Revista Colombiana de Matemáticas*, v. XIX, n. 19, p. 77-94, 1985.

[15] D'OTTAVIANO, I. M. L., The model extension theorems for J3-theories. Methods in Mathematical Logic. *Lecture Notes in Mathematics*, p. 157-173, 1985.

[16] D'OTTAVIANO, I. M. L. Definability and quantifier elimination for J_3-theories. *Studia Logica*, v. XLVI, v. 46, n. 1, p. 37-54, 1987.

[17] D'OTTAVIANO, I. M. L.; da COSTA, N. C. A. Sur un problème de Jáskowski. *Comptes Rendus de l'Académie de Sciences de Paris* (A-B), v. 270, p. 1349-1353, 1970.

[18] D'OTTAVIANO, I. M. L.; EPSTEIN, R. L. A many-valued paraconsistent logic. *Reports on Mathematical Logic*, Wydawnictwo U Jagiell., Krakow, v. 22, p. 89-103, 1988.

[19] DUNN, J. M.; HARDEGREE, G. M. *Algebraic methods in philosophical logic*. Oxford: Oxford University Press, 2001.

[20] ENDERTON, H. B. *A mathematical introduction to logic*. San Diego: Academic Press, 1972.

[21] EPSTEIN, R. L. *The semantic foundations of logic. Volume 1: propositional logics*. Dordrecht: Kluwer Academic Publishers, 1990.

[22] FEITOSA, H. A.; CRUZ, G. A.; GOLZIO, A. C. J. Um novo sistema de axiomas para a lógica paraconsistente J3. *C.Q.D.- Revista Eletrônica Paulista de Matemática*, v. 4, p. 16-29, 2015.

[23] FEITOSA, H. A.; PAULOVICH, L. *Um prelúdio à lógica*. São Paulo: Editora, UNESP, 2005.

[24] JAŚKOWSKI S. Propositional calculus for contradictory deductive systems. *Studia Logica*, v. XXIV, p. 143-157, 1969. (English version of [25]).

[25] JAŚKOWSKI S. Rachunek zdán dla sustemów dedukcyjnych sprzecznych. *Studia Societatis Scientiarum Torunensis*, Sec. A, I, n. 5, p. 55-57, 1948.

[26] HERRLICH, H.; HUSEK, M. Galois connections categorically. *Journal of Pure and Applied Algebra*, v. 68, p. 165-180, 1990.

[27] ŁUKASIEWICZ, J.; TARSKI, A. Untersuchungen über den Aussangenkalküls. *Comptes Rendus des séances de la Societé des Sciences et des Lettres de Varsovie*, Classe III, v. 23, p. 30-50, 1930.

[28] MALINOWSKI, G. *Many-valued logics*. Oxford: Clarendon Press, 1993.

[29] RASIOWA, H. *An algebraic approach to non-classical logics*. Amsterdam: North-Holland, 1974.

[30] MIRAGLIA, F. *Cálculo proposicional: uma interação da álgebra e da lógica*. Campinas: UNICAMP/CLE, 1987. (Coleção CLE, v. 1)

[31] ORE, O. Galois connections. *Transactions of the American Mathematical Society*, v. 55, p. 493-513, 1944.

[32] ORLOWSKA, E.; REWITZKY, I. Algebras for Galois-style connections and their discrete duality. *Fuzzy Sets and Systems*, v. 161, p. 1325-1342, 2010.

[33] SILVA, H. G.; FEITOSA, H. A.; CRUZ, G. A. Um sistema de tableaux para a lógica paraconsistente J3. *Kínesis*, v. 9, n. 20, p. 126-150, 2017.

[34] SMITH, P. *The Galois connection between syntax and semantics*. Technical report. Cambridge: Univisity of Cambridge, 2010.

Undecidability, incompleteness and beyond: an adventure.*

N. C. A. da Costa
NCACOSTA@TERRA.COM.BR

In 1985 I received a letter[1] from someone who presented himself as a mathematical physicist, and a former PhD student of my friend Leopoldo Nachbin. Boiling down to essentials, Francisco Antonio Doria, the letter's author, asked me an intriguing question, does Gödel incompleteness matter to physics? Here started a (so far) 35–year collaboration that has produced nearly 50 papers, book chapters, and a book that made the *Scientific American* best–seller lists, *Gödel's Way*, also coauthored by Gregory Chaitin.

In December 1987 the author and F. A. Doria formulated a research program on foundational issues in mathematics, and their consequences to the empirical sciences. Main tool was to be the axiomatization of some theory through Suppes predicates; and as a first goal they decided to formalize as much as possible of classical physics. The original motivation was P. Cohen's remark, that non–Cantorian (forcing) models in set theory might be as useful in physics as non–Euclidean geometries had been a century before.

In May 1988 the authors drew up a first list of problems in that direction; in May 1991 they conceived a second list, which was based on the experience they obtained in their work on the first one. A selection and reappraisal of those problems is presented

---

*Partially supported by FAPESP and CNPq (Brazil, Philosophy Section).
[1]Remember: this happened before e–mail...

here, with new comments and an updated list of references with the results obtained.

References:

1. N. C. A. da Costa and F. A. Doria, "Structures, Suppes Predicates and Boolean–Valued Models in Physics," in J. Hintikka, ed., *Festschrift in Honor of Prof. V. Smirnov on his 60th Birthday* (1991).

2. N. C. A. da Costa, F. A. Doria and J. A. de Barros, "A Formally Undecidable Statement in Electromagnetic Theory," in *Encontro de Física Teórica do Rio de Janeiro — Homenagem Póstuma ao Professor Carlos Marcio do Amaral*, C. A. Bertulani and J. Lopes Filho, eds., pp. 145–155, Editora da UFRJ (1990).

1 Generic universes in gravitation theory

The Einstein equations describe the gravitational field. Local solutions are patched up to characterize universe models, that is, space–time manifolds whose geometry directly stems out of known solutions for the gravitational equations. We can also proceed the other way round: given a differentiable manifold, we can examine the space of all metric tensors on it. That space is a Fréchet manifold, a rather nasty structure. Research in the late sixties and early seventies by J. Marsden, J. Arms and others has shown that symmetric solutions for the Einstein equations belong to a first–category set in that manifold, and it is known that non–symmetric solutions may exhibit some kind of chaotic behavior.

Now think of the set of all spacetimes. It can be given (at worst) a Fréchet manifold topology. So we are left with the following questions: can we identify the spaces of topologically generic spacetimes, set–theoretically generic spacetimes, with the set of chaotic spacetimes? (At least modulo some reasonable, small set.)

Can we identify "topologically generic spacetimes" with "set–theoretically generic spacetimes" in a convenient set–theoretic model? Is there some kind of physically meaningful predicate that characterizes the 'interesting' generic metric tensors over a fixed differentiable manifold? (By 'physically meaningful' we think of a predicate which explicitly involves 'physical' concepts.)

We conjectured that almost anything obtained in that direction would turn out to be undecidable in ZF.

What we have learned: first, we require Martin's Axiom in order to talk about topologically and measure–theoretically generic sets of manifolds. Then we must deal with an unexpected phenomenon, the plethora of exotic smooth 4–manifolds These have been shown to (at least in some examples) originate in the underlying physics. But how does a set–theoretical, exotic, spacetime look like? It can be argued that the typical spacetime is precisely an exotic, set–theoretically generic object which can only be known through "local" depictions, that is, the novelty we reach through our model–theoretic construction is lost.

References:

1. N. C. A. da Costa, F. A. Doria and J. A. de Barros, "A Suppes Predicate for General Relativity and Set–Theoretically Generic Spacetimes," *International Journal of Theoretical Physics*, **29**, 935–961 (1990).

2. N. C. A. da Costa and F. A. Doria, "Suppes Predicates for Classical Physics," in J. Echeverría et al., eds., *The Space of Mathematics*, Walter de Gruyter, Berlin–New York (1992).

2 Generic economies, generic social structures

Population dynamics is described by (mainly) nonlinear differential equations. So, when we turn to the space of their solutions we can ask the same kind of questions we asked about general relativity. The questions on chaotic behavior turn out to be very difficult questions about the characterization of strange attractors and the like, but we can still ask about the overall relation between chaos and genericity, both topological and set–theoretic.

Also, given the results of Arrow and Debreu on the topological genericity of equilibria in competitive markets, we can also ask: well, there certainly *exist* equilibria in competitive markets. But, given sufficient computational power can we compute those equilibria beforehand?

What we have learned: The most interesting result we have obtained is a consequence of a Rice–like theorem we proved in our work on the undecidability of chaos. The fact that Nash games always have solutions doesn't imply that those solutions are *computable*, that is, that *we cannot algorithmically check whether a Nash game has reached equilibrium values*. When we go from Nash games to Arrow–Debreu equilibria, our result (the sometimes called Tsuji–da Costa–Doria theorem) means that we cannot check whether a given market configuration is an equilibrium configuration.

References:

1. M. Tsuji, N. C. A. da Costa, F. A. Doria, "The Incompleteness of Theories of Games," *J. Philosophical Logic*, **27**, 553–568 (1998).

2. N. C. A. da Costa and F. A. Doria, "Suppes Predicates and the Construction of Unsolvable Problems in the Axiomati-

zed Sciences," in P. Humphreys, ed., *Patrick Suppes, Mathematician, Philosopher*, Kluwer (1994).

3. N. C. A. da Costa and F. A. Doria, "Gödel Incompleteness in Analysis, with an Application to the Forecasting Problem in the Social Sciences," *Philosophia Naturalis* **31**, 25–62 (1994).

3 Is stability decidable? Is chaos decidable?

That problem was originally formulated by M. Hirsch in 1983:

> An interesting example of chaos — in several senses — is provided by the celebrated *Lorenz System*. [...] This is an extreme simplification of a system arising in hydrodynamics. By computer simulation Lorenz found that trajectories seem to wander back and forth between two particular stationary states, in a random, unpredictable way. Trajectories which start out very close together eventually diverge, with no relationship between long run behaviors.
>
> But this type of chaotic behavior has not been *proved*. As far as I am aware, practically nothing has been proved about this particular system. Guckenheimer and Williams proved that there do indeed exist many systems which exhibit this kind of dynamics, in a rigorous sense; but it has not been proved that Lorenz's system is one of them. It is of no particular importance to answer this question; but the lack of an answer is a sharp challenge to dynamicists, and considering the attention paid to this system, it is something of a scandal.

The Lorenz system is an example of (unverified) *chaotic dynamics*; most trajectories do not tend to stationary or periodic orbits, and this feature is persistent under small perturbations. Such systems abound in models of hydrodynamics, mechanics and many biological systems. On the other hand experience (and some theorems) show that many interesting systems can be expected to be nonchaotic: most chemical reactions go to completion; most ecological systems do not oscillate unpredictably; the solar system behaves fairly regularly. In purely mathematical systems we expect heat equations to have convergent solutions, and similarly for a single hyperbolic conservation law, a single reaction–diffusion equation, or a gradient vectorfield.

A major challenge to mathematicians is to determine which dynamical systems are chaotic or not. Ideally one should be able to tell from the form of the differential equations. The Lorenz system illustrates how difficult this can be.

S. Smale formulated a particular version of that question: is Lorenz chaos due to the the Guckenheimer attractor?

Another related set of questions: can stability be algorithmically decided for polynomial dynamical systems over Z? Such is the content of V. Arnol'd's 1974 Hilbert Symposium problems:

> *Is the stability problem for stationary points algorithmically decidable?* The well–known Lyapounov theorem solves the problem in the absence of eigenvalues with zero real parts. In more complicated cases, where the stability depends on higher order terms in the Taylor series, there exists no *algebraic* criterion.
>
> *Let a vector field be given by polynomials of a fixed degree, with rational coefficients. Does an algorithm exist, allowing to decide, whether the stationary point is stable?*

A similar problem: Does there exist an algorithm to decide, whether a plane polynomial vector field has a limit cycle?

A more detailed statement of those problems was recently communicated to da Costa and Doria by Arnol'd himself in a private exchange:

> In my problem the coefficients of the polynomials of known degree and of a known number of variables are written on the tape of the standard Turing machine in the standard order and in the standard representation.
>
> The problem is whether there exists an algorithm (an additional text for the machine independent of the values of the coefficients) such that it solves the stability problem for the stationary point at the origin (i.e., always stops giving the answer "stable" or "unstable").
>
> I hope, this algorithm exists if the degree is one. It also exists when the dimension is one. My conjecture has always been that there is no algorithm for some sufficiently high degree and dimension, perhaps for dimension 3 and degree 3 or even 2. I am less certain about what happens in dimension 2.
>
> Of course the nonexistence of a general algorithm for a fixed dimension working for arbitrary degree or for a fixed degree working for an arbitrary dimension, or working for all polynomials with arbitrary degree and dimension would also be interesting.
>
> Integers were introduced in the formulation only to avoid the difficulty of explaining the way the data are written on the machine's tape. The more realistic formulation of the problem would require the definition of an analytic algorithm working with real numbers and functions (defined as symbols). The algorithm should permit arithmetical operations, modulus,

differentiation, integration, solution of nondifferential equations (also for implicit functions in situations where conditions for the implicit function theorems are violated), exponentiation, logarithms, evaluation of 'computable' functions for 'computable' arguments.

The conjecture is that with all those tools one is still unable:

1. To solve the general stability problem starting from the right hand side functions as symbols with which one may perform the preceding operations.
2. To solve the above problem for polynomial vectorfields with real or complex coefficients.
3. To solve them with integer coefficients.

However as far as I know there are no words in logic to describe the above problem and I have thus preferred to stop at the level of algorithms in the usual sense rather than to try to explain to logicians the meaning of the impossibility of the solution of differential equations of a given type by quadratures (e.g., in the Liouville case in classical mechanics or in the theory of second order ordinary differential equations). The main difficulty here is that the solvability or unsolvability should be defined in a way that makes evident the invariance of this property under admissible changes of variables defined by functions that one can construct from the right hand side of the equations in a given coordinate system. In other terms we should explicitly describe the structure of the manifold where the vectorfield is given, with respect to which the equation is nonintegrable.

In the usual approach this structure is a linear space structure, and I think it is too restrictive.

In any case I would like to know whether you think you have proved my conjectures on polynomial vectorfields with integer coefficients:

- For some pair (degree, dimension);
- For some dimension;
- For some degree,

the polynomials being given on the tape of the machine in the standard form. If one of those undecidability conjectures is proved, it would be interesting to know for which pair (degree, dimension), or value of the dimension or value of the degree is the undecidability proven.

What we have learned: it came as a surprise. Doria and I discovered that there is a Rice–like theorem in the language of classical analysis that implies a whole infinite family of undecidability and incompleteness theorems. For instance, chaos is undecidable, *no matter which definition we use for chaos*, we just have to require it to be nontrivial (either everything is chaotic, or nothing is chaotic). We have used the Rice–like result to show the undecidability/incompleteness of chaotic systems, and a related result to show the undecidability of the Arnol'd problems. Actually it is easier to prove a result for dynamical systems in the language of analysis than to restrict it to polynomial dynamical systems over the rationals.

References:

1. N. C. A. da Costa and F. A. Doria, "Undecidability and Incompleteness in Classical Mechanics," *International Journal of Theoretical Physics* **30**, 1041–1073 (1991).

2. N. C. A. da Costa, F. A. Doria and A. F. Furtado–do–Amaral, "A Dynamical System where Proving Chaos is Equivalent to

Proving Fermat's Conjecture," *International Journal of Theoretical Physics*, **32**, 2187–2206 (1993).

3. N. C. A. da Costa, F. A. Doria, A. F. Furtado–do–Amaral and J. A. de Barros, "Two Questions on the Geometry of Gauge Fields," *Foundations of Physics* **24**, 783–799 (1994).

4. N. C. A. da Costa and F. A. Doria, "'On Arnol'd's Hilbert Symposium Problems," in G. Gottlob, A. Leitsch, D. Mundici, eds., *Proceedings of the 1993 Kurt Gödel Colloquium: Computational Logic and Proof Theory*, Lecture Notes in Computer Science **713**, Springer (1993), pp. 152–158. (1993).

5. N. C. A. da Costa and F. A. Doria, "An Undecidable Hopf Bifurcation with an Undecidable Fixed Point," *International Journal of Theoretical Physics* **32**, 2187–2206 (1993).

4 $P = NP$?

We conjecture that it is undecidable in ZF. The main intuition is that, in order to talk about the set of all Boolean satisfiability problems we may require much more than arithmetic, and there seems to be room in the axiomatic framework where we embed the satisfiability problem for lots of wild–behaving sets. For example, there is a set $\mathcal{M}$ of intuitively poly Turing machines so that the sentence "$\mathcal{M}$ is a set of poly Turing machines" is undecidable in, say, ZFC.

That was our starting point. Actually the theory we are dealing with is marred by all sorts of undecidability phenomena. But there is more to be considered.

What we have learned: around mid–1994 we received an e–mail from a top ranking mathematician who shared with us some folklore–like, so far unpublished, results. The chief fact stated in that message was: "if $P < NP$, the counterexample function to $P = NP$ grows in its peaks faster than any total recursive function." (The hypothesis we make is essential to ensure that the counterexample function is a total function.)

It took us six months to prove that result — no proof was hinted at in the e–mail message; the proof we gave was checked by several researchers which acted as informal referees for us.[2]

One of the scholia in our discussion led to the following consequence: if total, the counterexample function cannot be proved to be such in ZFC. Follows that the counterexample function is noncomputable. It is in reality an intractable monster. A subsidiary consequence can be derived: "if ZFC is consistent, then the Σ_1–extension of ZFC proves that there is no proof of the totality of the counterexample function in ZFC." Doria pictured that result as follows: if you stand in the Σ_1–extension, and look down into ZFC, you'll see no such proof.

The problem is, isn't that proof 'masked' within ZFC by the extra stuff added by the extension?

References:

1. N. C. A. da Costa, F. A. Doria, E. Bir, "On the metamathematics of the P vs. NP question," *Applied Mathematics and Computation,*" **189**, 1223–1240 (2007).

2. F. A. Doria, "El Aleph, or a monster lurks in the belly of computer science," in S. Wuppuluri and F. A. Doria, eds., *The Map and the Territory*, 403–417, Springer (2018).

---

[2] Among them, Marcel Guillaume, who checked lots of arguments for us, and in general helped us in keeping safe our math.

Appendix

A The counterexample function f

Since this fact isn't well–known, and is lost within a plethora of folklore results, I repeat in this Appendix the proof we have given for it.

SAT is the set of all Boolean expressions in conjunctive normal form (cnf) that are satisfiable, and BGS is a recursive set of poly Turing machines that contains emulations of every conceivable poly Turing machines.

The full counterexample function f is defined as follows; let ω be also a set of codes for an enumeration of the Turing machines. Similarly we code by an analogous standard code SAT onto ω:

- If $n \in \omega$ isn't a poly machine, $f(n) = 0$.

- If $n \in \omega$ codes a poly machine:

    - $f(n) =$ first instance x of SAT so that the machine fails to output a satisfying line for x, plus 1, that is, $f(n) = x + 1$.
    - Otherwise $f(n)$ is undefined, that is, if $P = NP$ holds for n, $f(n) =$ undefined.

As defined, f is non computable. It will also turn out to be at least as fast growing as the Busy Beaver function, since in its peaks it dominates all intuitively total recursive functions.

- Use the s–m–n theorem to obtain Gödel numbers for an infinite family of "quasi–trivial machines" (see below). The table for those Turing machines involves very large numbers, and the goal is to get a compact code for that value in each quasi–trivial machine so that their Gödel numbers are in a sequence $g(0), g(1), g(2), \ldots$, where g is primitive recursive.

- Then add the required clocks as in the BGS sequence of poly machines, and get the Gödel numbers for the pairs machine + clock. We can embed the sequence we obtain into the sequence of all Turing machines.

- Notice that the subsets of poly machines we are dealing with are (intuitive) recursive subsets of the set of all Turing machines. More precisely: if we formalize everything in some theory S, then the formalized version of the sentence "the set of Gödel numbers for these quasi–trivial Turing machines is a recursive subset of the set of Gödel numbers for Turing machines" holds of the standard model for arithmetic in S, and vice versa.

  However S may not be able to prove or disprove that assertion, that is to say, it will be formally independent of S.

- We can thus define the counterexample functions over the desired set(s) of poly machines, and compare them to fast–growing total recursive functions over similar restrictions.

Definition A.1 *For* $f, g : \omega \to \omega$,

$$f \textbf{ dominates } g \leftrightarrow_{\text{Def}} \exists y \, \forall x \, (x > y \to f(x) \geq g(x)).$$

We write $f \succ g$ *for f dominates g.* □

Quasi–trivial machines

Recall that the operation time of a Turing machine is given as follows: if M stops over an input x, then the operation time over x,

$$t_M = |x| + \text{number of cycles of the machine until it stops.}$$

Example A.2

- **First trivial machine.** Note it O. O inputs x and stops.

  $t_O = |x| +$ moves to halting state $+$ stops.

  So, operation time of O has a linear bound.

- **Second trivial machine.** Call it O'. It inputs x, always outputs 0 (zero) and stops.

  Again operation time of O' has a linear bound.

- **Quasi–trivial machines.** A *quasi–trivial machine* Q operates as follows: for $x \leq x_0$, x_0 a constant, fixed value, Q $=$ R, R an arbitrary total machine. For $x > x_0$, Q $=$ O or O'.

  This machine has also a linear bound. □

Remark A.3 Now let H be any fast–growing, superexponential total machine. Also let H' be a total Turing machine. Form the following family $Q_{...}$ of quasi–trivial Turing machines with subroutines H and H':

1. If $x \leq H(n)$, $Q^{H,H',n}(x) = H'(x)$;
2. If $x > H(n)$, $Q^{H,H',n}(x) = 0$. □

Proposition A.4 *There is a family* $R_{g(n,|H|,|H'|)}(x) = Q^{H,H',n}(x)$, *where g is primitive recursive, and* $|H|, |H'|$ *denotes the Gödel number of H and of H'.*

Proof: By the composition theorem and the s–m–n theorem. □

Remark A.5 *Very important!* We are interested in quasi–trivial machines where H' $=$ T, the standard truth–table exponential algorithm for SAT. □

Notice that, for the counterexample function when defined over all Turing machines (with the extra condition that the counterexample function $= 0$ if M_m isn't a poly machine), we have:

Proposition A.6 *If $g(n)$ is the Gödel number of a quasi–trivial machine as in Remark A.3, then $f(g(n)) = H'(n) + 1$.*

Proof: Use the machines in Proposition A.4 and Remark A.5. □

The main result

Our goal here is to prove the following: *no total recursive function dominates f.*

We sketch here the idea of the proof. Suppose that there is a total recursive function $h(n)$ that dominates f. Get a total recursive $k(n)$ that dominates h and so that the relative growth speed of k with respect to h is faster that any primitive recursive function.

Why do we need such a condition? We use the quasi–trivial machines to reproduce k within f, that is, we replicate function $\langle n, k(n) \rangle$ within f by a sequence of machines with Gödel numbers $N(n), n = 0, 1, 2, \ldots$ (see above Proposition A.6), where N is primitive recursive, so that we have that k becomes the sequence of machines $N(n), n = 0, 1, 2, \ldots$, and we get the value of f at k as $\langle N(n), k(n) + 1 \rangle$ with $f(N(n)) = k(n) + 1$.

As N can be taken to be primitive recursive, monotonic increasing on n, it slows down the growth of k by a primitive recursive function. Given our construction — which is trivially fulfilled — we have that f still overtakes h infinitely many times, as k grows faster than h, and we are done.

Proposition A.7 *For no total recursive function h does $h \succ f$.*

Proof: Suppose that there is a total recursive function h such that $h \succ f$. Notice:

- Given such a function h, we can obtain another total recursive function h′ which satisfies:

    1. h′ is strictly increasing.
    2. For $n > n_0$, $h'(n) > h(g(n))$, with g as in Prop. A.6. □

- Given a total recursive h, there is a total recursive h′ that satisfies the previous conditions.

For given h, we obtain out of that total recursive function by the usual constructions a strictly increasing total recursive h*. Then if, for instance, F_ω is Ackermann's function, $h' = h^* \circ F_\omega$ will do. (The idea is that F_ω dominates all primitive recursive functions, and therefore h* composed with it dominates $g(n)$.)

We have that the Gödel numbers of the quasi–trivial machines Q are given by $g(n)$. Choose adequate quasi-trivial machines, so that $f(g(n)) = h'(n) + 1$, from Proposition A.6. We now conclude our argument. If we make explicit the computations, for $g(n)$ (as the argument holds for any strictly increasing primitive recursive g):

$$f(g(n)) = h'(n) + 1 = h^*(F_\omega(n)) + 1,$$

and

$$h^*(F_\omega(n)) > h^*(g(n)).$$

For $N = g(n)$,

$$f(N) > h^*(N) \geq h(N), \text{ all } N.$$

Therefore no such h can dominate f. □

Corollary A.8 *No total recursive function dominates f.* □

Notice that within theories like the ones we have been using here, we cannot prove that f is total.[3]

---
[3] This Appendix was based on the chapter that Doria wrote for his book (with S. Wuppuluri), *The Map and the Territory*, Springer (2018). For more details please check the reference.

Logic in Times of Big Data*

Marcelo Finger[†]
Department of Computer Science
Institute of Mathematics and Statistics
University of São Paulo, Brazil
mfinger@ime.usp.br

Abstract

Artificial Intelligence has become the area of prominence in Computer Science, and within it the so called area of Big Data, with its emphasis on processing large quantities of data via artificial neural networks. The collection of methods called automated learning has gained visibility both in science and in the media, including fierce discussions on social media.

In this paper, we discuss the role and possibilities for Logic in an environment where most, if not all, the attention is focused on the Big Data fever.

I would like to dedicate this article to Prof. Doria, with whom I had the immense pleasure of several remote discussions, and a few in-person ones. The breadth of his interests have always been a reason for wonder, and I hope the following piece makes justice to that wide view of Logic and Science.

---

*This study was financed in part by the Coordenação de Aperfeiçoamento de Pessoal de Nível Superior - Brasil (CAPES) - Finance Code 001

†Partially supported by Fapesp projects 2019/07665-4 (C4AI) and 2014/12236-1 (Animals) and CNPq grant PQ 303609/2018-4

Preamble

Science is a human activity. And so is Computer Science, thus it is subject to typically human phenomena, like concentrated focus on what is temporarily fashionable. Recently, Artificial Intelligence has become the area of prominence in Computer Science, and within it the so called area of Big Data, with its emphasis on processing large quantities of data via artificial neural networks. The collection of methods called automated learning has gained visibility both in science and in the media, including fierce discussions on social media.

When an area of science has this type and prominence it ends up affecting other areas as well. It was like that at the beginning of the last century, with the development of quantum physics, and likewise with Goedel's incompleteness theorems.

The Beginning of Modern Logic

The development of modern Logic started in the middle of the 19th century, notably with Boole's proposal for an algebra of truth values, in his significantly titled book "On the laws of thought" (Boole 1854), and also with the works of Charles Saunders Peirce in the United States (Peirce 1958). But the real thrust for the creation of modern Logic occurred with the project on formalizing mathematics, initially with Frege's works in Germany (Frege 1879), and with the publication of Whitehead and Russell's Principia Mathematica (Whitehead and Russel 1910). In this trajectory, it is important to note a change in objectives, where the goal was no longer to formulate the laws of thought, but to formalize the practice of Mathematics. In this way, modern Logic has crystallized as a tool for the analysis of rational thought and as the foundation of mathematical methods and proofs.

The formalist project, as proposed by Hilbert among others, began by seeking the mirage of automating all of mathematics (Hilbert 1925), a goal that proved impossible to be achieved with the results of incomplete arithmetic (Kleene 1952). However, the secondary results of this effort proved to be extremely fruitful, giving rise to the theory of recursive functions, to Turing Machines and finally to computer science (Turing 1936; Turing 1937).

Mathematical Logic itself has borne fruit in the field of Mathematics, serving as a basis for the modern Set Theory, and even today it generates new mathematical challenges, such as Voevodsky's homotopic type theory (HoTT) (Univalent Foundations Program 2013) and contemporary efforts to automate the proof verification in mathematics (Avigad 2020).

The Beginning of Artificial Intelligence as a Discipline

The first computers were built shortly after the Second World War (Augarten 1984), and it was not long before the idea of automating human reasoning returned to academic circles, this time renamed as *common sense reasoning*. Early on, programs were created in an attempt to perform a series of activities associated with human intelligence, such as automatic translation from Russian to English (a Cold War influenced choice), the first automatic theorem provers, and the creation of a model which was initially called Perceptron, and which later led to the creation of modern artificial neural networks (Buchanan 2005).

From the 1970s on, there occurred the development of what became known as Symbolic Artificial Intelligence, driven by the Japanese Fifth Generation Project (Pollack 1992). Both Lambda Calculus and First Order Logic, formalisms proposed for the study of computability in the 1930s (Church 1932; Barendregt 1981; Boolos and Jeffrey 1989), served as motivation to LISP and Prolog programming languages, respectively. Then came the first expert systems, which consisted of a large set of logical rules that addressed a subject in a specific domain (Brownston, Farrell, Kant, and Martin 1985). These rule-based systems showed a number of deficiencies in their handling, notably there were two main problems. First rules are categorical, and do not deal adequately with uncertain or fuzzy information. Second it was patently hard to come up with rules that were well designed for the goals of the expert systems, since they demanded the treatment of a very large number of cases, each of which was addressed not by one, but by a set of rules; and all these rules ended up interacting with each other, often in an unpredictable way (Widom and Ceri 1996).

The exploration of first of these deficiencies generated an entire sub-area of artificial intelligence known as Reasoning under Uncertainty, giving rise to formalisms such as fuzzy logics and fuzzy reasoning (Goodman and Nguyen 1988), probabilistic logics (Nilsson 1986), Markovian networks and Bayesian networks (Pearl 1988). The second deficiency has been also explored, leading to the appearance of an area called inductive logical reasoning; despite important advances, this area still struggles with difficulties in dealing with the combinatorial explosion associated with the learning of rules (Muggleton 1992), even when these rules incorporate probabilistic elements (De Raedt and Kersting 2008).

Although there has been a change in the focus for most works in artificial intelligence to other areas, such as probabilistic reasoning and neural architectures, we cannot say that logic-based systems met their end in artificial intelligence. On the contrary, several areas are undergoing a strong expansion, among which reasoning based on techniques of logical satisfiability, since the basic NP-problem, the SAT problem, had had significant computational advances (Eén and Sörensson 2006) and has successfully been used in formal hardware and software verification (De Moura and Bjørner 2008). The profusion of data systems and metadata over the internet also enabled the development of description logic as an interconnection and interface tool for these knowledge-based systems (Baader, Calvanese, McGuinness, Patel-Schneider, Nardi, et al. 2003; Baader, Horrocks, Lutz, and Sattler 2017).

However, large and rule-based systems undeniably ceased to be the focus of the scene in the field of artificial intelligence.

The Big Data Explosion and the Eclipse of Logic

It is important to note that neural networks aim to remedy several deficiencies presented by the first expert systems, both enabling the treatment of uncertain information and incorporating the learning of a very large number of data in a compact form, such as the weights of a neural network. In the beginning, when problems with the first expert systems were detected, neural networks did not have the capacity to address these problems. Neural networks have their own internal deficiencies, such as the problem of overfitting, which causes networks to memorize the input data and fail to generalize to unseen data; similarly, neural networks suffer from the problem of underfitting, when due to the scarcity of data the system cannot identify the relevant patterns in the examples provided, also leading to a lack of generalization in unseen data. In addition, the high computational cost of training such a network can be too heavy to be successfully used in practice.

These problems only began to be solved at the beginning of the 21st century. First, with the spread of the internet, a profusion of data began to appear in digital format, which could then be used by networks, reducing the underfitting problem. On the other hand, the availability of highly parallel hardware has enabled the efficient training of large-sized neural networks; paradoxical as it may sound, the development of this highly parallel hardware was driven by the need for very fast video processors for rendering images in computer games[1]. Finally, the problem of overfitting can be addressed through the development of new and sophisticated neural network architectures, no longer restricted to multilayer perceptrons.

Big Data is characterized by the use of vast amounts of information that is used by supervised or unsupervised learning methods, and is employed in various tasks, such as the classification of data, texts or images; medical diagnosis; prediction of numerical values,

---

[1] https://techcrunch.com/2017/10/27/how-video-game-tech-makes-neural-networks-possible/, last visited September 2020

generalizing the logistic regression methods; the elaboration of recommendation systems; the translation of texts between an increasing number of languages; among many others.

On the other hand, the inherent complexity of traditional logic-based techniques led to a situation in which data explosion had the effect of eclipsing the use of logical methods. The requirements of an extremely expensive process of handcrafting rules can be compared to jewel-making techniques; their inadequacies remain the same, whether one constructs rules in first order logic, or in Prolog, or whether one focuses in developing systems based on fuzzy logic, or on probabilistic logic, or on one of several families of description logic.

Expressiveness versus Complexity

Modern logic was created to formalize and express all mathematical activities. Right away, this task proved to be infeasible, because Russell's paradox showed that when aiming for "total" expressiveness, logic collapses in contradiction. The first attempt to control expressivity created first-order logic which, however, proved to be a non-recursive logic and, by the Church-Turing hypothesis, not computable as well. It is noteworthy that the classic propositional logic, which was already considered as "trivial" by Hintikka (Hintikka 1973), was the logic in which the first NP-complete problem was shown to go, in Cook's 1971 result (Cook 1971).

In this way, Complexity Theory had two of its key moments defined within the scope of classical Logic: the non-computability of first order logic, and the NP-completeness of classic propositional Logic. In addition, the decision of quantified propositional logic (best known as QBF, quantified boolean formulas) was shown to be PSPACE-complete, and most logics mentioned above have their decision complexity in the polynomial hierarchy located between NP and PSPACE. Thus, most fuzzy and probabilistic logics, as well as modal logics, which encompass description logics, have their decision problem in this range. Moreover combining logics, for example, by means of the product of three or more of those logics, almost always results in a system whose decision problem is not computable. That is, individually each of these logics is intractable and when composed they become non-computable (Gabelaia, Kurucz, Wolter, and Zakharyaschev 2005).

With this degree of complexity, it is not surprising that those logics are barred at the Big Data party.

Logics Both Interesting and Tractable

The emergence of the Big Data phenomenon forces us to rethink the goal of finding highly expressive formalisms. Instead, we are proposing to search for formal systems whose decision problem complexity allows their efficient application to a data repository that keeps growing.

One possible search for low complexity systems started with the research project in the area of logic approximations (Selman and Kautz 1991; Schaerf and Cadoli 1995; Finger and Wassermann 2006). Although almost all logics in this approach were tractable ones, the nature of systems was almost always ad hoc, looking for artificial ways to impose syntactic or semantic restrictions in order to avoid a combinatorial explosion. As a consequence, very few among those approximating systems have been generated to describe specific situations.

An exception to the ad hoc nature of tractable logics is found in intelim logic (D'Agostino and Floridi 2009), which arose from a critique of the eminently intuitionist nature of deduction rules in the Natural Deduction method. This analysis led to a reformulation of the set of natural deduction rules in two subgroups. In a subgroup, we have the rules for inserting and eliminating connectors whose application does not require the use of extra hypotheses; this set of rules is called intelim. The second group contains the rules that require the formulation of hypotheses that must later be discarded in the deduction, called hypothetical rules.

It turns out that the use of classical propositional calculus with only intelim rules produces a logic whose decision problem is tractable, and which can be used to approximate classical propositional logic as the number of hypotheses introduced in a deduction grows. This method is capable of polynomially approximating any classic decision. The complexity of the basic intelim formalism is only quadratic (D'Agostino, Finger, and Gabbay 2013).

There are other well-established logics with tractable fragments, such as Horn's logics, which are complete over the polynomial complexity class, and the clause fragment with only two literals, 2-SAT, whose complexity is LOGSPACE-complete .

However, there are no widespread methods for learning formulas in none of these logics. There are two reasons for this. First learning involves searching for theories, and such space is of size exponential in the number of variables. Second even with an efficient learning method, the problem of dealing with uncertain information is not yet addressed.

In both cases, the fundamental problems of rule learning remain, and until efficient methods are found that provide adequate solutions to these problems, we will not see a revival of rule-based methods.

Logic and Some Problems Created by Big Data

The arrival of a new technology brings new solutions to existing problems, but invariably it also unveils new problems that were latent before. Among the problems brought about by Big Data technology, we will focus on two, namely, the ethical problems of the use of personal information, and the problem of lack of transparency due to the fact that automated learning methods in general, and neural networks in particular, do not allow direct inspection of the information that was automatically learned. This is called the *black box method problem*. In a way, we can say that the second problem aggravates the first one.

The lack of transparency of neural networks brought the new

challenge of creating an artificial intelligence capable of presenting explanations, which was called *explainable artificial intelligence*. A neural network is a set of weights that are modified over countless learning interactions; at no point in this process is there a justification for the reason a set of weights has one specific set of values. After all, the method does not present any semantics, in the sense that it does not allude to a mapping of its values in any other domain in which they can be inspected (Bender and Koller 2020).

Looking at the core of a neural network, it is nothing more than a continuous function in a multidimensional space. And as such, this function can be approximated by a piecewise linear function. This approximation can be as precise as needed, increasing the number of linear parts of the approximation and, thus, reducing the error of the approximation.

That property reveals a way in which logic can be used to approximate any continuous function by means of a logical sentence. We are not referring here to an approach by classical logic, which is known to have only two truth values; on the contrary, the representation of this property requires a non-classical fuzzy logic, which has in its semantics any values in the closed interval $[0, 1]$. There are countless logics with this property, but a famous result of McNaughton's Theorem is that in Łukasiewicz's Logic with infinite truth values, any linear function by parts with integer coefficients has a formula whose true value corresponds to this function (McNaughton 1951; Cignoli, d'Ottaviano, and Mundici 2000). Note that there is an unavoidable restriction on *integer coefficients*; to approximate *any* continuous function we need a slightly more expressive logic, capable of capturing piecewise linear functions with any real or rational coefficients. Several logics that are capable of capturing this expressiveness have already been proposed in the literature, many of which were logics with expressiveness much higher than necessary to capture only this fragment (Amato and Porto 2000; Esteva, Godo, and Montagna 2001; Gerla 2001). Above all, the literature only focused on proving that logics had the necessary expressiveness, without providing efficient algorithms capable of producing a formula given a linear function by parts that one wishes to represent (Mundici 1994; Aguzzoli 1998; Aguzzoli and Mundici 2003).

We note, however, that such a method has been recently proposed for a logic that is capable of expressing all and exclusively piecewise linear functions with rational coefficients. It is the Łukasiewicz Logic modulo satisfiability (MODSAT), a logic that allows expressing piecewise linear functions with a pair of formulas, in which the models of the first formula are restricted by those values that satisfy the second formula with truth value 1 (thus, modulo 1-satisfiability, or simply MODSAT) (?). In addition, a tractable method has been proposed that is capable of producing a pair of formulas as described above for each piecewise linear function provided as input (Preto and Finger 2020).

As it is a well-known logic for which there are already several proof methods implemented (Finger and Preto 2018), by transforming a neural network into a pair of Łukasiewicz Logic MODSAT formulas, we can perform typical analyzes of program model checkers, such as reachability of a given state, safety and liveness properties that guarantee that some undesirable state will never be reached. It is true that these methods involve NP-complete decision problems, but it is also true that almost no proof-based verification method has tractable methods, and since we are assuming that we will be checking critical systems, the investment in advanced off-line verification methods ends up paying off, providing proof of system security or the disclosure of a failure.

Will logic be able to participate in the Big Data party this way?

Acknowledgement

The author would like to thank Sandro Preto for proof reading an earlier version of his paper.

References

Aguzzoli, S. (1998). The complexity of McNaughton functions of one variable. *Advances in Applied Mathematics 21*(1), 58–77.

Aguzzoli, S. and D. Mundici (2003). *Weierstrass Approximation Theorem and Lukasiewicz Formulas with one Quantified Variable*, pp. 315–335. Heidelberg: Physica-Verlag HD.

Amato, P. and M. Porto (2000). An algorithm for the automatic generation of a logical formula representing a control law. *Neural Network World 10*(5), 777–786.

Augarten, S. (1984). *Bit by Bit: An Illustrated History of Computers*. Ticknor & Fields. Available at http://ds-wordpress.haverford.edu/bitbybit/bit-by-bit-contents/.

Avigad, J. (2020). Reliability of mathematical inference. *Synthese*.

Baader, F., D. Calvanese, D. McGuinness, P. Patel-Schneider, D. Nardi, et al. (2003). *The description logic handbook: Theory, implementation and applications*. Cambridge university press.

Baader, F., I. Horrocks, C. Lutz, and U. Sattler (2017). *An Introduction to Description Logic*. Cambridge University Press.

Barendregt, H. P. (1981). *The Lambda Calculus: Its Syntax and Semantics*. Number 103 in North-Holland Studies in Logic and the Foundation of Mathematics. Elsevier Science Publishers.

Bender, E. M. and A. Koller (2020, July). Climbing towards NLU: On meaning, form, and understanding in the age of data. In *Proceedings of the 58th Annual Meeting of the Association for Computational Linguistics*, Online, pp. 5185–5198. Association for Computational Linguistics.

Boole, G. (1854). *An Investigation on the Laws of Thought*. London: Macmillan. Available on project Gutemberg at http://www.gutenberg.org/etext/15114.

Boolos, G. S. and R. C. Jeffrey (1989). *Computability and Logic* (third ed.). Cambridge University Press.

Brownston, L., R. Farrell, E. Kant, and N. Martin (1985). *Programming Expert Systems in OPS5: An introduction to rule-based programming*. Addison-Wesley.

Buchanan, B. G. (2005). A (very) brief history of artificial intelligence. *AI Magazine 26*(4), 53–53.

Church, A. (1932). A set of postulates for the foundation of logic. *Annals of Mathematical Studies 33*(2), 346–366.

Cignoli, R., I. d'Ottaviano, and D. Mundici (2000). *Algebraic Foundations of Many-Valued Reasoning*. Trends in Logic. Springer Netherlands.

Cook, S. A. (1971). The complexity of theorem-proving procedures. In *Conference Record of Third Annual ACM Symposium on Theory of Computing (STOC)*, pp. 151–158. ACM.

D'Agostino, M., M. Finger, and D. M. Gabbay (2013). Semantics and proof-theory of depth bounded boolean logics. *Theor. Comput. Sci. 480*, 43–68.

D'Agostino, M. and L. Floridi (2009, March). The enduring scandal of deduction. is propositional logic really uninformative? *Synthese 167*(2), 271–315.

De Moura, L. and N. Bjørner (2008). Z3: An efficient smt solver. In *International conference on Tools and Algorithms for the Construction and Analysis of Systems*, pp. 337–340. Springer.

De Raedt, L. and K. Kersting (2008). Probabilistic inductive logic programming. In *Probabilistic Inductive Logic Programming*, pp. 1–27. Springer.

Eén, N. and N. Sörensson (2006). Translating pseudo-boolean constraints into sat. *Journal on Satisfiability, Boolean Modeling and Computation 2*(1-4), 1–26.

Esteva, F., L. Godo, and F. Montagna (2001, Jan). The ŁΠ and ŁΠ$\frac{1}{2}$ logics: two complete fuzzy systems joining Łukasiewicz and product logics. *Archive for Mathematical Logic 40*(1), 39–67.

Finger, M. and S. Preto (2018). Probably half true: Probabilistic satisfiability over łukasiewicz infinitely-valued logic. In D. Galmiche, S. Schulz, and R. Sebastiani (Eds.), *Automated Reasoning - 9th International Joint Conference, IJCAR 2018, Held as Part of the Federated Logic Conference, FloC 2018, Oxford, UK, July 14-17, 2018, Proceedings*, Volume 10900 of *Lecture Notes in Computer Science*, pp. 194–210. Springer.

Finger, M. and R. Wassermann (2006). The universe of propositional approximations. *Theoretical Computer Science 355*(2), 153–166.

Frege, G. (1879). Begriffsschrift. In *(van Heijenoort 1982)*, pp. 1–82. Harvard University Press.

Gabelaia, D., A. Kurucz, F. Wolter, and M. Zakharyaschev (2005). Products of 'transitive' modal logics. *Journal of Symbolic Logic 70*(3), 993?1021.

Gerla, B. (2001). Rational Łukasiewicz logic and DMV-algebras. *Neural Network World 11*(6), 579–594.

Goodman, I. and H. Nguyen (1988). Conditional objects and the modeling of uncertainties. In M. G. et al (Ed.), *Fuzzy Computing*, pp. 119–138. North Holland.

Hilbert, D. (1925). On the Infinite. In *(van Heijenoort 1982)*, pp. 367–392. Harvard University Press.

Hintikka, J. (1973). *Logic, language games and information. Kantian themes in the philosophy of logic*. Oxford: Clarendon Press.

Kleene, S. C. (1952). *Introduction to metamathematics*. Bibl. Matematica. Amsterdam: North-Holland.

McNaughton, R. (1951). A theorem about infinite-valued sentential logic. *Journal of Symbolic Logic 16*, 1–13.

Muggleton, S. (1992). *Inductive logic programming*. Number 38. Morgan Kaufmann.

Mundici, D. (1994). A constructive proof of mcnaughton's theorem in infinite-valued logic. *The Journal of Symbolic Logic 59*(2), 596–602.

Nilsson, N. (1986). Probabilistic logic. *Artificial Intelligence 28*(1), 71–87.

Pearl, J. (1988). *Probabilistic reasoning in intelligent systems: networks of plausible inference*. San Francisco, CA, USA: Morgan Kaufmann Publishers Inc.

Peirce, C. S. (1931–1958). *Collected Papers of Charles Sanders Peirce*. Harvard University Press. Volumes 1–8, edited by Charles Hartshorne, Paul Weiss and Arthur Burks.

Pollack, A. (1992). 'Fifth Generation' became japan's lost generation. *The New Tork Times*. June 5, Section D, Page 1.

Preto, S. and M. Finger (2020). An efficient algorithm for representing piecewise linear functions into logic. In *LSFA2020 – Proceedings of the 15th International Workshop on Logical and Semantic Frameworks, with Applications*, pp. 1–15.

Schaerf, M. and M. Cadoli (1995). Tractable reasoning via approximation. *Artificial Intelligence 74*(2), 249–310.

Selman, B. and H. Kautz (1991, July). Knowledge compilation using horn approximations. In *Proceedings AAAI-91*, pp. 904–909.

Turing, A. (1936). On computable numbers, with an application to the entscheidungsproblem. *Journal of Math 58*(345-363), 5.

Turing, A. (1937). Computability and λ-definability. *Journal of Symbolic Logic 2*, 153–163.

Univalent Foundations Program (2013). *Homotopy Type Theory: Univalent Foundations of Mathematics*. Institute for Advanced Study (Princeton, N.J.).

van Heijenoort, J. (Ed.) (1982). *From Frege to Gödel: A Source*

Book in Mathematical Logic, 1879–1931. Cambridge, Massachussets: Harvard University Press.

Whitehead, A. N. and B. A. W. Russel (1910). *Principia Mathematica.* Cambridge University Press.

Widom, J. and S. Ceri (Eds.) (1996). *Active Database Systems — Triggers and rules for Advanced Database Processing.* Morgan Kaufmann.

Epistemology of quasi-sets

Adonai S. Sant'Anna

Abstract

I briefly discuss the epistemological role of quasi-set theory in mathematics and theoretical physics. Quasi-set theory is a first order theory, based on Zermelo-Fraenkel set theory with *Urelemente* (ZFU). Nevertheless, quasi-set theory allows us to cope with certain collections of objects where the usual notion of identity is not applicable, in the sense that $x = x$ is not a formula, if x is an arbitrary term. Basically, quasi-set theory offers us some sort of logical apparatus for questioning the need for identity in some branches of mathematics and theoretical physics. I also use this opportunity to discuss a misunderstanding about quasi-sets due mainly to Nicholas J. J. Smith, who argues, in a general way, that sense cannot be made of vague identity.

Key words: quasi-sets, identity, indistinguishability, epistemology

1 Introduction

Francisco "Chico" Doria was my doctoral co-advisor. More than that, he is one of the greatest influences on my professional and even personal life. The most important lesson I learned from him is this: be honest with yourself! If you want to publish papers, you go ahead and publish papers. But if you want to find unconventional results in science, you should not submit yourself to what other people expect from you. He never said that to me in so many words. But his actions through theoretical physics and mathematics said that to me in a very clear way. I found this hidden message in his successful works on undecidability and incompleteness in physical theories and in his "exotic" results on the P versus NP problem in computer science as well, both in collaboration with Newton da Costa, my doctoral advisor. So, that is what I try to do since then. All I want from this life is to be honest with myself. If all my efforts fail, that will be my sole responsibility. But this journey I started almost three decades ago, thanks to da Costa and Doria, is truly worthwhile in itself. So, let's talk about quasi-sets and their curious epistemological role.

Quasi-set theory $\mathcal{Q}$ [1] [10] [11] is a first order theory without identity which allows the existence, among its terms (the objects of $\mathcal{Q}$), of collections (sets and q-sets) and atoms (*Urelemente*). Some of those collections correspond to ZFU sets, in the sense that a binary predicate letter of *extensional equality* $=_E$ is explicitly defined in $\mathcal{Q}$ and a given translation from ZFU to $\mathcal{Q}$ guarantees that every translated axiom of ZFU is a theorem in $\mathcal{Q}$ (where ZFU equality is translated as the extensional equality in $\mathcal{Q}$). In other words, extensional equality in quasi-set theory has all the usual properties of standard identity in ZFU. Nevertheless, the axioms of $\mathcal{Q}$ do not allow that $x =_E x$ is necessarily a formula, for any term x. Concerning atoms in $\mathcal{Q}$, there are two kinds of *Urelemente*, termed m-atoms and M-atoms, which are identified by two unary predicates $m(x)$ and $M(x)$, respectively. M-atoms correspond (in a precise sense) to standard atoms of ZFU theory, while m-atoms are something else (or less, if the reader allows our poetic view). A weaker binary relation of "indistinguishability" (denoted by $\equiv$), is used instead of identity, and it is postulated that $\equiv$ has the standard properties of an equivalence relation. The defined binary predicate letter of extensional equality $=_E$ cannot be applied to m-atoms, since no expression of the form $x =_E y$ is a formula if either x or y denote m-atoms. Hence, there is a precise sense in saying that m-atoms can be indistinguishable without being identical. In standard mathematics, when we say that $x = y$ (x is identical to y) we are talking about the very same object, with two different labels: x and y. In quasi-set theory $\mathcal{Q}$, the formula $x \equiv y$ does not entail we are necessarily talking about the very same object. Axioms of quasi-set theory are a very natural extension of the axioms of Zermelo-Fraenkel set theory with Urelemente (ZFU). So, no one here is abandoning standard mathematics. Actually, quasi-set theory provides a specific methodology for a better understanding of identity and its role in mathematics. Geometry was better understood when its postulates were questioned in the last two centuries. Something similar happens here, concerning identity.

The development of quasi-set theory was motivated mostly by certain striking phenomena in quantum mechanics, all related to the well known *problem of non-individuality* among elementary particles [1]. According to any standard statistical mechanics textbook [3], Maxwell-Boltzmann (MB) statistics gives us the most probable distribution of N *distinguishable* objects into, say, boxes with a specified number of objects in each box. By "box", in this context, we mean "state". Thus, MB statistics can be easily described through a classical mathematical picture for ensembles of individual particles. Nonetheless, in quantum mechanics the very notion of state of a particle differs considerably from that one in classical mechanics. Besides, quantum statistics are supposed to give us the most probable distribution of N *indistinguishable* objects into distinguishable boxes with a specified number of objects in each box. The usual and physically meaningful quantum statistics are Bose-Einstein and Fermi-Dirac.

Hence, it has been argued that classical particles are "individuals" of some sort. Even when classical particles share the very same set of intrinsic properties, their individuality must be ascribed by something which "transcends" such intrinsic properties [1] [19] [22]. In quantum statistics, on the other hand, the Indistinguishability Postulate (IP) asserts that "If a permutation is applied to any state for an assembly of particles, then there is no way of distinguishing the resulting permuted state function from the original one by means of any observation at any time" [2]. IP is one of the most basic principles of quantum mechanics and it implies that permutations of quantum particles are not usually regarded as observable.

The non-individuality problem in quantum mechanics is not limited to quantum statistics. The helium atom, e.g., is probably the simplest realistic situation where the problem of individuality plays an important role. With the non-individuality question put aside, the wave function of the helium atom would be just the product of two hydrogen atom wave functions with Z = 1 changed to Z = 2. Nevertheless, the space part of the wave function for the case where one of the electrons is in the ground state (100) and the other one is in excited state (nlm) is:

$$\phi(x_1, x_2) = \frac{1}{\sqrt{2}}[\psi_{100}(x_1)\psi_{nlm}(x_2) \pm \psi_{100}(x_2)\psi_{nlm}(x_1)]$$

where the + and − signs are designated for the spin singlet and spin triplet, respectively, while x_1 and x_2 denote the vector positions of both electrons.

For the ground state, however, the space function must necessarily be symmetric. In that case, the problems regarding non-individuality have no physical effect. The most interesting case, however, is the excited state of helium. The equation stated above reflects our ignorance about which electron is in position x_1 and which one is in position x_2. Nevertheless, in the same equation there are terms like $\psi_{100}(x_1)$, which corresponds to a specific physical property of an individual electron.

A unified quasi-set-theoretic approach to quantum distributions and the helium atom is introduced in [12]. And a comprehensive quasi-set-theoretic approach to quantum statistics is presented in [25]. Notwithstanding, a quasi-set theoretic approach to the quantum interference produced by two light beams has not yet been developed, despite the fact that a profound mathematical link between indistinguishability and coherence was presented almost three decades ago by Leonard Mandel [16].

From an epistemological and even methodological point of view, quasi-set theory presents some peculiar features, if we compare it to other mathematical tools usually developed and employed in theoretical physics. And that is a good reason to raise some questions about the theoretical and philosophical role of quasi-sets.

Mostly, mathematical approaches to cope with physical problems are developed with one simple purpose in mind: to solve problems! In order to do that, mathematical tools are created and developed. But what is the usual procedure to develop a mathematical tool which is useful in theoretical physics? The standard answer to this question emerges from *stronger* concepts, *stronger* models, and *stronger* theories.

Usually, the cornerstone of any physical theory is either a differential equation or a class of differential equations. Newtonian mechanics is based on Newton's Second Law, classical electromagnetism is based on Maxwell's equations, quantum mechanics is based on Schrödinger's equation, and so on. And differential equations simply establish the *limits* to what extent physical phenomena are supposed to be, at least from a theoretical point of view.

Quasi-set theory does not seem to work that way! Quasi-set theory is not supposed to be a problem solver, in the usual sense. And that is why I wrote this paper (in honor to my friend and former Ph.D. co-advisor Francisco Doria). Quasi-set theory allows us to discuss about a mathematical universe which is, in a sense, weaker than ZFU (since $x \equiv y$ does not entail that x and y denote the very same object). And such a feature may cause some surprise to many people. A good example to illustrate this last claim is Nicholas J. J. Smith's misunderstandings concerning quasi-sets and their role in the scientific enterprise.

So, let's elaborate the main points of this paper.

2 Epistemological character of $\mathcal{Q}$

Most of mathematics used in theoretical physics is based on set theories. Despite the fact that physicists in general are not usually concerned with set theories *per se*, there seems to be no major problem if I claim that any common mathematical scenario for theoretical physics is based on Zermelo-Fraenkel-like axioms, including those found in ZFC and ZFU. An excellent defense for such a claim may be found in Patrick Suppes' magistral book *Representation and Invariance of Scientific Structures* [30]. Other mathematical approaches are possible and even necessary in theoretical physics, like category theory [4] [15]. But ZFU set theory is powerful enough for dealing with plenty of important and well-known applications.

But, what is set theory after all? Since there is a huge myriad of set theories in the literature (ZF, von Neumann's, NBG, NF, among many others), someone could argue that the common element among them is the formal study of collections of objects, in some intuitive way which partially rescues the main original ideas due to Georg Cantor (regarding *Mengenlehre*). But that is hardly an accurate description. The most popular set theories in literature (ZF, ZFU, and NBG) are simply the formal study of two predicative letters, namely, membership ($\in$) and identity ($=$). Of course, any set theory seems to be associated to an intended interpretation of collection, as it was originally proposed by Cantor. But, from a purely formal point of view, the standard axioms of popular set theories simply state how those standard binary predicative letters ($\in$ and $=$) behave, in a very general way.

Nevertheless, physicists need much more than a simple knowledge concerning two binary predicative letters (I am not suggesting that either ZF or other formal set theories are simple theories!). Physicists need mostly functions rather than sets, and whatever they can do with such functions [27]. A function, in extensional set-theoretic terms, is usually a particular case of a set (a very specific set of ordered pairs). And that works as some sort of mathematical constraint: physicists need specialized sets called functions. But, physicists do need specialized functions as well, namely, those functions which are solutions of very specific differential equations submitted to very specific boundary conditions. And so we have one more level of mathematical constraint.

By mathematical constraints we mean, in the present context, mathematical formulas that somehow can be intuitively translated into either physical laws or physical principles. Such a specialized mathematical apparatus provides the necessary impulse to raise and develop physical theories. After all, it is far from enough to say the position of a given particle in a given space can be described by a function. Physicists do need to assert special conditions for a function in order to describe position: like the Second Law of Classical Mechanics, namely, a specific differential equation!

On the other side of this mathematical spectrum, however, there are some few works in the literature which follow a different philosophical approach. It is something that has more to do with a *revisionist perspective* than with plain mathematical applications. For example, it is easy to show that in many natural axiomatic formulations of physical and even mathematical theories, there is a substantial list of superfluous concepts usually assumed as primitive [27], like time, space and spacetime. That happens mainly when those theories are formulated in a set-theoretic language, such as Zermelo-Fraenkel's. On the other hand, in 1925, John von Neumann created a set theory where sets are definable as special cases of functions. And in [27] it is provided a reformulation of von Neumann's set theory where it is demonstrated that such an axiomatic framework can be used to formulate physical and mathematical theories with less primitive concepts in a very natural fashion.

Another historical example is Hertz's mechanics [6]. Despite the fact that Padoa's method guarantees that force is an indispensable concept in certain axiomatic formulations for non-relativistic classical particle mechanics [17], nothing prevents us from introducing a formal framework for non-relativist classical particle mechanics without any notion of force [24]. Even Kepler's Laws for planetary orbits may be derived within this "forceless" framework [26]. Concerning the standard concepts of force and mass, a fascinating and enlightening discussion may be found in [8].

All this means that mathematics may play a twofold epistemological role in theoretical physics: application and revision. Applications are achieved by means of stronger assumptions than those naturally offered by formal set theories axioms (like specialized sets called functions, and specialized functions which are solutions of specific differential equations under specific boundary conditions). Revision is achieved when we use foundational tools in order to answer to the following natural question: Do we really need all standard mathematical apparatus in order to do mathematical physics? More specifically: Do we really need the mathematical counterpart which describes space? Do we need time? Do we need spacetime? Are we doomed to always use identity in all theoretical physics? What do we really mean by usual concepts like space, time, spacetime, mass, force, and identity?

Quasi-set theory is just a useful mathematical tool which allows us to answer rather important questions concerning mathematical methods in physics.

From a formal point of view, identity is usually associated to the predicate letter $=$ in standard set theories. One of the major mathematical applications of identity is the statement of symmetry principles. How can we say, for example, the angular momentum of the Earth-Moon system is invariant? The usual answer is provided by a formula grounded on the standard notion of identity:

$$\mathbf{H} = \mathbf{H}_E + \mathbf{H}_M,$$

where $\mathbf{H}$ is the total angular momentum of the Earth-Moon system, $\mathbf{H}_E$ and $\mathbf{H}_M$ are the angular momenta of Earth and Moon, respectively, and

$$\frac{d\mathbf{H}}{dt} = \mathbf{0},$$

where t is the time parameter and $\mathbf{0}$ is the null vector. Recall that functions $\mathbf{H}$, $\mathbf{H}_E$, $\mathbf{H}_M$, $d\mathbf{H}/dt$, and $\mathbf{0}$ are sets of ordered pairs and, so, equality in the last two equations is indeed an identity between objects.

Nevertheless, do we really need identity in order to state symmetry principles? That is a vague and, so, a tricky question! Therefore, allow me to rephrase my doubt. Can we rewrite the last two equations as

$$\mathbf{H} \equiv \mathbf{H}_E + \mathbf{H}_M$$

and

$$\frac{d\mathbf{H}}{dt} \equiv \mathbf{0}?$$

Quasi-set theory has been used for a better understanding (in a philosophical fashion) of some phenomena in quantum mechanics. Within this context, the indistinguishability predicate $\equiv$ has been mostly used among *Urelemente* of $\mathcal{Q}$. Nevertheless, the same predicate letter may be used among collections as well, including quasi-functions (the quasi-set theoretic counterpart of functions). One thing is to say the total angular momentum of the Earth-Moon system is identical to the vector sum of the angular momenta of Earth and Moon. Another thing is to say both functions are simply indistinguishable. That sort of philosophical perspective may inspire new and relevant ideas. In the specific case of the last two "equations", it seems reasonable to consider both functions $\mathbf{H}$ and $\mathbf{H}_E + \mathbf{H}_M$ as standard ZF-sets, even within $\mathcal{Q}$. Thus, as a consequence we have the following:

$$\mathbf{H} =_E \mathbf{H}_E + \mathbf{H}_M$$

and

$$\frac{d\mathbf{H}}{dt} =_E \mathbf{0},$$

where $=_E$ stands for the extensional equality mentioned in the Introduction. Nevertheless, quasi-set theory allows us to state symmetry principles in a more relaxed way.

By using category theory, Shahn Majid proposed a very general symmetry principle inspired on Einstein's formulation for equivalence between gravitation and acceleration [15]. His idea was to propose a mathematical tool for coping with quantum gravitation. Analogously, if macroscopic objects are, in some sense, composed of microscopic objects, why can't we abandon identity right at start?

After all, microscopic particles (usually described by quantum theories) seem to be devoid of any classical notion of identity.

In standard set theories, an arbitrary permutation of the elements of a set x is a classical example of an automorphism. And the automorphism group of x is the symmetric group on x. Notwithstanding, in quasi-set theory the automorphism group simply presents indistinguishable elements. That's all! Nevertheless, such indistinguishable elements offer a new way to examine quantum statistical mechanics. And that is a revisionist perspective that mathematics can effectively offer to theoretical physics.

That revisionist perspective offered by quasi-set theory may be responsible for some misunderstandings among physicists and even philosophers. We illustrate this point in the next Section.

3 Nicholas Smith's criticisms

Nicholas J. J. Smith published ten years ago a paper about the philosophical dispute concerning any clear notion of vague identity [28]. According to him:

> [T]o make clear sense of something, one must at least model it set-theoretically; but due to the special place of identity in set-theoretic models, any vague relation that one does model set-theoretically will not be identity, for real identity will already be there, built into the background of the model, and perfectly precise.

More specifically, Smith argues that Krause and collaborators [13] made no progress towards the problem of clarifying any notion of vague identity. According to Smith, Krause and colleagues

> "simply present quasi-set theory as an *axiomatic theory* − i.e. as a list of definitions and axioms involving a primitive relation of indistinguishability, symbolised as ≡. This approach leads directly to the dilemma that Priest and van Inwagen faced. If we understand this

axiomatic theory in the usual way — i.e. as picking out the class of (standard set-theoretic) models in which all the axioms are true — then we can only understand ≡ as a new relation in addition to the precise identity relation which inheres in all standard models. We have then done nothing to remove precise identity in favour of the vague relation of indistinguishability — i.e. we have gone no way towards achieving Krause et al's stated goal of developing a framework which does not treat objects as individuals in the standard sense."

The works cited by Smith are [21] and [7] (Priest and Inwagen, respectively).

Well, let's analyse Smith's arguments. After all, it seems to me that no one did this until now, at least from the perspective I wish to explore here.

First of all, it is easy to find a manifold of prejudices to the interests of philosophy, when someone reads a statement like this: "to make clear sense of something, one must at least model it set-theoretically". Is this statement supposed to be a critical one? If that is the case, what is set theory supposed to be? Smith argues he is not talking about a particular formulation for set theory, like ZFC or NBG (the most popular examples in specialized literature). He refers to set theory as "a way of thinking with which we need to be inculcated if we are to understand any mathematical theory". That is quite puzzling! After all, there are many well known mathematical theories that require no particular notion of set whatsoever: classical propositional calculus [18], non-classical propositional calculi [9], certain formulations for category theory [5], mereology [14], type theory [23], and, of course, set theory itself [18] [31]!

Besides, ZFC and NBG are first order theories whose primitive concepts are equality $=$ and membership $\in$. And that is all! There is no need to talk about sets (in the intuitive sense of a collection of objects), when we are dealing with ZFC and NBG. Both ZFC and NBG are well known examples of formal theories in which the main purpose is to deal with two binary predicate letters, namely, equality and membership. The objects of ZFC and NBG are simply represented by *terms*, from a first order language point of view. So, they are not necessarily sets of collections of any kind. If a mathematician refers to such terms as *sets*, that happens because both ZFC and NBG are defined through axioms somehow associated to an *intended and merely intuitive interpretation* of what a set (collection) is supposed to be. Somehow, ZFC and NBG were erected in order to capture an intended interpretation that started with the work of Georg Cantor in the late 19th century. Nevertheless, there is no way to guarantee which formulation is better suited for Cantor's ideas about sets.

Cantor said "a set is any collection of definite, distinguishable objects of our intuition or of our intellect to be conceived as a whole". That is a really vague notion about what a set is supposed to be! That is why there is a huge myriad of set theories in specialized literature!

In von Neumann set theory, e.g., there are in fact (in a very precise and specific sense) sets [27]! A set may be *defined* as a quite particular term in von Neumann's axiomatic framework. Not all terms stand for sets in von Neumann's theory. But in ZFC and NBG, *all* terms (by a simple abuse of language) are usually called sets (except in ZFU and its variations). For all that matters, those terms could be called *things*! Even if that was the case, still ZFC and NBG should be seen as first order theories whose primitive concepts are just equality and membership. In a more rigorous way, membership is the only non-logical primitive symbol in ZFC and NBG. So, what makes Smith think he knows for sure what a set should be?

At this point we need to make some further clarification. The notion of set in von Neumann's theory is not the same as the notion of set in either ZFC or NBG. In von Neumann's theory a set is a syntactic notion, definable by means of some sort of ampliation of the formal language. An *ampliative definition* is that one which adds a new symbol to a formal language. A well known example of ampliative theory of definitions is Leśniewski's theory [29].

For practical purposes, sets, in von Neumann's, work analogously to characteristic functions. On the other hand, sets in ZFC and NBG, as we emphasized, are simply intended interpretations, whose meanings depend on metalinguistic considerations.

During the early development of formal set theories, many mathematicians displayed serious prejudice against certain set-theoretic ideas. The best known example is the Axiom of Choice. Banach-Tarski theorem was used sometimes as an attempt to refute the Axiom of Choice [20]. The argument was simple (yet naive!): since the theorem is intuitively false, the Axiom of Choice cannot be true! That happened because many mathematicians of the early 20th century had their own ideas about what a set is supposed to be (like Smith is trying to do, with his own sense of what a clear sense is supposed to be).

Mathematics is not the science of clear sense! Mathematics is a social enterprise which fundamentally depends on fresh ideas and, of course, intense and critical discussion. Despite the fact that mathematics can be done in a formal way, formal languages cannot hide mathematics from social criticism. Yuri Manin proposed in 1974:

> We should consider the possibilities of developing a totally new language to speak about infinity. Classical critics of Cantor (Brouwer *et al.*) argued that, say, the general choice axiom is an illicit extrapolation of the finite case.
>
> I would like to point out that this is rather an extrapolation of common-place physics, where we can distinguish things, count them, put them in some order, etc. New quantum physics has shown us models of entities with quite different behavior. Even 'sets' os photons in a looking-glass box, or of electrons in a nickel piece are much less Cantorian than the 'set' of grains of sand. In general, a highly probabilistic 'physical infinity' looks considerably more complicated and interesting than a plain infinity of 'things'.

That point raised by Manin was one of the major motivations for developing quasi-set theory. The point here is not the concept of set *per se*, but it is rather a formal way to cope with infinities, specially infinite collections in some sense. If quasi set theory answers to Yuri Manin questions, that is naturally debatable. Nevertheless, quasi-set theory is as licit as any usual standard set theory, if we are trying to talk about infinite collections in a rigorous and formal way.

Smith argues quasi-set theory is simply an axiomatic theory. Well, we can obviously say exactly the same about any standard set theory, like ZFC and NBG, among others! So, why quasi-set theory should be seen in any different way?

The point raised by Smith seems to be directly connected to (standard) model theory. Nevertheless, what prevents us from developing a quasi-set-theoretic version of model theory? Few people in the world work with quasi-sets. And there is a lot to do if we really want to get some answers good enough to resist to qualified criticisms. On the other hand, some shy examples of quasi-set-theoretic models already exist in the literature, like the ones presented in [12] and [25] (apparently ignored by Smith).

Georg Cantor himself once said: "The essence of mathematics lies in its freedom". So, let us be free to create! Manin wrote [1] "In accordance with Hilbert's prophecy, we are living in Cantor's paradise. So, we are bound to be tempted." That non-Cantorian proposal is in a poetic agreement to the notion of freedom advocated by Cantor himself!

Allow me to extrapolate Smith's ideas (at my own risk) to another realm: geometry. What about a thesis like the following one? "To make clear sense of geometrical concepts, one must at least model it in Euclidian geometry; but due to the special place of the parallel postulate in geometry, any vague axiom that one does model geometrically will not be the parallel postulate, for real parallelism will already be there, built into the background of the geometrical model, and perfectly precise." Well, nowadays non-euclidian geometry is as good geometry as euclidian geometry!

So, what seems to be the problem?

There is no precise definition for what geometry is! Analogously, there is no precise definition for what set theory is supposed to be!

The real issue here, mostly from a philosophical point of view, is this: can we cope with infinite collections (in a precise sense) without appealing to identity? If the answer is negative, why? And if the answer to this question is positive, what is the advantage of this kind of formal approach? That is why I wrote this paper! Any discussion about the importance of quasi-set theory should take epistemological issues under consideration. And the epistemological character of quasi-set theory is quite unusual, from a general scientific perspective. That happens as a consequence of the revision power quasi-set theory provides. We do not need to live in a world enslaved into identity. One of the most valuable lessons of twentieth century mathematics was this: there is no room for prejudice.

4 Acknowledgement

I would like to thank Décio Krause for criticisms and suggestions made in a previous version of this paper.

References

[1] French, S., Krause, D.: *Identity in Physics: A Formal, Historical and Philosophical Approach* (Oxford Un. Press, Oxford, 2006)

[2] French, S., "Identity and Individuality in Quantum Theory", *The Stanford Encyclopedia of Philosophy* (Fall 2015 Edition), Edward N. Zalta (ed.)

[3] Garrod, C., *Statistical Mechanics and Thermodynamics* (Oxford University Press, 1995).

[4] Geroch, R., *Mathematical Physics* (University of Chicago Press, 2015).

[5] Hatcher, W. S., *Foundations of Mathematics* (W. B. Saunders, 1968).

[6] Hertz, H. R., *The Principles of Mechanics* (Dover, New York, 1956).

[7] Inwagen, P., "How to reason about vague objects", *Philosophical Topics* **16** 255 - 284 (1988).

[8] Jammer, M., *Concepts of Mass in Classical and Modern Physics* (Dover, New York, 2016).

[9] Karpenko, A. S.: "The classification of propositional calculi" *Studia Logica* **66** 253 - 271 (2000).

[10] Krause, D., 'On a quasi-set theory', *Notre Dame Journal of Formal Logic* **33** 402-411 (1992).

[11] Krause, D., "Logical aspects of quantum (non-)individuality", *Foundations of Science* **15** 79-94.

[12] Krause, D., Sant'Anna, A. S., Volkov, A. G.: "Quasi-set theory for bosons and fermions: quantum distributions", *Foundations of Physics Letters*, **12** 67-79 (1999).

[13] Krause, D., Sant'Anna, A. S., Sartorelli, A.: "On the concept of identity in Zermelo-Fraenkel-like axioms and its relationship with quantum statistics" *Logique et Analyse* **189 - 192** 231 - 260 (2005).

[14] Leonard, H. S., Goodman, N.: "The calculus of individuals and its uses" *Journal of Symbolic Logic* **5** 45 - 55 (1940).

[15] Majid S., "Some physical applications of category theory", In: Bartocci C., Bruzzo U., Cianci R. (eds) *Differential Geometric Methods in Theoretical Physics. Lecture Notes in Physics* **375** (Springer Verlag, Berlin, Heidelberg, 1991)

[16] Mandel, L., "Coherence and indistinguishability" *Optics Letters* **16** 1882 - 1883 (1991).

[17] McKinsey, J. C. C., Sugar, A. C., Suppes, P., "Axiomatic foundations of classical particle mechanics", *Journal of Rational Mechanics and Analysis* **2** 253 - 272 (1953).

[18] Mendelson, E.: *Introduction to Mathematical Logic* (Chapman & Hall, London, 1997).

[19] Post, H.: "Individuality and physics", *The Listener* **70** 534-537 (1963).

[20] Potter, M.: *Set Theory and Its Philosophy: A Critical Introduction* (Clarendon Press, 2004)

[21] Priest, G.: "Fuzzy identity and local validity", *Monist* **81** 331 - 342 (1998).

[22] Redhead, M., Teller, P.: "Particle labels and the theory of indistinguishable particles in quantum mechanics", *British Journal for the Philosophy of Science*, **43** 201-218 (1992).

[23] Russell, B., Whitehead, A., *Principia Mathematica* (3 volumes) (Cambridge University Press, 1910, 1912, 1913).

[24] Sant'Anna, A. S., "An axiomatic framework for classical particle mechanics without force", *Philosophia Naturalis* **33** 187 - 203 (1996).

[25] Sant'Anna, A. S., Santos, A. M. S.: "Quasi-set-theoretical foundations of statistical mechanics: a research program", *Foundations of Physics* **30** 101 - 120 (2000).

[26] Sant'Anna, A.S., Garcia, C., "Gravitation in Hertz Mechanics", *Foundations of Physics Letters* **16** 565 - 578 (2003).

[27] Sant'Anna, A. S., Bueno, O.: "Sets and functions in theoretical physics", *Erkenntnis* **79** 257-281 (2014).

[28] Smith, N. J. J.: "Why sense cannot be made of vague identity" *Noûs* **42** 1-16 (2008).

[29] Suppes, P.: *Introduction to Logic* (Van Nostrand, Princeton, 1957).

[30] Suppes, P.: *Representation and Invariance of Scientific Structures* (CSLI, Stanford, 2002).

[31] Tarski, A., S. Givant, *A Formalization of Set Theory without Variables* (AMS, 1987).

Recursively Enumerable Sets and Fourier Series

Bruno Scarpellini

Dedicated to Prof. F.A. Doria on the occasion of his seventieth birthday and his pioneering work, among many others, on the relationship and interplay between analog and digital computing

Abstract

In our digression below we focus on some questions pertaining to the intersection of Computable Analysis and Fourier series. They appear also in the context of the discussions on "Hypercomputation" (see [Sy], [Cp 1,2] for an extended list of references). Here we restrict ourselves to the discussion of some mathematical aspects of this notion. The starting point are the sets $\mathscr{F}_s^+$ of Fourier series $f = \sum a_{\underline{n}} e^{i\underline{n}\underline{\alpha}}$, $(\underline{n} = (n_1,\ldots,n_s))$, $\underline{\alpha} = (\alpha_1,\ldots,\alpha_s)$, $\underline{n}\underline{\alpha} = n_1\alpha_1 + \cdots + n_s\alpha_s$) subject to the constraints: (a) $a_{\underline{n}} = 0$ if $n_j < 0$ for some j, $1 \leq j \leq s$, (b) $a_{\underline{n}} \in \mathbb{R}_+ = \{\xi \mid \xi \geq 0\}$, (c) $\sum_{\underline{n} \geq 0} a_{\underline{n}} < \infty$, where $\underline{n} \geq 0$ means $n_j \geq 0$, $j = 1, \ldots, s$.

With any $f \in \mathscr{F}_s^+$ we associate an s-ary relation $\Gamma(f) \subset \mathbb{N}^s$ over the natural numbers $\mathbb{N} = \{0, 1, \ldots\}$ according to the setting: (*) $\underline{n} = (n_1,\ldots,n_s) \in \Gamma(f)$ iff $a_{\underline{n}} > 0$. Based on this frame we give a characterisation of the recursively enumerable sets in terms of Fourier series belonging to $\mathscr{F}_s^+$, and some open problems are discussed.

1 Preliminaries

(A) In what follows we consider some questions involving concepts from computability theory as well as concepts from the theory of Fourier series whose purpose is to clarify the relationship between digital and analogue computers. Here, however, we ignore the aspect of digital and analogue computing to a large extent and treat these questions as mathematical problems in their own right. Yet, at a few places analogue computation is taken into account.

(B) In order to put these notes into precise form we fix some notation. By $\mathbb{N}, \mathbb{Z}, \mathbb{Q}, \mathbb{R}, \mathbb{C}$ we denote, respectively, the sets of non-negative integers, integer, rational, real and complex numbers. By $\mathscr{F}_s$, $s \in \mathbb{N} \setminus \{0\}$, we denote the set of continuous, complex valued functions $f(\alpha_1,\ldots,\alpha_s)$ of s real arguments $\alpha_1,\ldots,\alpha_s$, that are 2π-periodic in each α_j and whose Fourier series

$$f(\alpha_1,\ldots,\alpha_s) = \sum_{n_1,\ldots,n_s \in \mathbb{Z}} a_{n_1\ldots n_s} e^{i(n_1\alpha_1+\cdots+n_s\alpha_s)}$$

satisfies
$$\sum_{n_1,\ldots,n_s\in\mathbb{Z}} |a_{n_1\ldots n_s}| < \infty. \tag{1.1}$$

By $\mathscr{F}_s^+$ we denote the subset of all $f \in \mathscr{F}_s$ whose Fourier series expansion satisfies the additional condition

$$\begin{aligned} a_{n_1\ldots n_s} &\geq 0 \text{ for all } n_1,\ldots,n_s \in \mathbb{Z}, \text{ and} \\ a_{n_1\ldots n_s} &= 0 \text{ if } n_j < 0 \text{ for some } j \leq s. \end{aligned} \tag{1.2}$$

Below, α_j ($j \in \mathbb{N}$), α, β, a, b, denote variables ranging over $\mathbb{R}/2\pi\mathbb{Z}$. Finite sequences, such as $\alpha_1,\ldots\alpha_s$, or $a_1,\ldots,a_t$ are denoted by the shorthands $\underline{\alpha}_s$, $\underline{a}_t$ or even by $\underline{\alpha}$, $\underline{a}$, provided that s, t are given by the context. If $f(\alpha_1,\ldots,\alpha_s)$ depends only on few variables, say α_j, α_k for some $j < k$, we also write $f(\alpha_j,\alpha_k)$, or even $f(\alpha,\beta)$. Finally, given lists of variables, say $\underline{\alpha}_s, \underline{\beta}_t, \underline{n}_s, \underline{k}_s$, we write $f(\underline{\alpha})$, $f(\underline{\alpha},\gamma)$, $f(\underline{\alpha},\underline{\beta})$ and $\underline{n}^{\underline{k}}$ instead of $f(\alpha_1,\ldots,\alpha_s,\gamma)$, $f(\alpha_1,\ldots,\alpha_s,\beta_1,\ldots,\beta_t)$, or $n_1^{k_1}\ldots n_s^{k_s}$, respectively.

Finally, recalling (1.1) we endow the complex vector space $\mathscr{F}_s$ with the norm $\|\ \|_s$ according to

$$\|f\|_s = \sum_{\underline{n}\in\mathbb{Z}^s} |a_{\underline{n}}|. \tag{1.3}$$

It is straightforward to show that $\mathscr{F}_s, \|\ \|_s$ is a complex Banach algebra.

2 Production Systems

We now come to the objects to be investigated in the sequel. The key definition relating the members of $\mathscr{F}_s^+$ with the s-ary relations over $\mathbb{N}$, i.e. the members of $\mathscr{N}_s = \{L \mid L \subset \mathbb{N}^s\}$ is

Definition 2.1. Let $f \in \mathscr{F}_s^+$ have the Fourier series expansion

$$f(\alpha_1,\ldots,\alpha_s) = \sum_{\underline{n}\geq 0} a_{n_1\ldots n_s} e^{in_1\alpha_1}\ldots e^{in_s\alpha_s}.$$

Then $\mathscr{R}_s(f)$ denotes the s-ary relation over $\mathbb{N}$ defined by

$$(n_1,\ldots,n_s) \in \mathscr{R}_s(f) \text{ iff } a_{n_1\ldots n_s} > 0. \tag{2.1}$$

Remarks. If $\mathscr{R}_s(f) = L$ for some $f \in \mathscr{F}_s^+$ and $L \subset \mathbb{N}^s$ we call f a *representative* of L.

A mapping $P : (\mathscr{F}_s^+)^k$ (some k) to $\mathscr{F}_s^+$ is called a k-ary *production* on $\mathscr{F}_s^+$.

Definition 2.2. Given a subset $\mathcal{U} \subset \mathcal{F}_s^+$ and a finite sequence $P_1, \ldots, P_l$ of productions on $\mathcal{F}_s^+$ we denote this couple by $\mathcal{U}; P_1, \ldots, P_l$, and call it a *production system based on $\mathcal{U}$*.

We conclude these preparations with

Definition 2.3. A function $f \in \mathcal{F}_s^+$ shall be called *attainable* by the production system $\mathcal{U}; P_1, \ldots, P_l$ if there are functions $F_1, \ldots, F_m \in \mathcal{F}_s$ such that $f = F_m$, and for each $j \leq m$ at least one of the following conditions (a), (b) holds.

(a) $F_j \in \mathcal{U}$;

(b) there are $j_1, \ldots, j_k < j$ and a k-ary production P_t $(t \leq l)$ with $P_t(F_{j_1}, \ldots, F_{j_k}) = F_j$.

We now can specify the questions announced in the introduction (Sect. 1). Recalling Def. 2.1 and given the production system $\mathcal{U}; P_1, \ldots, P_l$ on $\mathcal{F}_s^+$ we ask:

(Q*) Does it hold that for every recursively enumerable relation $\Gamma \subset \mathbb{N}^s$ there is $f \in \mathcal{F}_s^+$ such that $\mathcal{R}_s(f) = \Gamma$ and such that f is attainable by $\mathcal{U}; P_1, \ldots, P_l$?

We now focus on a family of production systems, already investigated in [Bs-Sc]. These systems are all based on the same productions and differ only in the underlying set $\mathcal{U}$ of functions.

We first consider the productions. To this end we let $C(\mathbb{R}^s)$ be the vector space of complex valued functions $f : \mathbb{R}^s \to \mathbb{C}$ that are continuous and 2π-periodic in their arguments $\alpha_1, \ldots, \alpha_s$. The space $C(\mathbb{R}^s)$ is a Banach algebra with the norm given by

$$|f|_s = \sup\{|f(\underline{\alpha})| \mid \underline{\alpha} \in \mathbb{R}^s\} = \sup\{|f(\underline{\alpha})| \mid \underline{\alpha} \in [0, 2\pi]^s\}.$$

Now let $f, g \in \mathcal{F}_s^+$ be given by their Fourier series expansions, i.e.:

(I) $f(\underline{\alpha}) = \sum_{\underline{n} \geq 0} a_{\underline{n}} e^{i\underline{n}\underline{\alpha}}$

$g(\underline{\alpha}) = \sum_{\underline{n} \geq 0} b_{\underline{n}} e^{i\underline{n}\underline{\alpha}}, \qquad \underline{\alpha} = (\alpha_1, \ldots, \alpha_s), \ \underline{n} = (n_1, \ldots, n_s).$

Since the coefficients $a_{\underline{n}}, b_{\underline{n}}$ satisfy conditions (1.1), (1.2) as of Sect. 1 we infer $f, g \in C(\mathbb{R}^s)$, whence the product fg given by $(fg)(\underline{\alpha}) = f(\underline{\alpha})g(\underline{\alpha})$ is well defined.

Proposition 2.1. *Let $f, g \in \mathcal{F}_s^+$. Then $fg \in \mathcal{F}_s^+$ and $\|fg\|_s \leq \|f\|_s \|g\|_s$.*

Remark. We omit the proof for reasons of space. In fact, we never make full use of Prop. 2.1 since the elements $f(\underline{\alpha})$ that shall appear in the course of the arguments are, by construction, automatically in $\mathscr{F}_s^+$.

Prop. 2.1 allows us to define the production $M : (\mathscr{F}_s^+)^2 \to \mathscr{F}_s^+$ via

(II)   for $f, g \in \mathscr{F}_s^+$, set $M(f,g) = fg$.

Next, fix $r \in \{1, \ldots, s\}$. Define $P_r : \mathscr{F}_s^+ \to \mathscr{F}_{s-1}^+$ by the stipulation

(III)   for $f \in \mathscr{F}_s^+$ set $(P_r f)(\alpha_1, \ldots, \alpha_{s-1}) = f(\alpha_1, \ldots, \alpha_{r-1}, 0, \alpha_r, \ldots, \alpha_{s-1})$.

We refer to the production P_r as a *projection*. We note the important property

Proposition 2.2. *Let $f \in \mathscr{F}_s^+$, then*

(IV) $(n_1, \ldots, n_{s-1}) \in \mathscr{R}_{s-1}(P_r f)$ iff

$$(\exists m)\{(n_1, \ldots, n_{r-1}, m, n_r, \ldots, n_{s-1}) \in \mathscr{R}_s(f)\}.$$

Proof. Assume, for simplicty, $r = s$. Let $f(\underline{\alpha}) \in \mathscr{F}_s^+$ have the Fourier series expansion

$$f(\underline{\alpha}) = \sum_{\underline{n} \geq 0} a_{\underline{n}} e^{i\underline{n}\,\underline{\alpha}} \tag{i}$$

subject to clauses (1.1) and (1.2) in Sect. 1. We use the shorthand $\underline{n}' = n_1, \ldots, n_{s-1}$, $\underline{\alpha}' = \alpha_1, \ldots, \alpha_{s-1}$, $\underline{n}'\underline{\alpha}' = n_1\alpha_1 + \cdots + n_{s-1}\alpha_{s-1}$. Furthermore, we do not mention anymore that all indices are ≥ 0. Thus, (i) can be rewritten as follows

$$f(\underline{\alpha}) = f(\underline{\alpha}', \alpha_s) = \sum_{\underline{n}', n_s} a_{\underline{n}'n_s} e^{i\underline{n}'\underline{\alpha}'} e^{in_s\alpha_s} \tag{ii}$$

whence

$$(P_r f)(\underline{\alpha}') = f(\underline{\alpha}', 0) = \sum_{\underline{n}', n_s} a_{\underline{n}'n_s} e^{i\underline{n}'\underline{\alpha}'} \tag{iii}$$

that is,

$$(P_s f)(\underline{\alpha}') = \sum_{\underline{n}'} e^{i\underline{n}'\underline{\alpha}'} A_{\underline{n}'}, \tag{iv}$$

where

$$A_{\underline{n}'} = \sum_{n_s} a_{\underline{n}'n_s}.$$

By virtue of (iv) and Def. 2.1 we infer

$$(n_1, \ldots, n_{s-1}) \in \mathscr{R}_{s-1}(P_s f) \iff A_{\underline{n}'} > 0 \tag{v}$$

i.e.

$$\iff \sum_{n_s} a_{\underline{n}' n_s} > 0. \tag{v*}$$

Since $a_{\underline{n}' n_s} \geq 0$, by virtue of (1.2) in Sect. 1, we infer from (v), (v*)

$$\underline{n}' \in \mathscr{R}_{s-1}(P_s f) \text{ iff there is } m \in \mathbb{N} \text{ such that } a_{\underline{n}' m} > 0. \tag{vi}$$

From (vi) and Def. 2.1 we conclude

$$\underline{n}' = (n_1, \ldots, n_{s-1}) \in \mathscr{R}_{s-1}(P_s f)$$
$$\iff (\exists m)\{(n_1, \ldots, n_{s-1}, m) \in \mathscr{R}_s(f)\} \tag{vii}$$

□

From a formal point of view functions like $e^{i\alpha_k}$ and $e^{i\alpha_l}$, for $k \neq l$ are not the same object. We add therefore the following productions to our system termed "relabelling". For any permutation π of the set of integers $\{1, \ldots, s\}$ a relabelling map $R_\pi : \mathscr{F}_s^+ \to \mathscr{F}_s^+$ is defined by the stipulation

(V) $(R_\pi F)(\alpha_1, \ldots, \alpha_s) = F(\alpha_{\pi 1}, \ldots \alpha_{\pi s})$.

While the productions introduced so far have elementary character the next one is already more powerful: it concerns convolutions $\star_t$. Let $f, g \in \mathscr{F}_s^+$, let $1 \leq t \leq s$, then we define

(VI) $(f \star_t g)(\alpha_1, \ldots, \alpha_s) = (2\pi)^{-t} \times$

$$\times \int_0^{2\pi} \cdots \int_0^{2\pi} f(\alpha_1 - \xi_1, \ldots, \alpha_t - \xi_t, \alpha_{t+1}, \ldots, \alpha_s) g(\xi_1, \ldots, \xi_t, \alpha_{t+1}, \ldots, \alpha_s) \underline{d\xi}^t$$

or in shorthand

$$(f \star_t g)(\underline{\alpha}) = (2\pi)^{-t} \int_0^{2\pi} \cdots \int_0^{2\pi} f(\underline{\alpha}_t - \underline{\xi}_t, \underline{\alpha}_{t+1}^s) g(\underline{\xi}_t, \underline{\alpha}_{t+1}^s) \underline{d\xi}^t,$$

where $\underline{\alpha}_{t+1}^s = \alpha_{t+1}, \ldots, \alpha_s$. We note

Proposition 2.3. For $f, g \in \mathscr{F}_s^+$ one has $(f \star_t g) \in \mathscr{F}_s^+$.

Proof. This follows straightforwardly from properties (1.1), (1.2) (Sect. 1) through computations with Fourier series. □

3 Theta-like Functions and Simple Functions

In [B-Sc] one had, among other things, to construct 2π-periodic functions of complicated nature out of simple ones, using only four methods of construction which, they too, were aimed at to be as simple as possible. Use was thereby made of Jacobi's theta function [Han]. Here a slight extension is used given by

Definition 3.1. (a) A recursive function $\mu : \mathbb{N}^t \to \mathbb{N}$, is called *regular* if

$$\mu(n_1, \ldots, n_t) \geq n_1 + \cdots + n_t$$

for all $n_j \in \mathbb{N}$, $1 \leq j \leq t$.
(b) Let $\lambda \in \,]0,1[$; a function $\theta_\lambda \in \mathscr{F}_s^+$ is called *theta-like* if it is of the form

$$\theta_\lambda(\alpha_1, \ldots, \alpha_s) = \sum_{n \geq 0} \lambda^{\mu(n)} e^{in\alpha_1} e^{in^2 \alpha_2}$$

for some regular recursive $\mu : \mathbb{N} \to \mathbb{N}$.

Remarks. We may also write $\theta_\lambda(\alpha_1, \alpha_2)$ instead of $\theta_\lambda(\alpha_1, \ldots, \alpha_s)$, or even $\theta_\lambda(\alpha, \beta)$ for distinct α, β. We apply rearrangement of variables justified by the application of "relabelling".

Definition 3.2. Fix $\lambda \in \,]0,1[\cap \mathbb{Q}$. A function $f \in \mathscr{F}_s^+$ is called λ-simple" or just "simple" if it is of the form (a) or (b) below.

(a) $f(\underline{\alpha}) = e^{ik_1\alpha_1 + \cdots + ik_s\alpha_s}$

(b) $f(\underline{\alpha}) = (1 - \lambda e^{ik_1\alpha_1 + \cdots + ik_s\alpha_s})^{-1}$, for some $k_1, \ldots, k_s \in \mathbb{N}$.

We denote by $\mathscr{U}_\lambda$ the family of λ-simple functions, and get $\mathscr{U}(\theta_\lambda) = \mathscr{U}_\lambda \cup \{\theta_\lambda\}$, the set of λ-basic functions or simply the set of basic functions.

4 Some Results

We list two results, proved in [B-Sc] which center around problem (Q*) but now for the production system defined above, i.e. $\mathscr{U}_\lambda(\theta_\lambda); M, P_r, R_\pi, \star_t$.

Theorem 4.1. *Let $\theta_\lambda(\alpha, \beta)$ be theta-like and let $\mathscr{U}(\theta_\lambda) = \mathscr{U}_\lambda \cup \{\theta_\lambda\}$ be the induced set of basic functions. Then there exists for every recursively enumerable set $\Gamma \subset \mathbb{N}^s$ a function $f \in \mathscr{F}_s^+$ attainable by the production system $\mathscr{U}_\lambda(\theta_\lambda); M, P_r, R_\pi, \star_t$, such that $\mathscr{R}_s(f) = \Gamma$.*

We postpone a discussion of Thm. 4.1 and focus instead on Thm. 4.2 which is an improvement of Thm. 4.1 provided a number theoretic conjecture due to R. Büchi [Bch], [PPV] is true. This conjecture is as follows :

Büchi's conjecture (BC):

There is an integer $K > 0$ with the following property: if a sequence $x_1, x_2, \ldots, x_K \in \mathbb{N}$ satisfies

(a) $0 \leq x_1 < x_2 < \cdots < x_K$,

(b) $x_{j-1}^2 + x_{j+1}^2 = 2(x_j^2 + 1)$, $2 \leq j \leq K$,

then $x_j = x_1 + j - 1$, $1 \leq j \leq K$.

Sequences of the form $k+1, k+2, \ldots, k+K$ satisfy (a), (b) and are called trivial; all solutions not of this form are termed "non trivial". One can show that there are infinitely many non trivial sequences of length four, but despite considerable efforts by computer, no nontrivial sequence of length five has ever been found. Thus, we assume for what follows:

(BC) *There is $K \geq 5$ such that no non-trivial sequence of length K exists.*

As for a reference see [PPV]. In order to apply (BC) we recall the Jacobi function

$$\vartheta_\lambda(\alpha) = \sum_{n \geq 0} \lambda^{n^2} e^{in^2 \alpha}$$

and substitute it for the theta-like $\theta_\lambda(\alpha, \beta)$ in the basic set $\mathscr{U}_\lambda \cup \{\theta_\lambda\}$ that underlies Thm. 4.1, so as to get a new basic set and hence a new production system $\mathscr{U}_\lambda \cup \{\vartheta_\lambda\}$ resp. $\mathscr{U}_\lambda \cup \{\vartheta_\lambda\}; M, P_r, R_\pi, \star_t$. The major step in which extensive use of (BC) is made is provided by

Lemma*. *Assume that (BC) holds. Then there exists a regular recursive function $\tilde{\rho} : \mathbb{N} \to \mathbb{N}$ such that the corresponding theta-like function*

$$\sum_{n \geq 0} \lambda^{\tilde{\rho}(n)} e^{in\alpha} e^{in^2 \beta}$$

is attainable by the production system $\mathscr{U}_\lambda \cup \{\vartheta_\lambda\}; M, P_r, R_\pi, \star_t$.

By combining Thm. 4.1 with Lemma* we get

Theorem 4.2. *Assume that (BC) holds. Let $\Gamma \subset \mathbb{N}^s$ be recursively enumerable. Then there is $f \in \mathscr{F}_s^+$ attainable by te production system $\mathscr{U}_\lambda \cup \{\vartheta_\lambda\}; M, P_r, R_\pi, \star_t$ such that $\mathscr{R}_s(f) = \Gamma$.*

Remark. Thm. 4.1, Thm. 4.2 and Lemma* are proved in [B-Sc] under the labels Thm. 3.12, Thm. 5.3, Thm. 5.2 respectively.

Comments We now come to the comparison of Thm. 4.1 with Thm. 4.2. We thereby assume that there is an integer $K \geq 5$ as requested by (BC).

The property with respect to which the underlying production system is tested is its capacity to generate functions $f \in \mathscr{F}_s^+$ that give rise to recursively enumerable non recursive predicates $\mathscr{R}_s(f)$.

Thus, the question of interest of a given production system is:

(Pr*) Given a recursiv enumerable predicate $\Gamma \subset \mathbb{N}^s$, is there $f \in \mathscr{F}_s^+$ attainable by the production system such that $\mathscr{R}_s(f) = \Gamma$?

The answer to (Pr*) is affirmative in the case of Thm. 4.1 where the set of basic functions has the form $\mathscr{U}_\lambda \cup \{\theta_\lambda\}$ with θ_λ theta-like. The answer is again affirmative in the case of Thm. 4.2 provided that (BC) holds: In this case the set of basic function is given by $\mathscr{U}_\lambda \cup \{\vartheta_\lambda\}$ with ϑ_λ the Jacobi theta function. In all other cases the answer is not known. E.g. if the set of basic functions coincides with the set of simple functions (and hence the production system of the form $\mathscr{U}_\lambda; M, P_r, R_\pi, \star_t$) the problem ((Pr*)) is open. The problem (Pr*) gets more difficult if we do not assume that that (BC) is true.

5 Concluding Remarks

In [Sc1] it was attempted to bring Fourier series into relation with concepts from recursion theory by proceeding as follows. Starting with a set of "elementary functions", new functions were generated by solving repeatedly ODEs or Fredholm equations. In [Bus] to this end a principle of analytic continuation was used. The aim was to generate in this way, in finitely many steps, periodic functions that had non recursive properties. The hope was that these periodic functions could be simulated by analog computers of the type GPAC (Shannon-Pour-El) or some improvements thereof but still in the range of the technically realisable.

However, this goal was not achieved since one could not avoid to construct theta-like functions , (i.e. $\sum_{n \geq 0} \lambda^{\rho(n)} e^{in\alpha} e^{in^2 \beta}$) by methods not accessible to the allowed class of computers. However, the proof of Theorem 5.3 in [B-Sc] and of Thm. 4.2 here avoids this difficulty thanks to the conjecture (BC) assumed to be true.

Appendix 1

Based on arguments in [Ric] we show that a production system may be considered as a system of termes provided with an interpretation. Our presentation applies to any production system $\mathscr{U}; P_1, \ldots, P_l$, but for reasons of space we focus attention on the systems as of Sect. 2 which are of the form

(i) $\mathscr{U}; M, P_r, R_\pi \star_t$

with $\mathscr{U}$ the set of basic functions. What counts here is that $\mathscr{U}$ is denumerable:

(ii) $\mathscr{U} = \{v_j \in \mathscr{F}_s^+, \, j = 1, 2, \dots\}$.

With every $v \in \mathscr{U}$ we associate a symbol $\hat{v}$, the "name" of v, considered as a term of length $\ell(\hat{v}) = 1$.

We also need finitely many symbols in order to build terms out of them:

(iii) $(\,)\,,\,[\,]\,\,|\,\,\hat{\pi}\,\,\hat{P}_r\,\,\star_t\,\,\langle\,\rangle$,

for each of the permutations π of $1, \dots, s$ and each of the productions P_r. We define terms inductively and define a mapping h that associates with every term ϕ an image $h(\phi) \in \mathscr{F}_s^+$:

Definition A.1. (a) For $v \in \mathscr{U}$, $\hat{v}$ is a term of length $\ell(\hat{v}) = 1$ and image $h(\hat{v}) = v$;

(b) if A, B, C are terms, $r, t \leq s$, and π a permutation, then

$$[A,B] \quad \langle A,B\rangle_t \quad \text{and} \quad (\hat{P}_r C) \quad (\hat{R}_\pi C)$$

are terms with lengths

$$\ell([A,B]) = \ell(\langle A,B\rangle_t) = \ell(A) + \ell(B) + 2$$

while

$$\ell((\hat{P}_r C)) = \ell((\hat{R}_\pi C)) = \ell(C) + 1.$$

Finally, we define $h(\ )$ according to the stipulation

$$h([A,B]) = Mh(A), h(B), \quad h(\langle A,B\rangle_t) = h(A) \star_t h(B)$$

and

$$h((\hat{P}_r C)) = P_r h(C), \quad h((\hat{R}_\pi C)) = R_\pi h(C).$$

We recall the notion of "attainability" introduced ins Sect. 2 and adapt it to the present context

Definition A.2. $f \in \mathscr{F}_s^+$ is attainable by the production system $\mathscr{U}; M, P_r, R_\pi, \star_t$, if there is an $n \in \mathbb{N}$ and a sequence $f_j \in \mathscr{F}_s^+$, $j \leq m$, such that $f_m = f$, and for $j \leq m$ one of the following conditions (a), (b) holds

(a) $f_j \in \mathscr{U}$;

(b) there exist $j_1, j_2 < j$ such that either $f_j = M f_{j_1} f_{j_2}$

or $f_j = f_{j_1} \star_t f_{j_2}$ for some $t \leq s$

or $f_j = P_r f_{J_1}$ for some $r \leq s$

or $f_j = R_\pi f_{j_1}$ for some permutation π of $(1,\ldots,s)$.

Proposition A.1. *If f, g are attainable, then so are Mfg, $f \star_t g$, $P_r f$, $R_\pi f$.*

Proof. Consider e.g. the case of Mfg, all other cases are treated in the same way.

By assumption, there are sequences $f_j \in \mathscr{F}_s^+$, $j \leq m$ (some m) and $g_k \in \mathscr{F}_s^+$, $k \leq n$ (some k) related to f respectively, g via Def. A.2. We define a sequence $h_l \in \mathscr{F}_s^+$, $l \leq m+n+1$ as follows: (1) for $j \leq m$ we set $h_j = g_j$; (2) for $k \leq n$ we set $h_{n+k} = f_k$. It is now easily seen that the sequence h_l, $l \leq m+n+1$ satisfies the condition for attainability stated in Def. A.2. Hence, Mfg is attainable. □

Proposition A.2.

(a) *If ϕ is a term, then $h(\phi)$ is attainable.*

(b) *If f is attainable, then $f = h(\phi)$ for some ϕ.*

Outline of proof. We proceed in both cases by induction. As to (a), let ϕ be a term. If ϕ is not composed, then necessarily $\phi = \check{v}$ for some basic function $v \in \mathscr{U}$. Hence, $h(\phi) = v$, where $h(\phi)$ is attainable by definition. Thus, let ϕ be composed, e.g. $\phi = [A,B]$ for some terms A, B, hence, $h(\phi) = Mh(A)h(B)$. Since $\ell(\phi) = \ell(A) + \ell(B) + 2$, by definition, we have $\ell(\phi) > \ell(A), \ell(B)$. By the induction hypothesis we have that $h(A), h(B)$ are attainable. It thus follows from Prop. A.1 that $h(\phi)$ is attainable. This proves (a) for $\phi = [A,B]$. All other cases are handled in the same way.

As to (b), let f be attainable. Then there is a sequence f_j, $j \leq n$, related to f via Def. A.1. We show by induction with respect to j that $f_j = h(\phi)$ for some term ϕ. Thus, let $j > 1$ and assume e.g.

(a*) $f_j = Mf_{j_1}f_{j_2}$ for some $j_1, j_2 < j$.

By the induction hypothesis there are terms A, B such that $h(A) = f_{j_1}$, $h(B) = f_{j_2}$, where by (a*) we have

(b*) $f_j = Mh(A)h(B)$.

Since $h([A,B]) = h(A)h(B)$, by definition, we infer

(c*) $h([A,B) = f_j$

proving (b) for this case. All other cases are treated in the same way. □

Appendix 2

We add some remarks concerning Thm. 4.1 resp. Thm. 3.12 in [B-Sc]. With $s \in \mathbb{N}$ fixed let a set Γ of ordered s-tuples of positive integers be given, i.e.

(I) $\Gamma \subseteq \mathbb{N}^s$.

A fundamental theorem on diophantine relations [DMR], [Smo, Chpt. 2] states:

(II) Γ *is recursively enumerable iff it is diophantine.*

This in turn implies

(III) $\Gamma \subseteq \mathbb{N}^s$ *is recursively enumerable iff there are polynomials $F(\underline{x}, \underline{z})$, $G(\underline{x}, \underline{z})$ whose coefficients are in $\mathbb{N}$ such that $\underline{m} \subset \Gamma$ iff there are $\underline{p}$ such that $F(\underline{m}, \underline{p}) = G(\underline{m}, \underline{p})$, where $\underline{m} = m_1, \ldots, m_s$, $\underline{p} = p_1, \ldots, p_t$, and $m_j, p_k \in \mathbb{N}$, while $\underline{x} = x_1, \ldots, x_s$ and $\underline{z} = z_1, \ldots, z_t$.*

We recall the δ-function:

(IV) $\delta(x,y) = 1$ iff $x = y$, $\delta(x,y) = 0$, if $x \neq y$.

Now to the proof of Thm. 4.1 (resp. Thm. 3.12 in [B-Sc]). In a first step one constructs a sequence of functions $f_j \in \mathscr{F}^+_{s+t}$, $j \leq l$, that satisfies conditions (a), (b) of Def. A.2 whose last member, i.e. f_l, looks as follows

(V) $f_l(\underline{\alpha}, \underline{\beta}) = \sum_{\underline{m},\underline{p} \geq 0} \lambda^{\mu(\underline{m},\underline{p})} e^{i\underline{m}\underline{\alpha}} e^{i\underline{p}\underline{\beta}} \delta(F(\underline{m},\underline{p}), G(\underline{m},\underline{p}))$,

where $\mu(\underline{\alpha}, \underline{\beta}) : \mathbb{N}^{s+t} \to \mathbb{N}$ is a regular recursive function emerging from the construction of the sequence $f_j, j \leq l$.

To $f_l(\underline{\alpha}, \underline{\beta})$ we apply the productions $P_{s+1}, \ldots, P_{s+t}$ so as to obtain the function $f_l(\underline{\alpha}, \underline{0})$ which, written more explicitly looks as follows;

(VI) $f_l(\underline{\alpha}, \underline{0}) = \sum_{\underline{m} \geq 0} e^{i\underline{m}\underline{\alpha}} A_{\underline{m}}$, with $A_{\underline{m}} = \sum_{\underline{p} \geq 0} \lambda^{\mu(\underline{m},\underline{p})} \delta(F(\underline{m},\underline{p}), G(\underline{m},\underline{p}))$.

From the construction of $f_l(\underline{\alpha}, \underline{0})$ via Def. A.2 and from (VI) one infers:

(1) $f_l(\underline{\alpha}, \underline{0})$ is attainable by $\mathscr{U}; M, P_r, R_\pi, \star_t$.

(2) $\mathscr{R}_s(f_l(\underline{\alpha}, \underline{0})) = \Gamma$.

(3) $\sum_{\underline{m} \geq 0} A_{\underline{m}} < \infty$, i.e. $f_l(\underline{\alpha}, \underline{0}) \in \mathscr{F}^+_s$.

That is, $f_l(\underline{\alpha}, \underline{0})$ has the properties required by Thm. 4.1.

Acknowledgement. We would like to thank Peter Buser for valuable discussions and for the typesetting in Latex.

References

[Bch] Büchi, R. *The collected works of J. Richard Büchi*. Edited and with a preface by Saunders Mac Lane and Dirk Siefkes. Springer-Verlag, New York, 1990.

[Bus] Buser, P. *Darstellung von Prädikaten durch analytische Funktionen*. Master thesis, Universität Basel, 1972, 41 pp.

[B-Sc] Buser, P. and Scarpellini, B. *Undecidability through Fourier series*. Ann. Pure Appl. Logic **167** (2016), no. 7, 507–524.

[Cp1] Copeland, J. *Hypercomputation*. Minds and Machines **12** (2002), 461–502.

[Dav] Davis, M. *Computability and unsolvability*. McGraw-Hill, New York, 1958.

[DMR] Davis, M., Matijasevič, Y. and Robinson, J. *Hilbert's tenth problem: Diophantine equations: positive aspects of a negative solution*; in: Mathematical developments arising from Hilbert problems. Proc. Sympos. Pure Math., Vol. XXVIII, Northern Illinois Univ., De Kalb, Ill., (1974), pp. 323–378.

[Han] Hancock, H. *Lectures on the theory of elliptic functions: Analysis*. Dover Publications,New York, 1958.

[Mat] Matiyasevich, Y.V. *Hilbert's tenth problem*. MIT Press, Cambridge, MA, 1993.

[PPV] Pasten, H., Pheidas, T. and Vidaux, X. *A survey on Büchi's problem: new presentations and open problems*. J. Math. Sci. (N. Y.) **171** (2010), 765–781.

[P-E] Pour-El, M.B. *Abstract computability and its relation to the general purpose analog computer*. Trans. Amer. Math. Soc. 199 (1974), 1–28.

[PrRi] Pour-El, M.B. and Richards, I. *Computability and noncomputability in classical analysis*. Trans. Amer. Math. Soc. **275** (1983), 539–560.

[Ric] Richardson, D. *Some undecidable problems involving elementary functions of a real variable*. J. Symbolic Logic **33** (1968), 514–520.

[Rub] Rubel, L.A. *The extended analog computer*. Adv. in Appl. Math. **14** (1993), 39–50.

[Sc1] Scarpellini, B. *Zwei unentscheidbare Probleme in der Analysis*. Zeitschr. f. math. Logik u. Grundlagen der Math., **9** (1963), 265–289.

[Sc2] Scarpellini, B. *Two undecidable problems of analysis*. Minds and Machines, **13** (2003), 49–77.

[Sc3] Scarpellini, B. *Comments on "Two undecidable problems of analysis"*. Minds and Machines, **13** (2003), 79–85.

[Smo] Smoryński, C. *Logical number theory. I. An introduction*. Universitext. Springer-Verlag, Berlin, 1991.

[Syr] Syropoulos, A. *Hypercomputation. Computing beyond the Church-Turing barrier*. Springer, New York, 2008.

[WW] Whittaker, E. T. and Watson, G. N. *A course of modern analysis*. Reprint of the fourth (1927) edition. Cambridge University Press, Cambridge, 1996.

Categories of Fuzzy Structures and Fuzzy Categories

Apostolos Syropoulos
Greek Molecular Computing Group
366, 28th October Str.
GR-671 33 Xanthi, Greece
asyropoulos@yahoo.com

Abstract

After a brief discussion of vagueness and its importance and short introductions to fuzzy sets and category theory, I discuss categories of fuzzy structures (i.e., categories whose objects are fuzzy structures) and I present two versions of fuzzy categories: one that is based on the theory of fuzzy graphs and one that is based on the idea that category theory is about arrows and their properties. The second approach is to define functors, natural transformations, and F-coalgebras.

1 Introduction

It is not easy to write or to talk about a highly abstract mathematical theory, in general. However, when your readers and/or your audience is supposed to be knowledgeable about many things, then things can become really tricky. For example, what happens when you are not sure whether they will really understand what you plan to write and/or say? I am not really interested in a general strategy. Instead, I wanted to know how to deal with people who think they know enough about *fuzzy logic* (or more precisely: fuzzy logic in the broad sense) but have no "respect" to generalities. On the other hand, it seems that people well versed in category theory have some sort of "allergy" to fuzzy logic. And this is yet another problem! To summarize: people with a background in fuzzy logic are usually snobbish to abstract ideas while people with solid background in logic and/or category theory are snobbish to fuzzy logics. Thus I expect that readers will not be prepared to simultaneously read and/or listen about mathematical universes and vagueness in nature! However, on a second thought, I am not really sure that people with a background in fuzzy logic really know what vagueness is, which is the essence of fuzzy mathematics. In fact, I am not getting tired telling the story about the submission of a paper of mine describing a fuzzy abstract chemical machine (see [32]) to a "prestigious" journal devoted to the international advancement of the theory and application of fuzzy sets. Unfortunately, the paper was rejected because three (!) reviewers insisted that it is a "fact that there is an equivalence between fuzzy set theory and probability theory." In different words, three "reviewers" insisted that a probabilistic abstract chemical machine and a fuzzy abstract chemical machine are identical just like probabilistic automata and fuzzy automata are identical (they are not!). Moreover, the editor of the journal agreed with them! And when I told him that his decision is wrong, he replied to me by saying that I am arrogant! Now try to talk to these people about vagueness and abstract mathematics... Most probably, they will think that you are insane. Fortunately, not everybody agrees with these "scholars". In fact, most scholars think this is one of the major problems of peer-review: many reviewers are not really peers.

Many great ideas in science are based on simple assumptions. For example, general relativity is based on the idea that time is not absolute and Euclidean geometry is based on the idea that given a line and a point outside the line, we can have only one line that goes over the point and is parallel to the first line. Similarly, category theory is based on the remark that in different settings we see the same constructs. On the other hand, fuzzy set theory is based on the remark that certain things, properties, attributes cannot be classified as either true or false. For example, under the current circumstance, the sentence *Trump will be the next president of the US* is such a statement while the sentence *Erdoğan will be the next president of Turkey* seems not to be such a statement! Thus it makes sense to talk about membership to a collection to a degree. Furthermore, there is a very general structure called the category of categories. This structure has been proposed as a foundation for mathematics by Francis William Lawvere [21]. In simple words, one can write down a system of "simple" *axioms* that suffice to define all usual mathematical objects and to prove their usual properties. Thus whatever we can do with fuzzy logic, we should be able to do with some fuzzy version of the category of categories. Of course we need to prove this equivalence once we have a fully developed theory. But before going there, we need to show how to fuzzify category theory.

Plan of the chapter The contents of this chapter reflect my worries as they have been outlined in the first paragraph. In particular, in what follows first I will try to explain what vagueness is and what is the relationship between fuzzy logic and vagueness. Then, I will introduce the notions of category, functor, and natural transformation. Other, concepts and properties will be introduced if and when it is necessary. Next, I will present categories of fuzzy structures as the first step to reconcile category theory and fuzzy logic. The next step will follow and it is about fuzzy categories and their properties. The chapter will conclude with a brief discussion of the ideas presented.

2 What is Vagueness?

The word vague is used to characterize objects that are not sharply determined. Alternatively, a vague object is one for which it is not clear if it has or does not have a specific property or characteristic. When we observe this, we say that this property has *fuzzy boundaries*. A noted objection to this idea was put forth by Jiri Benovsky [4] who claimed that all ordinary objects are vague and that ordinary objects have sharp boundaries, therefore vague objects have sharp boundaries! The Sorites Paradox, which was introduced by Eubulides of Miletus, is a typical example of an argument that demonstrates what fuzzy boundaries are. The term sorites derives from the Greek word soros, which means "heap." The paradox is about the number of grains of wheat that makes a heap. All agree that a single grain of wheat does not comprise a heap. The same applies for two grains of wheat as they do not comprise a heap, etc. However, there is a point where the number of grains becomes large enough to be called a heap, but there is no general agreement as to where this occurs. Although there is no precise definition of vagueness, still most people would agree that adjectives like tall, old, short, young, etc., express vague concepts. For example, there is no general agreement on what makes a person tall or short, young or old. Of course a baby is definitely young but can we say the same for a person that is 30 years old? Furthermore, there are objects that one can classify as vague. For example, a cloud is vague since its boundaries are not sharp. Also, a dog is a vague object since it loses hair all the time and so it is difficult to say what belongs to it.

I am sure there are many people who consider such arguments as sophisms. Others consider vagueness as a *linguistic* phenomenon, that is, something that exists only in the realm of natural languages and gives us greater expressive power. And there are others that think that vagueness is a property of the world. As far as I know, there are three views regarding vagueness: the *ontic* view, the *semantic* view, and the *epistemic* view. According to the ontic view, the world itself is vague and, consequently, language is vague so to describe the world. The semantic view asserts that vagueness exists only in our language and our thought. In a way, this view is similar to the mental constructions of intuitionism, that is, things that exists in our minds but not in the real world. On the other hand, the epistemic view asserts that vagueness exists because we do not know where the boundaries exist for a "vague" concept. So we wrongly assume they are fuzzy. Personally, I favor that onticims about vagueness is the right view. In different words, I believe that vague objects exist and that vagueness is a property of the real world. It seems that semanticism is shared by many people, engineers in particular who use fuzzy mathematics, while if epistemicism is true, then there is simply no need for fuzzy mathematics and you are wasting your time reading this text.

Countries, lakes, rivers, etc., are also objects that are vague. In 1967, Benoit Mandelbrot [24] argued that the measured length of the coastline of Great Britain depends on the scale of measurement. This means that the boundaries of Great Britain are not sharp and so the country is a vague object since it has vague characteristics. Of course one may argue that here that there is no genuine vagueness but this is a problem of representation. A response to this argument was put forth by Michael Morreau [25]. Obviously, if the existence of vague objects is a matter of representation, then there are obviously no vague objects including animals. Consider Koula, my dog. She has hair that she will lose tonight so it is a questionable part of her. Because she has many such questionable parts (e.g., nails, whisker, etc.), Koula is a vague dog. Assume that Koula is not a vague dog. Instead, assume that there are many precise dogs that differ around the edges of the hair. Then, there are many animals that should be dogs because they differ slightly when compared to Koula. All of these candidates are dogs and they have very small differences between them. If vagueness is a matter of representation, then whenever I own a dog, in fact I own at least a thousand dogs! Clearly, this is not true.

Gareth Evans [10] presented an argument that *proves* that there are no vague objects. Evans used the modality operator $\triangledown$ to express indeterminacy. Thus $\triangledown\phi$ is pronounced as *it is indeterminate whether* ϕ. The dual of $\triangledown$ is the $\triangle$ operator and $\triangle\phi$ is pronounced as *it is determinate that* ϕ. Evans started his argument with the following premise:

$$\triangledown(a = b) \qquad (1)$$

This means that it is true that it is indeterminate whether a and b are identical. Next, he transformed this expression to an application of some sort of λ-abstraction:

$$\lambda x. \triangledown (x = a)b \qquad (2)$$

Of course it is a fact that it is not indeterminate whether a is identical to a:

$$\neg \triangledown (a = a) \qquad (3)$$

Using this "trick" to derive formula (2), one gets

$$\neg \lambda x. \triangledown (x = a)a \qquad (4)$$

Finally, he used the *identity of indiscernibles* principle to derive from (2) and (4):

$$\neg(a = b)$$

meaning that a and b are not identical. So we started by assuming that it is indeterminate whether a and b are identical and concluded they are not identical. In different words, indeterminate identities becomes nonidentities, which makes no sense. Therefore the assumption makes no sense. The identity of indiscernibles principle (see [11] for a thorough discussion of this principle) states that if, for every property F, object x has F if and only if object y has F, then x is identical to y. This principle was initially formulated by Wilhelm Gottfried Leibniz.

A first response to this argument is that the logic employed to deliver this proof is not really adequate. Pelletier [28] points out that when one says that an object is vague, this means that there is a predicate that neither applies nor does not apply to it. Thus when you have a meaningful predicate Fa, it makes no sense to make it indeterminate by just prefixing it with the $\triangledown$ operator. Although this logic is not appropriate for vagueness, still this is not a response to Evan's argument.

Edward Jonathan Lowe [23] put forth an argument that is a response to Evans' "proof":

> Suppose (to keep matters simple) that in an ionization chamber a free electron a is captured by a certain atom to form a negative ion which, a short time later, reverts to a neutral state by releasing an electron b. As I understand it, according to currently accepted quantum-mechanical principles there may simply be no objective fact of the matter as to whether or not a is identical with b.

The idea behind this example is that "identity statements represented by '$a = b$' are 'ontically' indeterminate in the quantum mechanical context" [12] (for a thorough discussion of the problem of identity in physics see [13]). Lowe's argument prompted a series of responses, nevertheless, I am not going to describe them here and the interested reader can read a summary of these responses in [5]. In a way, these responses culminated to a revised à la Evans proof based on Lowe's initial argument:

1. At t_1, $\triangledown(a$ has been emitted).

2. So at t_1, $\lambda x. \triangledown (x$ has been emitted)a.

3. But at t_1, $\neg \triangledown (b$ has been emitted).

4. So at t_1, $\lambda x. \neg \triangledown (x$ has been emitted)b.

5. Therefore, $a \neq b$.

It is possible to provide a re-interpretation of quantum mechanics that does not use probabilities but possibilities instead [33]. This re-interpretation assumes that vagueness is a fundamental property of the world. Although the ideas involved are very simple, still they require a good background in quantum mechanics and ideas that I am not going to discuss in this chapter.

3 What are Fuzzy Sets?

It is widely accepted that there are at least three different expressions of vagueness:

Many-valued logics and fuzzy logics Borderline statements are assigned truth-values that are between absolute truth and absolute falsehood.

Supervaluationism The idea that borderline statements lack a truth value.

Contextualism The truth value of a proposition depends on its context (i.e., a person may be tall relative to American men but short relative to NBA players).

Here I am interested in fuzzy logic, in general, and fuzzy sets, in particular. Fuzzy sets are used to mathematically represent borderline cases and so elements of a fuzzy set belong to it to some degree. Usually, this degree is a number that belongs to the interval $[0,1]$. Fuzzy sets were introduced by Lotfi Askar Zadeh [35]:

Definition 3.1 *Let X be a set that we call a universe. A fuzzy subset A of X, is characterized by a function $A : X \to [0,1]$, which is called the membership function. For every x from X, the value $A(x)$ is called a degree to which element x belongs to the fuzzy subset A.*

Note that I have used the word *characterized* and not *is* mainly because Zadeh used it in his original formulation of fuzzy sets. However, we know that $\mathcal{P}(X)$, the powerset of X, is isomorphic to 2^X (i.e., the *function space*, that is, the set of all functions from X to $2 = \{0,1\}$.), the set of all functions from X to $\{0,1\}$, and similarly, $\mathcal{F}(X)$, the set of all fuzzy subsets of X, is isomorphic to the set of all fuzzy characteristic functions, $[0,1]^X$. Therefore, there is no need to distinguish between "is" and "characterized"! Also, it is common in the literature to talk about fuzzy sets when, in fact, the term refers to fuzzy subsets. Let me now provide the definition of some common set operations.

Definition 3.2 *Assume that $A : X \to [0,1]$ and $B : X \to [0,1]$ are two fuzzy subsets of X. Then*

- *their* union *is*
$$(A \cup B)(x) = \max\{A(x), B(x)\};$$

- *their* intersection *is*

$$(A \cap B)(x) = \min\{A(x), B(x)\};$$

- *the* complement *of A is the fuzzy subset*

$$\bar{A}(x) = 1 - A(x) \quad for \quad x \in X.$$

Usually one writes $a \wedge b$ for $\min(a,b)$ and $a \vee b$ for $\max(a,b)$.

Definition 3.3 *Given a lattice[1] L and a function $f : X \to Y$, the image $f^\to : L^X \to L^Y$ and the preimage $f^\leftarrow : L^X \to L^Y$ operators are defined by*

$$f^\to(A)(y) = \bigvee\{A(x) \mid x \in X \text{ and } f(x) = y\}$$

and

$$f^\leftarrow(B) = B \circ f,$$

respectively. In addition, it is a fact that these two operators form an adjunction $f^\to \dashv f^\leftarrow$, so for all $A \in L^X$ and all $B \in L^Y$ it holds that $A \subseteq f^\leftarrow(f^\to(A))$ and $f^\to(f^\leftarrow(B)) \subseteq B$.

4 What is Category Theory?

Category theory is a very general formalism whilst others view it as a *language* [20]. The theory was introduced by Samuel Eilenberg and Saunders MacLane [9]. Category theory is based on the remark that if we consider vector spaces and their linear transformations, groups and their homomorphisms, topological spaces and their continuous mappings, ordered sets and their order preserving transformations, etc., and focus on their common characteristics and properties, then we start seeing patterns that we could not see before. Tom Leinster's [22] picturesque description of category theory makes clear this remark:

> Category theory takes a bird's eye view of mathematics. From high in the sky, details become invisible, but we can spot patterns that were impossible to detect from ground level. How is the lowest common multiple of two numbers like the direct sum of two vector spaces? What do discrete topological spaces, free groups, and fields of fractions have in common?

---
[1] A partially ordered set P is a *lattice* if and only if every finite subset of P has both a greatest lower bound and a least upper bound.

John Baez and Mike Stay [1] described how category theory, as a new Rosetta Stone, can aid researchers attempting to go from one field of science to another. In particular, they described how to go from one field to another from a specific list of fields. This list includes quantum physics, topology, computation, and logic. The following table is the "pocket" version of their "Rosetta Stone":

Category Theory	Physics	Topology	Logic	Computation
object	system	manifold	proposition	data type
morphism	process	cobordism	proof	program

As was hinted above, a category is a mathematical universe that includes all objects of a particular form together with maps between them. These maps must obey a few basic principles. In particular,

Definition 4.1 *A category $\mathcal{C}$ is made up of*

1. *a collection of things that are called $\mathcal{C}$-objects (objects for simplicity);*

2. *a collection of "bridges" between $\mathcal{C}$-objects that are called $\mathcal{C}$-arrows (arrows fro simplicity) or $\mathcal{C}$-morphisms (or just morphisms for simplicity);*

3. *each arrow f has as* domain *the object $dom\,f$ and as* codomain *the object $cod\,f$. If $A = dom\,f$ and $B = cod\,f$, then we write $f : A \to B$;*

4. *an operation that assigns to each pair (g, f) of $\mathcal{C}$-arrows, such that $dom\,g = cod\,f$, a $\mathcal{C}$-arrow $g \circ f$, such that $dom(g \circ f) = dom\,f$ and $cod(g \circ f) = cod\,g$, that is, $g \circ f : dom\,f \to cod\,g$. In addition, given the arrows $f : A \to B$, $g : B \to C$, and $h : C \to D$, then $h \circ (g \circ f) = (h \circ g) \circ f$;*

5. *for each $\mathcal{C}$-object A there is a $\mathcal{C}$-arrow $id_A : A \to A$ called the* identity *arrow, such that for any $f : A \to B$ and $g : B \to C$, $id_B \circ f = f$ and $g \circ id_B = g$.*

Eilenberg and MacLane [9, p. 237] used the term *aggregates* but the term *collection* is more common nowadays. To keep things simple I do not use the term "set" to avoid paradoxes like the set of all sets. Also, Eilenberg and MacLane noted that since the last rule "provides a one-to-one correspondence between the set of all objects of the category and the set of all its identities", it is thus clear that "the objects play a secondary role, and could be entirely omitted from the definition of a category". A side effect of their omission would be that "the manipulation of the applications would be slightly less convenient were this done" [9, p. 238].

Example 4.1 *The collection of all sets and functions between them forms the category that is traditionally denoted by* **Set**.

In passing, I would like to mention that a purely algebraic definition of categories was given by Peter J. Freyd and Andre Scedrov [14].

In category theory *commutative diagrams* play the role equations play in algebra. In the simplest case, a commutative diagram can be identified with two different paths starting from the same object A and ending with the same object B in which the composition of the arrows that make up the first path and the composition of the arrows of the second path yield two arrows that have the same effect (i.e., when applied to the same object(s), they yield the same result). For example, the identity law can be expressed by the following commutative diagrams:

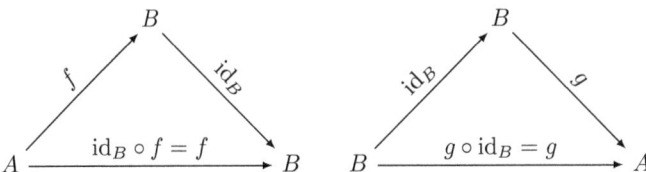

Note that a commutative diagram is actually a proof of some statement.

Categories were introduced in order to introduced functors which allowed the introduction of natural transformations (see [9, p. 247]). First, I will give the definition of functors:

Definition 4.2 *A functor F from a category $\mathcal{C}$ to a category $\mathcal{D}$ is a function that assigns*

1. *to each $\mathcal{C}$-object A, a $\mathcal{D}$-object $F(A)$;*

2. *to each $\mathcal{C}$-arrow $f : A \to B$ a $\mathcal{D}$-arrow $F(f) : F(A) \to F(B)$ such that*

    (a) $F(id_A) = id_{F(A)}$ *for all $\mathcal{C}$-objects A,*

    (b) $F(g \circ f) = F(g) \circ F(f)$ *if $g \circ f$ is defined.*

Think of a category whose objects are categories, its arrows will be functors. Indeed, it is possible to define functional composition for two functors $F : \mathcal{C} \to \mathcal{D}$ and $G : \mathcal{D} \to \mathcal{E}$ to get the functor $G \circ F : \mathcal{C} \to \mathcal{E}$. And since this operation is associative, we can easily define the category of categories of Lawvere. Having defined categories of categories, is it possible to define categories of functors? Surprisingly, the answer is: Yes! Consider two categories $\mathcal{C}$ and $\mathcal{D}$, then we are going to construct the category of all functors between these two categories. This new category will be denoted by $\mathcal{D}^{\mathcal{C}}$.

Definition 4.3 *A natural transformation from a functor $F : \mathcal{C} \to \mathcal{D}$ to a functor $G : \mathcal{C} \to \mathcal{D}$ is an assignment $\tau : F \dot{\to} G$ that provides, for each $\mathcal{C}$-object A, a $\mathcal{D}$-arrow $\tau_A : F(A) \to G(A)$, such that for any $\mathcal{C}$-arrow $f : A \to B$ we have $\tau_B \circ F(f) = G(f) \circ \tau_A$.*

At this point we are ready to proceed with the presentation of categories whose objects are fuzzy structures.

5 Categories of Fuzzy Structures

It is not difficult to define categories of fuzzy structures. For example, Carol Walker [34] and Siegfried Gottwald [17] have defined such categories. However, some other categories of fuzzy structures emerged from efforts to define fuzzy models of *linear logic* [15]. In particular, Michael Barr [3], Basil Papadopoulos and this author [26, 27, 31] have introduced such categories. A simple example of a category of fuzzy sets is the category **SET**(L):

Definition 5.1 *Let L be a frame.*[2] *Category* **SET**(L) *has as objects Goguen sets, that is, pairs (S, σ), where S is a set and $\sigma \in L^S$. Category* **SET**(L) *has as arrows maps between Goguen sets: Given two Goguen sets (S, σ) and (T, τ) a map $f : (S, \sigma) \to (T, \tau)$ is a function $f : S \to T$ such that $\sigma \leq f^{\leftarrow}(\tau)$.*

Clearly, composition of maps is function composition, which is associative, and the identity arrows are the identity functions. Now that we have defined a category of fuzzy structures, we can embed this category into one that has known properties. Here we will show how we can embed it into *Dialectica* and *Chu* categories.

In 1958, the Austrian logician Kurt Friedrich Gödel [16] published in the journal *Dialectica* an interpretation of intuitionistic arithmetic (i.e., the intuitionistic analogue of Peano arithmetic known as Heyting arithmetic) in a quantifier-free theory of functionals of finite type, which has come to be known as *Dialectica Interpretation*. Valeria de Paiva [6] presented a categorical version of the Dialectica interpretation. In her thesis, she presented two categories with arrows that correspond to the Dialectica interpretation of implication. Later on she presented one more categorical version of the Dialectica interpretation [7]. The term *Dialectica space* refers to objects of *Dialectica* categories. The Dialectica categories are models of linear logic. This is particularly interesting since categorical semantics model derivations (i.e., proofs) and are not just used to deduce whether a theorem is true or not. Another widely known categorical model of linear logic is based on the Chu construct described by Po-Hsiang Chu

---
[2] A partially ordered set P is a *frame* if and only if
1. every subset has a least upper bound;
2. every finite subset has a greatest lower bound; and
3. the operator $\wedge$ distributes over $\vee$:

$$x \wedge \bigvee Y = \bigvee \{x \wedge y \mid y \in Y\}.$$

in an appendix of [2]. Let me begin with Chu spaces and the representation of fuzzy structures as Chu spaces.

Chu Spaces A *Chu space* over an alphabet Σ (i.e., an arbitrary set whose structure is of no importance) is a triplet (X, r, A), where X and A are arbitrary sets and $r : X \times A \to \Sigma$ is a function. Function r relates the elements of X with the elements of A. For example, suppose that $\Sigma = \{0, 1\}$ and that A stands for the set of open subsets of X. Then, $r(x, a) = 1$ if x belongs to the open subset a, else $r(x, a) = 0$. Following a similar way of thinking, one can represent any relational structure (e.g., groups, vector spaces, categories, etc.) as a Chu space. Assume that $\mathcal{A} = (X, r, A)$ and $\mathcal{B} = (Y, s, B)$ are two Chu Spaces. Then, a transformation from $\mathcal{A}$ to $\mathcal{B}$ is just a pair of functions $(f, \bar{f})$, where $f : X \to Y$ and $\bar{f} : B \to A$, such that

$$s(f(x), b) = r(x, \bar{f}(b)), \text{ for all } x \in X \text{ and all } b \in B, \qquad (1)$$

or as a commutative diagram:

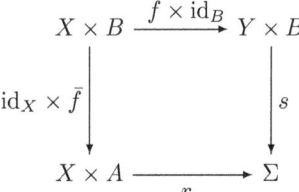

This condition is called the *adjointness condition*. We can build a category **Chu**(Σ) with objects all Chu spaces and with arrows pairs of functions that fulfill the adjointness condition. Arrow composition is the usual composition of functions in pairs. For any Chu space (X, r, A) it is easy to verify that the identity arrow is the pair of functions $(\text{id}_X, \text{id}_A)$.

Embedding Fuzzy Categories to Chu Categories We consider a subcategory of **SET**(L) that has the following property: for each pair of objects (S, σ) and (T, τ), a function $f : S \to T$ is an arrow between them if and only if $\sigma = f^\leftarrow(\tau)$. We call this subcategory **SET**(L)$_=$. Although the restriction imposed on the arrows of **SET**(L)$_=$ may seem too strong, the following proposition shows that there are enough arrows in **SET**(L)$_=$.

Proposition 5.1 *Suppose that (S, σ) and (T, τ) are Goguen sets and that $f : S \to T$ is an injective function satisfying $f^\to(\sigma) = \tau$. Then, $f : (S, \sigma) \to (T, \tau)$ is an arrow such that $\sigma = f^\leftarrow(\tau)$.*

This proposition is immediate from the following remarks:

Remark 1 *If $f : L \to M$, $g : L \leftarrow M$ are isotone maps (i.e., maps that are monotone increasing and therefore order-preserving) between partially ordered sets, then $f \dashv g$ implies that $f \circ g \circ f = f$. In addition, if f is injective, then $g \circ f = \mathrm{id}_L$.*

Remark 2 *If $f : X \to Y$ is injective, then $f^{\to}$ is injective, which implies that $f^{\leftarrow} \circ f^{\to} = \mathrm{id}_{L^X}$.*

Let me define a functor $\mathfrak{F}$ from the subcategory $\mathbf{SET}(L)_=$, for some fixed L, to the category $\mathbf{Chu}(L)$. First, the definition the object part:

Definition 5.2 *(Object Part) Let (S, σ) be an object of $\mathbf{SET}(L)_=$. Then, functor $\mathfrak{F}$ maps it to the Chu space $(S, r, \{\sigma\})$, where $r(s, \sigma) = \sigma(s)$.*

The following result is direct consequence of the previous definition:

Corollary 5.1 *Functor $\mathfrak{F}$ is injective on objects.*

And now the definition of the morphism part of the functor:

Definition 5.3 *(Arrow Part) Suppose that $\mathfrak{F}(S, \sigma) = (S, r, \{\sigma\})$ and $\mathfrak{F}(T, \tau) = (T, s, \{\tau\})$. Moreover, suppose that $f : (S, \sigma) \to (T, \tau)$ is an arrow of $\mathbf{SET}(L)_=$. Then $\mathfrak{F}(f) = (f, g)$, where $g(\tau) = \sigma$.*

The following result is a direct consequence of the above definitions:

Theorem 5.1 *Any subcategory $\mathbf{SET}(L)_=$ fully embeds[3] into a category $\mathbf{Chu}(L)$.*

Dialectica Spaces Goguen sets can be represented more naturally as Dialectica spaces. Generally speaking, Dialectica spaces are Chu spaces with more arrows between them. As was noted above, there three different families of Dialectica categories, but here I am interested only in the categories $\mathbf{Dial}_L\mathbf{Set}$, where L is a *lineale* [8], that is, a monoidal partially ordered set with additional structure.

Definition 5.4 *A monoidal partially ordered set is a partially ordered set $(L, \leq)$ with a given symmetric monoidal structure $(L, \circ, \mathbf{1})$. That is, a set L equipped with a binary relation $\leq$, together with a monoid structure $(\circ, \mathbf{1})$ consisting of a (order-preserving) multiplication $\circ : L \times L \to L$ and a distinguished object $\mathbf{1}$ of L. We write a monoidal partially ordered set as a quadruple $(L, \leq, \circ, \mathbf{1})$.*

---
[3]Assume that $F : \mathcal{S} \to \mathcal{C}$ is a functor, where $\mathcal{S}$ is a subcategory of $\mathcal{C}$ (i.e., $\mathcal{S}$ consists of a subcollection of the collections of objects and arrows of $\mathcal{C}$). Then, $\mathcal{S}$ is a *full subcategory* if for every pair of $\mathcal{S}$-objects A and B and to every $\mathcal{C}$-arrow $g : F(A) \to F(B)$, there is an $\mathcal{S}$-arrow $f : A \to B$ such that $g = F(f)$.

Operator 'o' is a logical conjunction operator, which is not necessarily idempotent. In addition, **1** is not necessarily the top element of L. Suppose now that L is a monoidal partially ordered set and a, b are elements of L. Then, if there is an $x \in L$, which is the largest element of L such that $a \circ x \leq b$, this element is denoted by $a \multimap b$.

Definition 5.5 *A lineale (or close partially ordered set) is a monoidal partially ordered set such that $a \multimap b$ exists for all $a, b \in L$. We write a lineale as a quintuple $(L, \leq, \circ, \mathbf{1}, \multimap)$.*

In addition, it holds that $z \circ x \leq y$ if and only if $z \leq (x \multimap y)$. Practically, this is a proof of the following:

Corollary 5.2 *Given a frame L, the quintuple $(L, \leq, \wedge, \mathbf{1}, \Rightarrow)$, where $\Rightarrow$ is the exponential, is a lineale.*

We are now ready to define the family of categories **Dial**$_L$**Set** [7]:

Definition 5.6 *Let $(L, \leq, \circ, \mathbf{1}, \multimap)$ be a lineale. Then, the objects of a category* **Dial**$_L$**Set** *are triplets (X, r, A), where X and A are arbitrary sets and $r : X \times A \to L$ is function. Suppose that (X, r, A) and (Y, s, B) are two objects. Then, an arrow between them is a pair (f, g), where $f : X \to Y$ and $g : B \to A$, such that*
$$r(x, g(b)) \leq s(f(x), b), \quad \forall x \in X, \forall b \in B.$$

Note that any category **Dial**$_L$**Set** is a symmetric monoidal closed category with involution and products and it is a categorical model of intuitionistic linear logic.

By following a line of arguments similar to those employed for Theorem 5.1, we can prove the following theorem:

Theorem 5.2 *Any category* **SET**(L) *embeds fully into a Dialectica category* **Dial**$_L$**Set**.

6 Fuzzy Categories

Not so surprisingly, a category of fuzzy structures is not a fuzzy structure itself. Indeed, Alexander Šostak [29, 30] was the first researcher who had realized this. In order to remedy this situation, Šostak introduced a new structure that mimics the way fuzzy graphs are defined. However, these new structures make use of GL-monoids:

Definition 6.1 *A GL-monoid is a triple $(L, \leq, \circledast)$, where L is a complete lattice and $\circledast$ is a binary operator such that:*

1. $a \leq b$ implies $a \circledast c \leq b \circledast c$, for all $a, b, c \in L$;
2. $a \circledast b = b \circledast a$, for all $a, b \in L$;
3. $a \circledast (b \circledast c) = (a \circledast b) \circledast c$, for all $a, b, c \in L$;
4. $a \circledast \mathbf{1} = a$, for all $a \in L$;
5. $a \circledast \mathbf{0} = \mathbf{0}$, for all $a \in L$;
6. $a \circledast (b_1 \vee \cdots \vee b_n) = (a \circledast b_1) \vee \cdots \vee (a \circledast b_n)$, for all $a, b_1, \ldots, b_n \in L$;
7. when $a \leq b$, then there is a $c \in L$ such that $a = b \circledast c$.

Let us proceed with the definition of L-fuzzy categories:

Definition 6.2 *An L-fuzzy category (where L is a GL-monoid) is a quintuple*

$$\mathcal{C} = (\mathrm{Ob}(\mathcal{C}), \omega, M(\mathcal{C}), \mu, \circ),$$

where

1. $\mathcal{C}_\perp = (\mathrm{Ob}(\mathcal{C}), M(\mathcal{C}), \circ)$ *is a usual (classical) category called the bottom frame of the fuzzy category $\mathcal{C}$;*
2. $\omega : \mathrm{Ob}(\mathcal{C}) \to L$ *is an L-fuzzy subclass of the class of objects $\mathrm{Ob}(\mathcal{C})$ of $\mathcal{C}_\perp$;*
3. $\mu : M(\mathcal{C}) \to L$ *is an L-subclass of the class of morphisms $M(\mathcal{C})$ of $\mathcal{C}_\perp$.*

In addition, ω and μ must satisfy the following conditions:

1. *if $f : X \to Y$, then $\mu(f) \leq \omega(X) \wedge \omega(Y)$;*
2. *$\mu(g \circ f) \geq \mu(g) \circledast \mu(f)$ whenever composition $g \circ f$ is defined;*
3. *if $e_X : X \to X$ is the identity morphism, then $\mu(e_X) = \omega(X)$.*

Given an L-fuzzy category $\mathcal{C}$ and $X \in \mathrm{Ob}(\mathcal{C})$, then $\omega(X)$ is the *degree* to which a potential object X of the L-fuzzy category $\mathcal{C}$ is indeed its object. Similarly, if $f \in M(\mathcal{C})$, then $\mu(f)$ is the degree to which the map f is indeed a morphism $\mathcal{C}$.

Definition 6.3 *Assume that $\mathcal{C} = (\mathrm{Ob}(\mathcal{C}), \omega, M(\mathcal{C}), \mu, \circ)$ is a fuzzy category. Then, an L-fuzzy subcategory of $\mathcal{C}$ is an L-fuzzy category*

$$\mathcal{C}' = (\mathrm{Ob}(\mathcal{C}), \omega', M(\mathcal{C}), \mu', \circ),$$

where $\omega' \leq \omega$ and $\mu' \leq \mu$. $\mathcal{C}'$ is a full subcategory of $\mathcal{C}$ if

$$\mu'(f) = \mu(f) \wedge \omega'(X) \wedge \omega'(Y),$$

for all $f \in M_\mathcal{C}(X, Y)$ and all $X, Y \in \mathrm{Ob}(\mathcal{C})$.

For some time, I have been working on a slightly different approach to the fuzzification of category theory [32]. In particular, since objects are supporting "actors" in this play, it is not really useful to assign membership degrees to them. The idea is that all objects and arrows belong to a category but only arrows should be associated with a plausibility degree since arrows are possible bridges between objects. It is instructive to think of a category as a graph to which each edge is associated with a plausibility degree that indicates the degree to which one can go from one node to another.

Definition 6.4 *A fuzzy category $\mathcal{C}$ is an ordinary category $\mathcal{C}$ but in addition:*

1. *There is an operation p that assigns to each arrow a plausibility degree $\rho = \mathrm{p}f \in [0,1]$. Thus an arrow that starts from A and ends at B with plausibility degree ρ is written as:*

$$A \xrightarrow[\rho]{f} B \quad or \quad f : A \xrightarrow{\rho} B;$$

2. *For the composite $g \circ f$ it holds that $\mathrm{p}(g \circ f) = (\mathrm{p}\,f) \wedge (\mathrm{p}\,g)$. The associative law holds since $\wedge$ is an associative operation.*

3. *an assignment to each $\mathcal{C}$-object B of a $\mathcal{C}$-arrow $\mathbf{1}_B : B \xrightarrow{1} B$, called the identity arrow on B, such that the following identity law holds true:*

$$\mathbf{1}_B \circ f = f \quad and \quad g \circ \mathbf{1}_B = g$$

for any $\mathcal{C}$-arrows $f : A \xrightarrow{\rho_f} B$ and $g : B \xrightarrow{\rho_g} A$.

Example 6.1 *Let me give a relatively simple example of a fuzzy category that has as objects sets. Assume that $f : X \to Y$ is function. Then we say that f is an arrow from X to Y with plausibility degree λ if there are fuzzy subsets that are described by the functions $A : X \to \mathrm{I}$ and $B : Y \to \mathrm{I}$ such that $B(f(x)) - A(x) \geq \lambda$, for all $x \in X$. Assume that $f : X \xrightarrow{\lambda_1} Y$ and $g : Y \xrightarrow{\lambda_2} Z$ are two arrows. Then since $B(f(x)) - A(x) \geq \lambda_1$ and $C(g(f(x))) - B(f(x)) \geq \lambda_2$, for all $x \in X$, one concludes that $C(g(f(x))) - A(x) \geq \lambda_3$, for all $x \in X$ and where $\lambda_3 = \min\{\lambda_1, \lambda_2\}$. Thus, the composite arrow $g \circ f : X \xrightarrow{\lambda_3} Z$ exists and can be defined form its constituents. It is not difficult to see that the associative law holds also. Clearly, the identity arrow for some object X is the identity function of this set. Also, it is almost trivial to see that the identity law holds. The resulting fuzzy category will be called **FSet**.*

Example 6.2 *Assume that $R : X \times X \to [0,1]$ is a fuzzy relation such that $R(x,x) = 1$ for all $x \in X$ and $R(x,y) * R(y,z) \leq R(x,z)$ for all $x, y, z \in X$ and where $*$ is a t-norm. A fuzzy relation with these properties is called a $*$-fuzzy* preorder. When the t-norm is function min, then it will be called just *fuzzy preorder*. Any fuzzy preorder relation R determines a fuzzy preorder category P in which the arrows $p \xrightarrow{\rho} p'$ are exactly those pairs $\langle p, p' \rangle$ for which $R(p, p') = \rho$. The fuzzy relation is transitive, which implies that there is a unique way of composing arrows. Also, the fuzzy relation is reflexive and so there are the necessary identity arrows.

Example 6.3 *Let $\wedge$ be the only object of a category and let us identify all arrows of this category with their plausibility degrees. For example $\wedge \xrightarrow[\lambda]{\lambda} \wedge$ is the arrow λ whose plausibility degree is obviously λ. Given two arrows λ_1 and λ_2, and assuming that $\wedge$ denotes the minimum, as is usually the case, then $\lambda_1 \circ \lambda_2 = \lambda_1 \wedge \lambda_2$. Assume there is an arrow $1_\wedge$ such that $1_\wedge \circ \lambda = \lambda$ and $\lambda \circ 1_\wedge = \lambda$ for all arrows λ, then according to the definition of arrow composition this arrow is the identity arrow, that is, $1_\wedge = 1$. Note that it is quite possible to have more than one arrow that has plausibility degree equal to one, nevertheless, for our purposes these arrows will behave exactly like 1 does. In different words, they will be isomorphic, but we will say more about isomorphisms in a while.*

Example 6.4 *A category is a* deductive system *(for example, see [19] for a thorough description of this categories-as-deductive-systems paradigm). In this paradigm, objects are seen as* formulas, *arrows as* proofs *(or* deductions*), and an operation on arrows as a* rule of inference. *In particular, each arrow $f : A \to B$ is thought of as the "reason" why A entails B. Thus, the identity law is the reason why A entails A, for all A objects (formulas) and the associative law becomes the following rule of inference:*

$$\frac{f : A \longrightarrow B \quad g : B \longrightarrow C}{g \circ f : A \longrightarrow C}$$

Similarly, a fuzzy category is a fuzzy deductive system *in which objects may be fuzzy formulas (remember, the objects of any fuzzy category are not necessarily "crisp"), arrows are fuzzy deductions, and the associative law is the following fuzzy inference:*

$$\frac{f : A \xrightarrow{\rho_f} B \quad g : B \xrightarrow{\rho_g} C}{f \circ g : A \xrightarrow{\rho_{g \circ f}} C}$$

Commutative diagrams Assume that

$$A \xrightarrow[\lambda_n]{f_n} \ldots \xrightarrow[\lambda_1]{f_1} B$$
$$A \xrightarrow[\rho_m]{g_m} \ldots \xrightarrow[\rho_1]{g_1} B$$

are two paths. Then, these paths form a *strong* commutative diagram provided that
$$\bigwedge\{\lambda_1,\ldots,\lambda_n\} = \bigwedge\{\rho_1,\ldots,\rho_m\}.$$
Otherwise, we write
$$f_n \circ \cdots \circ f_1 =_\omega g_m \circ \cdots \circ g_1$$
and say that the two paths commute with plausibility degree ν, where
$$\omega = \left[\bigwedge\{\lambda_1,\ldots,\lambda_n\}\right] \wedge \left[\bigwedge\{\rho_1,\ldots,\rho_m\}\right].$$

Fuzzy Functors and Fuzzy Natural Transformations Let me now introduce fuzzy functors and fuzzy natural transformations. Since functors and natural transformations are actually arrows, they should be fuzzy too.

Definition 6.5 *A fuzzy functor F from a fuzzy category $\mathcal{C}$ to a fuzzy category $\mathcal{D}$ is a triple (F_0, F_1, F_2) where*

1. *$F_2 : [0,1] \to [0,1]$ is a complete lattice homomorphism;*
2. *F_0 maps a $\mathcal{C}$-object A to a $\mathcal{D}$-object $F_0(A)$;*
3. *F_1 maps a $\mathcal{C}$-arrow $f : A \xrightarrow{\lambda} B$ to a $\mathcal{D}$-arrow $F_1(f) : F_0(A) \xrightarrow{F_2(\lambda)} F_0(B)$ such that*
   (a) *$F_1(id_A) = id_{F_0(A)}$ for all $\mathcal{C}$-objects A,*
   (b) *$F_1(g \circ f) =_\nu F_1(g) \circ F_1(f)$ provided $g \circ f$ is defined.*

Definition 6.6 *A fuzzy natural transformation from a fuzzy functor $F : \mathcal{C} \to \mathcal{D}$ to a fuzzy functor $G : \mathcal{C} \to \mathcal{D}$ is an assignment $\tau^\nu : F \xrightarrow[\nu]{} G$ that provides, for each $\mathcal{C}$-object A, a $\mathcal{D}$-arrow $\tau_A^\nu : F_1(A) \xrightarrow{\nu} G_1(A)$, such that for any $\mathcal{C}$-arrow $f : A \xrightarrow{\mu} B$ we have $\tau_B^\nu \circ F(f) =_\nu G(f) \circ \tau_A^\nu$.*

Fuzzy Coalgebras Coalgebras [18] is a mathematical theory that aims to describe computational dynamics. Roughly, coalgebras are used to describe the behavior of a computer system. Although it is impossible to develop a fuzzy version of the theory of coalgebras in a few pages, still it is possible to give a (fuzzy) categorical definition of the fuzzy analog of coalgebras. Assume that $\mathcal{C}$ is a fuzzy category and $F : \mathcal{C} \to \mathcal{C}$ a fuzzy functor. Then, a fuzzy F-*coalgebra* consists of an object X and an arrow $c : X \xrightarrow{\lambda} F(X)$. Usually, X is called the state space and c the transition. Given coalgebras $c : X \xrightarrow{\lambda} F(X)$ and $d : Y \xrightarrow{\lambda'} F(Y)$, a map from c to d consists of an arrow $f : X \xrightarrow{\kappa} Y$ such that
$$F(f) \circ c =_\omega d \circ f.$$

7 Conclusions

My goal was to show that it is possible to devise a fuzzy category theory. I explained why vagueness matters, what fuzzy set theory is, and what category theory is. Although I demonstrated that it is possible to marry category theory and fuzzy set theory by defining categories of fuzzy structures, im my humble opinion the end result is not... vague. Thus only a really fuzzy version of category theory would allow one to describe ideas related to vagueness in a natural way. I presented two approaches that solve the problem of fuzzification of category theory. The first one is more... algebraic while the second more close to the true essence of category theory. Of course, the theory is far from being complete. However, it is the first step towards the real fuzzification of category theory.

8 Acknowledgements

I would like to thank Costas Drossos and Georgios O. Papadopoulos for their comments and suggestions. Also, I would like to thank the editors of this volume for inviting me to contribute to this Festschrift.

References

[1] J. Baez and M. Stay. Physics, Topology, Logic and Computation: A Rosetta Stone. In B. Coecke, editor, *New Structures for Physics*, pages 95–172. Springer, Berlin, 2011.

[2] M. Barr. *∗-Autonomous Categories*. Number 752 in Lecture Notes in Mathematics. Springer-Verlag, Berlin, 1979.

[3] M. Barr. Fuzzy models of linear logic. *Mathematical Structures in Computer Science*, 6(3):301–312, 1996.

[4] J. Benovsky. Vague Objects with Sharp Boundaries. *Ratio: An international journal of analytic philosophy*, 28(1):29–39, 2015.

[5] G. Darby. Vague Objects in Quantum Mechanics? In K. Akiba and A. Abasnezhad, editors, *Vague Objects and Vague Identity*, pages 69–108. Springer, Dordrecht, The Netherlands, 2014.

[6] V. de Paiva. The Dialectica Categories. Technical Report 213, Computer Laboratory, University of Cambridge, UK, 1991.

[7] V. de Paiva. Dialectica and Chu Constructions: Cousins? Talk presented at the Workshop on *Chu Spaces: Theory and Applications*, Santa Barbara, California, USA, 2000.

[8] V. de Paiva. Lineales: Algebraic Models of Linear Logic from a Categorical Perspective. In D. Barker-Plummer, D. I., Beaver, J. van Benthem, and P. S. di Luzio, editors, *Words, Proofs and Diagrams*, pages 123–142. CLSI Publications, Stanford, CA, USA, 2002.

[9] S. Eilenberg and S. MacLane. General theory of natural equivalences. *Transactions of the American Mathematical Society*, 58(2):231–294, 1945.

[10] G. Evans. Can there be vague objects? *Analysis*, 38(4):208, 1978.

[11] P. Forrest. The identity of indiscernibles. In E. N. Zalta, editor, *The Stanford Encyclopedia of Philosophy*. Metaphysics Research Lab, Stanford University, winter 2016 edition, 2016.

[12] S. French and D. Krause. Quantum Vagueness. *Erkenntnis*, 59:97–124, 2003.

[13] S. French and D. Krause. *Indentity in Physics: A Historical, Philosophical and Formal Analysis*. Oxford University Press, Oxford, UK, 2006.

[14] P. J. Freyd and A. Scedrov. *Categories, Allegories*, volume 39 of *North-Holland Mathematical Library*. Elsevier, Amsterdam, 1990.

[15] J.-Y. Girard. Linear Logic. *Theoretical Comput. Sci.*, 50:1–102, 1987.

[16] K. Gödel. Über eine bisher noch nicht benützte erweiterung des finiten standpunktes. *Dialectica*, 12(3–4):280–287, 1958.

[17] S. Gottwald. Universes of Fuzzy Sets and Axiomatizations of Fuzzy Set Theory. Part II: Category Theoretic Approaches. *Studia Logica*, 84:23–50, 2006.

[18] B. Jacobs. *Introduction to Coalgebra: Towards Mathematics of States and Observation*, volume 59 of *Cambridge Tracts in Theoretical Computer Science*. Cambridge University Press, Cambridge, 2016.

[19] J. Lambek and P. J. Scott. *Introduction to higher order categorical logic*. Cambridge University Press, Cambridge, U.K., 1994.

[20] E. Landry. Category theory: The language of mathematics. *Philosophy of Science*, 66:S14–S27, 1999.

[21] F. W. Lawvere. The Category of Categories as a Foundation for Mathematics. In S. Eilenberg, D. K. Harrison, S. MacLane, and H. Röhrl, editors, *Proceedings of the Conference on Categorical Algebra*, pages 1–20. Springer Berlin, Berlin, 1966.

[22] T. Leinster. *Basic Category Theory*. Number 143 in Cambridge Studies in Advanced Mathematics. Cambridge University Press, Cambridge, 2014.

[23] E. J. Lowe. Vague Identity and Quantum Indeterminacy. *Analysis*, 54(2):110–114, 1994.

[24] B. Mandelbrot. How Long Is the Coast of Britain? Statistical Self-Similarity and Fractional Dimension. *Science*, 156(3775):636–638, 1967.

[25] M. Morreau. What Vague Objects Are Like. *The Journal of Philosophy*, 99(7):333–361, 2002.

[26] B. K. Papadopoulos and A. Syropoulos. Fuzzy Sets amd Fuzzy Relational Structures as Chu Spaces. *International Journal of Uncertainty, Fuzziness and Knowledge-Based Systems*, 8(4):471–479, 2000.

[27] B. K. Papadopoulos and A. Syropoulos. Categorical relationships between Goguen sets and "two-sided" categorical models of linear logic. *Fuzzy Sets and Systems*, 149:501–508, 2005.

[28] F. J. Pelletier. The not-so-strange modal logic of indeterminacy. *Logique et Analyse*, 27(108):415–422, 1984.

[29] A. Šostak. Fuzzy categories related to algebra and topology. *Tatra Mountains Mathematical Publications*, 16(1):159–185, 1999. Available from the web site of the Slovak Academy of Sciences.

[30] A. Šostak. Fuzzy Functions and an Extension of the Category $\mathbf{L-TO}P$ of Chang-Goguen L-Topological Spaces. In P. Simon, editor, *Proceedings of the Ninth Prague Topological Symposium*, pages 271–294. arXiv:math/0204165 [math.GN], 2002.

[31] A. Syropoulos. Yet Another Fuzzy Model for Linear Logic. *International Journal of Uncertainty, Fuzziness and Knowledge-Based Systems*, 14(1):131–135, 2006.

[32] A. Syropoulos. Fuzzy Categories. `arXiv:1410.1478v1 [cs.LO]`, 2014.

[33] A. Syropoulos. On Vague Computers. In A. Adamatzky, editor, *Emergent Computation: A Festschrift for Selim G. Akl*, pages 393–402. Springer International Publishing, Cham, 2017.

[34] C. L. Walker. Categories of fuzzy sets. *Soft Computing*, 8(4):299–304, 2004.

[35] L. A. Zadeh. Fuzzy Sets. *Information and Control*, 8:338–353, 1965.

The Philosophical Significance of Incompleteness

Otávio Bueno[*]

Abstract

Incompleteness is a widespread trait of several cognitive systems, such as theories, models, databases, experimental setups, and clinical trials. It is a ubiquitous feature of the sciences and of scientific practice more generally. It is also a multifaceted phenomenon, involving inferential, informational and representational traits as well as semantic features. Inferential incompleteness involves true statements that cannot be derived from consistent axiomatizations satisfying certain conditions. Informational incompleteness concerns the unavailability, or only the partial availability, of certain bits of information given a particular context and the available representational devices for a certain domain of inquiry. Semantic incompleteness concerns vagueness and referential indeterminacy. In this paper, I discuss the importance of the first two forms of incompleteness—inferential and informational/representational incompleteness—and some aspects of their philosophical significance.

1 Introduction

Scientific information should ideally have a number of features; in particular, accuracy, empirical support, and relevance. But should it be complete as well? In this paper, I argue that the gap between the ideal and the actual—especially in the case of completeness—is significant and wide. For a variety of reasons, scientific information is not always accurate, sometimes it is less than properly empirically supported, and not as relevant as it should be. And although there are good reasons for scientists still to aim at yielding accurate, empirically supported, relevant information in the sciences, the case for completeness is different. Here, I argue, incompleteness is not only ubiquitous, but inevitable. It is, thus, crucial to learn how to live with and cherish this salient feature of scientific practice. I examine different kinds of incompleteness in scientific reasoning and examine some aspects of their philosophical significance.

[*]Department of Philosophy, University of Miami, Coral Gables, FL 33124, USA. E-mail: otaviobueno@mac.com.

Three kinds of incompleteness can be highlighted: (a) *Inferential incompleteness* involves situations in which certain statements cannot be proven from axiomatic systems that are presumed consistent and that satisfy certain conditions (Gödel's incompleteness theorems provide the key source of such incompleteness). (b) *Informational and representational incompleteness*, ubiquitous in the sciences, concerns the partiality of data and of information about particular domains and what can be represented about them given available representational devices. (c) *Semantic incompleteness*, in turn, involves cases of vagueness and referential indeterminacy. In what follows, I focus on inferential as well as informational and representational incompleteness.

2 Inferential Incompleteness

Inferential incompleteness emerges primarily from a gap between true statements expressible in a given theory and what can be derived from the theory. As noted, a significant source of inferential incompleteness, related to unprovability, emerges from Gödel's incompleteness results. Roughly speaking, in light of Gödel's first incompleteness theorem, given a consistent theory T, which is expressive enough to formulate arithmetic in it, there is a sentence that is true but cannot be proven in T. With Gödel's second incompleteness theorem, also roughly speaking, if T is consistent and allows for arithmetic to be formulate in it, T's consistency cannot be proven in T. We have here the typical form of inferential incompleteness: true statements cannot be proven in certain theories, given their content and the underlying logic, which is typically taken to be classical. (In what follows, I will focus primarily on the first incompleteness result.)

In the case of arithmetic, Gödel's result requires an understanding of arithmetical concepts that is not given fully by the axioms of the theory, since there are arithmetical sentences—namely, Gödel sentences—that are recognized to be true but which cannot be proven from the axioms in question. It is a significant fact about Gödel's result that it presupposes such intuitive understanding of arithmetic for the implementation of the incompleteness proof. Without such understanding, one would be unable to recognize the truth of the Gödel sentences, and thus prevented from asserting the result.

It could be argued that it is enough that a Gödel sentence G—which asserts its own unprovability—be true; it need not be *recognized* to be true. But given G's function in the incompleteness theorem, G's truth ultimately needs to be recognized. Let T be a consistent theory that extends minimal arithmetic ([1]). G is provably equivalent to its Gödel number not being in the set of Gödel numbers of sentences in T. In order to determine that G's Gödel number is not a member of that set, an understanding of arithmetic *prior* to T's axiomatization is required. How else could G's truth be settled but by invoking the previously accessible arithmetical concepts used to express G and the set of Gödel numbers of the sentences in question? This means that the recognition of the truth of Gödel sentences is indeed needed: otherwise, the incompleteness—which calls for the unprovability of *true*, rather than *presumed* true, sentences—would not go through. Recognition of G's truth is crucial for that.

To illustrate this point, consider the key reasoning invoked to establish that the set of Gödel numbers of theorems of T—a consistent extension of minimal arithmetic—is not definable in T ([1, pp. 222-223]). If the Gödel sentence G is not a theorem of T, its Gödel number is not in the set of Gödel numbers of sentences in T. Since G is (provably) equivalent to its Gödel numeral *not* being in the set of Gödel numbers of sentences in T, it follows that G is a theorem of T. However, if this is indeed the case, G's Gödel number *is* in the set of Gödel numbers of sentences in T. Hence, G is not a theorem of T. It follows that T is inconsistent, contrary to the assumption that was made. As this reasoning makes transparent, the determination that G's Gödel number is (or is not) in the set of Gödel numbers of sentences in T is crucial for the argument. Decisive for this determination is an intuitive understanding of arithmetic, in terms of which the truth of the Gödel sentence can be recognized.

There is also an inevitability to Gödel's result. As is well known, even adding the Gödel sentence as a new axiom to T will not result in a complete theory, since a new Gödel sentence is then obtained. Central to this argument are, once again, an intuitive understanding of arithmetic and the fact that the Gödel sentence is *recognized* to be true. As Michael Dummett points out:

> In attempting to characterise the totality of natural numbers—e.g. for the purpose of pointing out that any model of our formal system in which U [a particular Gödel sentence for the system] is false will contain elements not in this totality—we should use some expression such as 'set' or 'finitely often'. If we then tried to give an account of the meaning of this expression by means of a stipulation of the assertions we wish to make involving it, this account could in turn be embodied in a formal system within which our definition of 'natural number' could be given. In view of the fact that Gödel's theorem applies to any system which contains arithmetic, there would again be an arithmetical statement expressible but not provable in the system, which we could recognize to be true: we should thus not have succeeded by this means in giving a complete characterization of the concept 'natural number' [11, pp. 186-187].

On this account of Gödel's theorem, our understanding of arithmetical concepts is not fully characterized by the axioms of arithmetic, since any attempt to codify this understanding in terms of explicit axioms will satisfy the conditions for Gödel's theorem, thus generating true but unprovable statements. In turn, the implementation of Gödel's incompleteness result itself requires, as noted, the recognition of the truth of the Gödel sentence, which can only be done by invoking previously understood arithmetical concepts.

The incompleteness result, thus, as is well known, indicates a significant limitation in any purely formal account of arithmetic—and, by extension, any mathematical theory that includes arithmetic—given that the formal account will always leave part of the formal content unreachable by inferential (deductive) means. Interestingly, the proof of the incompleteness result also crucially relies on an intuitive understanding of arithmetical concepts. This indicates the centrality of the intuitive conception for both the incompleteness of arithmetic and the implementation of the incompleteness result. We need to grasp arithmetical concepts intuitively in order to understand and eventually establish that any axiomatization of these concepts, satisfying the usual conditions, is ultimately incomplete.

Does the incompleteness identified in Gödel's theorem extend to other areas that include arithmetic but go beyond what is explicitly considered in the theorem's proof? Is our understanding of the concepts in question, in these cases, not fully characterized by the axioms under consideration? The answer seems to be affirmative. Understanding a concept and formulating it in a particular language are related but distinct enterprises. In the case of arithmetical concepts, their understanding is prior to the axiomatization of the theories that invoke them. The process of axiomatization may change, refine or distort the concepts in question, but the understanding of such concepts need to be already in place *prior to* the axiomatization. Understanding arithmetical concepts is needed both to suggest suitable axioms and, as noted above, to secure the adequacy of Gödel's result.

It could be argued that Gödel sentences are not only very artificial, but they also have no genuine mathematical content. So, perhaps, the incompleteness that emerges from Gödel's theorem will not affect the content of mathematics. This is not so. Gödel sentences with perfectly significant mathematical content have been found. Consider, for instance, the following well-known cases: (a) Suppose that S is a Borel set (a set in a topological space that can be obtained from open sets through the operations of countable union, countable intersection, and relative complement). In this case, there exists a Borel function f such that the graph of f is either included in or disjoint from S. Harvey Friedman established that this theorem cannot be proved in full second-order arithmetic. Its proof requires full ZFC ([18, p. 23] and [16]). (b) Here is another example: the statement to the effect that all projective sets are Lebesgue measurable cannot be proven in full ZFC ([20] and [16]). Its proof requires the existence of extremely large cardinals: Woodin cardinals ([22], [13], [14] and [16]). In both cases, true mathematical statements—with significant mathematical content—cannot be proven in certain theories.

Newton da Costa, Francisco Doria and their collaborators found significant extensions of the Gödel phenomena in a number of additional areas. (i) In classical mechanics, they proved that it is impossible to decide whether a given Hamiltonian can be integrated by quadratures, which leads to a formulation of Gödel's incompleteness theorem for Hamiltonian mechanics ([5] and [6]). (ii) In dynamic systems and chaos theory, they established that there is no algorithmic procedure to test whether a given dynamical system is chaotic ([8]), and given a dynamical system that has an equilibrium point at the origin, one cannot algorithmically decide whether that equilibrium is stable or not ([7]). (iii) In economics, they showed that there are finite games whose equilibria cannot be proven to be computable ([21]).

As is well known, inferential incompleteness crucially depends on the underlying logic. (i) If the logic is infinitary and includes an ω-rule, Gödel's theorems do not go through ([19]). (ii) If the logic is paraconsistent, and the consistency of the theories in question is no longer assumed, Gödel's theorems also do not go through ([15, pp. 236-237]). Inferential incompleteness crucially relies on the constraints provided by the language that is used to express and formulate the results in question. This should not surprise us: what follows or not from certain statements depends on what can or cannot be stated in the language in use. First-order logic is complete given its limited expressive resources. If quantification over properties, rather than only over objects, is allowed for, as is the case with second-order languages, true statements can be formulated in the higher-order language that cannot be proved. As a result, second-order logic, richer in expressive resources than its first-order counterpart, is incomplete. Hence, the now familiar trade-off between inconsistency and incompleteness emerges. What is more valuable: a consistent but incomplete theory, or an inconsistent but complete one? To answer this question, broader commitments to the underlying logic need to be assessed. Those committed to classical logic may favor the former; those who adopt non-classical logics have additional avenues for research by exploring the domain of the inconsistent in search of better, consistent alternatives ([9]).

However these issues end up being resolved, the sheer scope of inferential incompleteness is impressive. As Francisco Doria rightly emphasized:

> Gödel incompleteness affects *every interesting predicate we can think about*. Think for instance that given a complicated expression which represents a real number, there is no algorithm to check whether it is algebraic or transcendental, and there will be some expressions (in ZFC) for real numbers whose transcendentality is formally undecidable! [10, p. 179].

In light of these considerations, it is inevitable that inferential incompleteness is widespread and affects large portions of the scientific enterprise.

3 Informational and Representational Incompleteness

Inferential incompleteness emerges primarily from what cannot be established inferentially (deductively) from certain theories. In contrast, informational incompleteness need not be concerned with inferential capabilities, but involves partiality of data or information about particular domains, whether scientific or not. In fact, it is often indeterminate whether certain relations among objects under investigation hold or not. The relations are, therefore, *partial* ([9], [2] and [3]). Given a domain D of objects and a family of relations R_i among these objects, a relation is partial provided that it has three components: (i) some objects in D are known to be in the relation (R_1); (ii) some objects in D are known *not* to be in the relation (R_2), and (iii) some objects in D are not known whether they are, or are not, in the relation (R_3). A structure S is partial provided that the relations among the objects in S are partial. A partial structure expresses and accommodates the partiality of information in the relevant domain.

Such partiality of information provides a significant source of informational incompleteness. It is a ubiquitous feature of scientific practice that available information is hardly ever complete: from gaps in the fossil record to unknowns in the molecular organic composition of airborne particles and their dependence on "formation, growth and aging processes that occur in air" ([12]), there is much to be completed in current understanding of empirical (and mathematical) phenomena. Informational incompleteness, similarly to its inferential counterpart, is widespread.

Consider, for instance, the challenges involved in modeling and predicting weather and climate extremes ([17]). As noted by Jana Sillmann *et al.*:

> The development of an extreme event depends on a favorable initial state, the presence of large-scale drivers, and positive feedbacks, as well as stochastic processes (noise). The relative importance of these factors varies for different types of extremes. For example, feedbacks for short lived events [...] like convective storms are typically associated with unstable atmospheric dynamics, whereas longer duration events [...] like heatwaves or droughts typically involve soil moisture-atmosphere interaction. External factors like global warming can influence extremes through these various factors. For example, the increased water vapor in a warmer atmosphere can enhance convective feedbacks, or increased surface evaporation might amplify heat waves and droughts [17, p. 66].

Significant informational incompleteness is found among all of these factors— an incompleteness expressed in the climate models that are developed to accommodate the phenomena under study. The problem, Sillmann *et al.* continues, is that:

> climate models can have large biases in some regions and may not be able to simulate key dynamical patterns such as atmospheric blocking or other weather regimes, jet stream position and intensity, tropical dynamics and teleconnections, or stratosphere-troposphere connections. A key challenge is to evaluate and improve models by targeting key processes that are relevant for a realistic, or at least a sufficient, representation of extremes [17, p. 66].

The limitations inherent in these models, given what can be accounted for by the available representational devices, are also salient. As a result, informational incompleteness in such models involves both the limited availability of relevant bits of information about a given domain, but also limitations in the modes of representation regarding that domain, which can be accommodated in terms of suitable partial structures, partial relations and partial mappings among such structures ([4]). Climate models' inability to simulate certain dynamical patterns, such as atmospheric blocking, clearly illustrates these representational limitations. This offers a glimpse of the way in which representational incompleteness contributes to informational incompleteness, given the importance that representing the relevant phenomena, *via* suitable models, has to obtaining and conveying information about the domain under investigation.

4 The Status of Incompleteness

Certain phenomena cannot but be incomplete. (i) It is an inherent feature of fiction that certain traits of fictional objects and fictional characters are left unspecified. One cannot specify every single aspect of a fictional object, and simply to stipulate that such object is complete without determining which features it has, or lacks, and what these features ultimately are, does not guarantee the completeness of the object. No stipulation of the completeness of Sherlock Holmes will decide every feature of the detective, including the color of the socks Holmes was wearing at any moment, or whether he was wearing socks at all. (This is a form of informational/representational incompleteness.)

(ii) The lack of denotation of certain expressions in the sciences provides an additional source of (informational) incompleteness. The attempt by Newtonian physicists to account for the disparity of the perihelion of Mercury by trying to find an unknown planet—Vulcan—that would be disrupting Mercury's orbit failed, leading to the conclusion that no such planet existed. An (informational) incompleteness inherent in Newtonian physics resulted.

(iii) Scientific idealizations provide an additional source of (representational) incompleteness. When certain parameters are adjusted to describe more easily a given object—such as the representation of Earth as a perfect sphere, when, in fact, it is more oblong—one ends up accurately describing not the target in question, but an abstract counterpart to it. As a result, relevant parameters regarding the target are left unspecified: the accurate representation of Earth's shape is not part of the idealized account, and is left incomplete.

(iv) Some expressions are vague and referentially indeterminate, and these are forms of semantic incompleteness. It is often indeterminate what the referents of certain terms are. Consider relativistic and Newtonian mass: one depends on the speed, the other does not. Which of them are we referring to when we refer to mass? Moreover, some terms are vague. Consider, for instance, the color spectrum. The lack of sharp boundaries between adjacent color hues indicates the incompleteness of color determination. It is unclear, and highlight controversial, whether it is possible to overcome this form of incompleteness in a principled way.

These considerations suggest three different ways of characterizing the status of incompleteness: (a) It can be an *epistemic* state, depending on the information and representational devices available to us in particular domains. This suggests that, with different and better developed devices, such incompleteness could be overcome, although a price will always be involved, as illustrated by the attempts, mentioned above, to undermine Gödel's incompleteness theorems by changing the underlying logic. (b) But perhaps incompleteness is an *ontic* trait, a feature of the world, rather than of our representations of it. On this conception, the incompleteness is unavoidable and ineliminable—unless the world itself changes. However, it is less clear, disregarding the cases of fiction and vagueness, in which way the world itself can be incomplete. Additional articulation of this conception is called for before it can be properly assessed. (c) Finally, taken as an heuristic device, incompleteness highlights where additional information and better inferential and representational mechanisms are needed. Understood in this way, incompleteness indicates and provides a rich avenue for potentially new discoveries. As a result, as illustrated by Gödel's theorems, incompleteness and novel pursuits often go hand in hand.

5 Conclusion

As opposed to inaccuracy, empirical inadequacy, and irrelevance, which typically should be corrected, incompleteness is a pervasive trait of scientific activity (and ordinary life). As noted above, it can be accommodate via suitable partial structures that encode partial information about the relevant domains, and it often suggests new avenues for research. Some of the incompleteness may decrease as new information is uncovered and better representational devices are developed. But given the heuristic power that incompleteness typically has, it is not something that can, and perhaps even should, be fully avoided.

6 Dedication

I am delighted to dedicate this paper to Professor Francisco Doria's 75th birthday. His unflagging search for understanding the multiple aspects and widespread reach of incompleteness has been an endless source of inspiration, and his tireless curiosity about every aspect of the world has been a model to everyone who has had the pleasure of getting to know him. His are giant shoulders; his generosity and insight, even larger.

My thanks also go to Professor Newton da Costa, for illuminating discussions about incompleteness over the decades that I have known him. Most of the ideas discussed above I learned from him.

References

[1] Boolos, G., Burgess, J. and Jeffrey, R. (2002). *Computability and Logic.* (Fourth edition.) Cambridge: Cambridge University Press.

[2] Bueno, O. (1997). "Empirical Adequacy: A Partial Structures Approach", *Studies in History and Philosophy of Science* 28: 585–610.

[3] Bueno, O. (2000). "Empiricism, Mathematical Change and Scientific Change", *Studies in History and Philosophy of Science* 31: 269–296.

[4] Bueno, O. and French, S. (2018). *Applying Mathematics: Immersion, Inference, Interpretation.* Oxford: Oxford University Press.

[5] da Costa, N.C.A. and Doria F.A. (1991a). "Undecidability and Incompleteness in Classical Mechanics", *International Journal of Theoretical Physics*, 30: 1041–1073.

[6] da Costa, N.C.A. and Doria F.A. (1991b). "Classical Physics and Penrose's Thesis", *Foundations of Physics Letters*, 4: 363–373.

[7] da Costa, N.C.A. and Doria F.A. (2008). *On the Foundations of Science.* Rio de Janeiro: E-papers.

[8] da Costa, N.C.A., Doria F.A., and Furtado do Amaral, A.F. (1993). "A Dynamical System where Proving Chaos is Equivalent to Proving Fermat's Conjecture", *International Journal of Theoretical Physics*, 32: 2187–2206.

[9] da Costa, N.C.A. and French, S. (2003). *Science and Partial Truth: A Unitary Approach to Models and Scientific Reasoning.* New York: Oxford University Press.

[10] Doria, F.A. (1996). "Some New Incompleteness Theorems and their Import to the Foundations of Mathematics", *Logique et Analyse*, 153-154: 165–182.

[11] Dummett, M. (1978). *Truth and Other Enigmas*. London: Duckworth.

[12] Finlayson-Pitts, B., Wingen, L., Perraud, V. and Ezell, M. (2020). "Open Questions on the Chemical Composition of Airborne Particles", *Communications Chemistry*, 3: 1–5.

[13] Martin, D. and Steel, J. (1988). "Projective Determinacy", *Proceedings of the National Academy of Sciences*, 85: 6582–6586.

[14] Martin, D. and Steel, J. (1989). "A Proof of Projective Determinacy", *Journal of the American Mathematical Society*, 2: 71–125.

[15] Priest, G. (2006). *In Contradiction*. (Expanded edition.) Oxford: Oxford University Press.

[16] Raatikainen, P. (2020). "Gödel's Incompleteness Theorems", Edward N. Zalta (ed.), *The Stanford Encyclopedia of Philosophy* (Winter 2020 Edition), URL = <https://plato.stanford.edu/archives/win2020/entries/goedel-incompleteness/>.

[17] Sillmann, J., Thorarinsdottir, T., Keenlyside, N., Schaller, N., Alexander, L., Hegerl, G., Seneviratne, S., Vautard, R., Zhang, X. and Zwiers, F. (2017). "Understanding, Modeling and Predicting Weather and Climate Extremes: Challenges and Opportunities", *Weather and Climate Extremes*, 18: 65–74.

[18] Simpson, S. (1999). *Subsystems of Second Order Arithmetic*. Berlin: Springer.

[19] Smorynski, C. (1977). "The Incompleteness Theorems", in Jon Barwise (ed.), *Handbook of Mathematical Logic*. Amsterdam: North-Holland, pp. 821–865.

[20] Solovay, R. M. (1970). "A Model of Set Theory in which Every Set of Reals is Lebesgue Measurable", *Annals of Mathematics* 92: 1–56.

[21] Tsuji, M., da Costa, N.C.A. and Doria, F.A. (1998). "The Incompleteness of Theories of Games", *Journal of Philosophical Logic*, 27: 553–568.

[22] Woodin, H. (1988). "Supercompact Cardinals, Sets of Reals, and Weakly Homogeneous Trees", *Proceedings of the National Academy of Sciences*, 85: 6587–6591.

Consciousness and Information, classical, quantum or algorithmic?*

Gregory Chaitin
University of Buenos Aires
https://uba.academia.edu/GregoryChaitin

Abstract

In Chapter 8 of his book *The Conscious Mind*, David Chalmers speculates on using information theory as the basis for a fundamental theory of consciousness. Building on his work, we attempt to flesh out an updated version of the Chalmers proposal by taking into account more recent developments including algorithmic information theory, quantum information theory, the holographic principle and the Bekenstein bound, and digital philosophy as sketched in two little-known monographs in Italian: *Introduzione alla filosofia digitale* and *Bit Bang: La nascita della filosofia digitale*.

Dedicated to Francisco Doria on his 75th birthday, a great friend, with admiration for his pathbreaking work with Newton da Costa on undecidability in physics.

---
*Paper based on talk at https://www.youtube.com/watch?v=RlotKFaHECU

1 Introduction — The Chalmers Proposal

We shall be concerned with the extremely stimulating book *The Conscious Mind: In Search of a Fundamental Theory* by David Chalmers [1]. In particular we shall consider Chapter 8, "Consciousness and Information: Some Speculation," and Chapter 10, the final chapter, "The Interpretation of Quantum Mechanics."

Work on information theory done in the quarter-century since Oxford University Press published Chalmers' book, suggests a significantly different version of the original Chalmers proposal.

What is this proposal? In a nutshell, Chalmers advocates a version of panpsychism identifying information with consciousness, according to which any physical system that is able to represent and process N bits of information contains N bits of consciousness. He also supports Wheeler's "It from bit" doctrine [1, p. 302] according to which information is the ontological substratum of the world. Furthermore Chalmers links quantum mechanics, which marks a shift from a materialist to an idealist fundamental theory of the physical world, with consciousness.

In the twenty-five years since the Chalmers book, consciousness has fortunately become a respectable subject. The author has only a tangential awareness of most of this recent work, including, for example, de Barros, Montemayor [2].

The goal of this paper is instead to suggest a few ideas on consciousness connected with digital philosophy and digital physics—which propose building the world out of information and computation rather than matter and energy—and to point out the synergy between these two research programs.

We should admit from the start, as does Chalmers, that these are but speculations intended to stimulate thinking on these extremely difficult topics.

2 Multiple Consciousness in Humans

Let us spell out a consequence of Chalmers' panpsychic stance. First of all, combining digital philosophy [3–5] with Chalmers, if the universe is built out of information, and if all information is conscious, then the entire universe would be conscious, and might perhaps be identified with the mind of God, an immanent not a transcendent God. This also connects with Leibniz's *Monadology* [6], an idealist ontology according to which the universe is composed of monads, individual minds with different degrees of perception, up to a maximal monad, God, with perfect perception.

This is analogous to Chalmers' persuasive argument for panpsychism, an argument by continuity. Clearly humans and dogs are conscious, and proceeding downwards, so must cells and viruses be, continuing on down to non-living physical systems like thermostats and light switches. As we proceed down this complexity scale, it is most implausible that consciousness suddenly cuts off, instead of merely diminishing in degree. And, we would argue, Chalmers' continuity argument should also be pursued in the opposite direction, attributing consciousness to families, companies, and nations—a collective consciousness distinct from an individual's consciousness—as well as to computers, computer networks and indeed the entire World Wide Web.

Furthermore, according to the Chalmers thesis, in an individual person, not only the top-level thinking being is conscious, but so is the immune system, each cell, and a person's "unconscious" mind, which is actually more powerful computationally than the conscious mind. In fact, we would argue [7] that the conscious mind operates at the cellular or neuronal level, and is serial, while the unconscious mind operates at the molecular level—as in DNA computers—and is highly parallel.

Some suggestive references: See Hadamard [8] on the psychology of creation in the mathematical field and Nørretranders' *Maerk Verden* (original title in Danish) [9] on how football players react faster than their conscious minds can. These two works suggest that the unconscious processes more bits of information than the conscious mind does. Consciousness is a narrow funnel. On the other hand, see Quiroga [10] on how people with photographic memory have immense storage capacity, e.g., von Neumann, who could reputedly recite backwards an arbitrary page of any book he had ever read.

Some unusual first-person experiences with consciousness: The author once fell asleep while driving at high speed one night on an empty highway, then woke up and realized he had briefly fallen asleep. Who was driving while the author was asleep?! This entity must also be conscious if Chalmers panpsychism holds. And sometimes one is momentarily distracted while reading and then returns to the text only to find that someone else had continued reading.

3 Which Information Theory To Use?

Chalmers follows the Shannon version of information theory [11] which contemplates discrete information measured in bits, and also a more subtle continuous information which, except for the presence of noise, would theoretically contain, as do infinite precision real numbers, an infinite amount of information. We shall discuss, and dismiss, the latter kind of information in the next section.

In a nutshell, we propose that instead of Shannon's version of information, in thinking about identifying consciousness with information it is better to employ algorithmic information theory (AIT) [12]. For processing information in physical systems is algorithmic and corresponds to the rate of thought. E.g., computers—if they are conscious—think very quickly, faster than the human brain.

Also DNA is a discrete algorithmic programming language of sorts, again closer to the way of thinking in algorithmic information theory than to the statistical way of thinking about information in Shannon information theory.

Possibly however, physical systems process qubits, not classical bits. Quantum information theory (QIT) teaches us to measure qubits of information in a physical system and to regard the time-evolution of every physical system as algorithmic information processing, i.e., as quantum computation. Hence the title of Nielsen and Chuang's standard treatise *Quantum Computation and Quantum Information* [13].

Please see the Appendix for more information on the relationship between classical bits and quantum qubits.

4 Dismissing Continuous Information

Chalmers remarks that due to noise, continuous physical systems can only represent a finite amount of information.

This is most fortunate, because algorithmic information theory cannot deal with real numbers, which contain an infinite amount of information. How then could we measure the amount of information and therefore the amount of consciousness?!

Going beyond the Chalmers argument, the holographic principle and the Bekenstein bound [14] imply that any physical system can contain only a finite number of (classical, not quantum) bits of information, which moreover grows only as the surface area of the system, not as the volume of the system—implying that in some sense the physical world is actually two-dimensional, not three-dimensional.

Roughly speaking the reasoning behind the Bekenstein bound, which comes from attempts to apply quantum mechanics to black holes and more generally to elaborate a quantum-mechanical version of general relativity, is like this: According to quantum mechanics, to represent an infinite amount of information, you have to go to extremely high energies (since particles and waves are identified and precise localization corresponds to extremely small wave-lengths and therefore extremely high energies).

But well-localized extremely high energy corresponds, following $E = mc^2$ to extremely high mass, which would provoke collapse to a black hole. For more on this, please see Smolin [14].

Furthermore, even in classical physics, no physical quantity has ever been measured to twenty digits (approximately sixty bits) of accuracy, as Rolf Landauer would always point out to the author [15].

5 Consciousness and Quantum Mechanics

Quantum information theory's successes in the past years (see Vedral [16], Neilsen and Chuang [13]) strengthens Chalmers' intuition linking consciousness with information theory and quantum mechanics (QM). Why? Because according to the current view, QM is about a new kind of information, so information and the quantum are in a sense the same.

QIT is not new physics, but it is a radically different way of perceiving QM. Before QM courses began with the Schrödinger equation, now many begin with the qubit.

6 Current State of Digital Philosophy

For the sake of completeness, we should remark that in the quarter-century since Chalmers published his panpsychism thesis [1, Chapter 8] which refers to "It from bit," this doctrine has been an active research area sometimes referred to as digital physics or digital philosophy [3–5].

Some of the workers in this area were indeed inspired by John Wheeler and his remarks on "It from bit." But others, such as Stephen Wolfram and Edward Fredkin, who are not cited by Chalmers, have had other sources of inspiration. Please see their respective websites at https://www.wolframphysics.org and https:/www.digitalphilosophy.org, and Wolfram's new book [17].

Digital philosophy and digital physics remain active, but still highly speculative and inconclusive. The most spectacular recent work in this area is gravity considered as an entropic force with space-time emerging from qubit entanglement, associated, among others, with the name of Erik Verlinde. See especially his crucial 2016 paper at https://arxiv.org/abs/1611.02269. This work has been well covered online in *Quanta Magazine* at https://www.quantamagazine.org/.

One of the interesting features of the Verlinde proposal is that it provides an alternative to the mysterious dark matter.

On the other hand, Randell Mills suggests that dark matter consists of hydrinos, a form of atomic hydrogen below the standard ground state (i.e., with the electron closer to the proton than is usually deemed possible). He provides experimental evidence [18] that the high energies in the solar corona—that is much hotter than the photosphere at the sun's surface—are due to hydrino formation in the corona, and he plans to use conversion of atmospheric hydrogen into dark matter to generate energy here on planet earth! So there is no shortage of fascinating new ideas, none at all.

7 Concluding Discussion — Tononi's Φ

In the quarter-century since Chalmers published his panpsychism thesis that any physical system that processes information is conscious, it has become possible to put more meat on the bones of the Chalmers proposal.

Unfortunately, it is possible to argue that as a consequence, the simplicity which was the principal argument in Chalmers' favor has become somewhat diluted. Nevertheless, it remains straightforward, at least in comparison with what is perhaps becoming its chief contender, the increasingly elaborate Tononi integrated information theory of consciousness [19, 20].

Note especially how much easier it is to measure consciousness according to Chalmers as opposed to the extreme computational difficulties of the Tononi Φ proposal, which requires considering all possible partitions of a physical system.

In conclusion, we believe that Chalmers panpsychism remains vigorous and stimulating, especially if taken in the context of digital philosophy, which attempts to build the world out of information and computation rather than matter and energy.

Appendix. What Are Qubits?

Should the Chalmers proposal be couched in terms of qubits rather than bits? I.e., should consciousness be measured in classical bits or quantum qubits?

What precisely are qubits? They are quantum mechanical superpositions of classical 0/1 bits. Consider a traditional N-bit string. According to quantum mechanics, to describe this N-bit string one has to associate a probability amplitude with each of the 2^N possibilities, a complex number $x + iy$ which has a phase as well as a magnitude. The probability associated with a probability amplitude $x + iy$ is given by $x^2 + y^2$.

A classical bit string is a special kind of qubit string in which the 2^N possibilities each have probability amplitude $\sqrt{(1/2^N)}$ and therefore probability $1/2^N$.

Note that probability amplitudes are vectors, they are probabilities which have a direction, which is why they can cancel instead of always adding as they do in classical probability theory. In other words, there can be destructive as well as constructive interference.

By the way, so-called entangled bits can occur, for instance, when 01 and 10 are possible, but not 00 and 11.

Cautionary note: If Randell Mills is correct and hydrinos actually exist—and he has amassed considerable experimental evidence in their favor—then perhaps we have to take seriously the theory that led Mills to the hydrino. This theory is nothing less than a proposal to replace quantum physics by a new, classical approach to microphysics. Recall that quantum mechanics was invented to explain the stability of the Bohr model of the hydrogen atom, a miniature solar system with an electron orbiting a proton. However, Mills believes that this was actually unnecessary. He has a stable classical model of the hydrogen atom involving something new he calls electron orbispheres. This could conceivably pull the rug out from under quantum mechanics.

Going from AIT to QIT: With classical bits as studied in AIT, computations are done via Turing machines, whereas in QIT we deal with quantum circuits, which are quantum mechanical versions of traditional boolean circuits involving and, or and not.

Bibliography

1. David Chalmers, *The Conscious Mind: In Search of a Fundamental Theory*, Oxford University Press, 1996.

2. J. Acacio de Barros, Carlos Montemayor, *Quanta and Mind: Essays on the Connection between Quantum Mechanics and Consciousness*, Synthese Library, vol. 414, Springer, 2019.

3. James Gleick, *The Information: A History, A Theory, A Flood*, Pantheon Books, New York, 2011.

4. Ugo Pagallo, *Introduzione alla filosofia digitale. Da Leibniz a Chaitin*, Giappichelli, 2005.

5. Giuseppe Longo, Andrea Vaccaro, *Bit Bang. La nascita della filosofia digitale*, Maggioli, 2013.

6. Lloyd Strickland, *Leibniz's Monadology: A New Translation and Guide*, Edinburgh University Press, 2014.

7. Gregory Chaitin, "Conceptual complexity and algorithmic information," *La Nuova Critica* 61–62 (2015), pp. 9–28. Reprinted in Shyam Wuppuluri, Francisco Doria, *Unravelling Complexity: The Life and Work of Gregory Chaitin*, World Scientific, 2020.

8. Jacques Hadamard, *The Psychology of Invention in the Mathematical Field*, Princeton University Press, 1945.

9. Tor Nørretranders, *The User Illusion: Cutting Consciousness Down to Size*, Penguin, 1999.

10. Rodrigo Quiroga, *Borges and Memory: Encounters with the Human Brain*, MIT Press, 2012.

11. Claude Shannon, Warren Weaver, *The Mathematical Theory of Communication*, University of Illinois Press, 1949.

12. Gregory Chaitin, *Algorithmic Information Theory*, Cambridge University Press, 1987.

13. Michael Nielsen, Isaac Chuang, *Quantum Computation and Quantum Information*, Cambridge University Press, 2000.

14. Lee Smolin, *Three Roads to Quantum Gravity*, Weidenfeld & Nicolson, 2000.

15. Rolf Landauer, "Information is physical," *Physics Today*, May 1991, pp. 23–29.

16. Vlatko Vedral, *Decoding Reality: The Universe as Quantum Information*, Oxford University Press, 2012.

17. Stephen Wolfram, *A Project to Find the Fundamental Theory of Physics*, Wolfram Media, 2020.

18. Brett Holverstott, *Randell Mills and the Search for Hydrino Energy*, KRP History, 2016.

19. Giulio Tononi, *Phi: A Voyage from the Brain to the Soul*, Pantheon, 2012.

20. Christof Koch, *Consciousness: Confessions of a Romantic Reductionist*, MIT Press, 2012.

Information and the Hard Problem of Consciousness

Carlos Montemayor[†] and J. Acacio de Barros[⋆]

[†]Department of Philosophy, San Francisco State University, San Francisco, CA, email: cmontema@sfsu.edu
[⋆]School of Humanities and Liberal Studies, San Francisco State University, San Francisco, CA, email: barros@sfsu.edu

Abstract

Information plays a vital role in our conscious experience. After all, consciousness is about how we experience the world, how we feel. Thus, it comes as no surprise that information theory is at the foreground of any attempts to define consciousness scientifically. In this paper, we argue that if we think of (phenomenal) consciousness as different from attention, information theory as we know it can be used for the latter but not for the former.

1 Introduction

Consciousness has an information problem. On the one hand, we seem to receive copious and uniquely valuable information about the world and ourselves through our conscious experiences. Many philosophers, perhaps a substantial majority, think that conscious awareness is the source of knowledge and moral reasoning. Conscious perception provides first-hand evidence about properties and objects in the immediate environment that seem fundamental for grounding epistemic justification. With respect to information capacity, consciousness seems, in fact, to be so rich that some authors think that it overflows cognitive access to information [3, 4, 5], thereby implying that consciousness has more information than any other cognitive process, including working memory and action-control.

On the other hand, there is the hard problem of consciousness [7, 8]. Succinctly stated, the problem that phenomenal consciousness (or what it is like for you to experience the intense and fragrant smell of a flower [24]) cannot be reduced to any functional or structural explanation [7]. This impossibility of reduction is quite a strong and striking claim considering scientific method's enormous success. This "impossibility" claim follows by entailment from arguments concerning cases, guided by intuition, which allegedly demonstrate that no descriptive, third-personal, scientific account could even come close to closing the "explanatory gap" between conscious awareness and physical reality (for opposing views, see [10, 14]).

For the science of consciousness, this conclusion is the equivalent of a "proof" of the incommensurability of conscious information. This is because science can only produce explanations about the structure and function of psychological or neural states. Therefore, there cannot be a science of phenomenal consciousness and its intrinsic subjectivity. In other words, we cannot have a scientific theory of what we consider to be the most informative cognitive state.

This situation is, to be sure, paradoxical. We have a relatively clear understanding of different kinds of information (e.g., semantic, algorithmic, syntactic) and a rigorous mathematical theory of information capacity and complexity. We also have a good understanding of how to define the main characteristics of phenomenal consciousness and various ways of demonstrating why phenomenal consciousness is very informative. Nevertheless, providing a clear characterization of how consciousness is informative seems extremely difficult and, according to the arguments just mentioned, perhaps impossible, particularly because of the hard problem. But what would it mean for consciousness to be uninformative? What would consciousness be without information?

Our purpose in this paper is twofold. We believe that there are problems concerning the definition of cognitive states that are informative. We identify these ambiguities by distinguishing different levels of information processing relevant to human psychology. We also offer a classification of information in terms of different types of consciousness and attention because we believe this more exact categorization may help move the debate on phenomenal information forward.

2 Layers of information processing

Cognitive information comes in many varieties. Some seem to be non-linguistic, map-like, or analog [18]; some seem to be formatted in both "digital" and "analog" formats, such as time cognition [21]. Empathic forms of communication also seem to concern non-linguistic cognition and can be found in many species [9]. Linguistic communication and attention to language might be a uniquely human feature because of syntax processing [22], but a substantial portion of our linguistic skills are shared with other species [13].

An obvious question is whether phenomenal consciousness is in a specific format or whether this issue is even relevant. Is analog formatting necessary? Or is language and more digital/symbolic formatting necessary? Because of the difficulties described above, there is no conclusive way of answering these questions. A more fruitful approach focuses on the possible relation and dynamics between consciousness and attention [23], independently of the issue of information encoding or formatting—and related difficulties concerning the content and the vehicle or medium of information.

It seems evident that many information-processing mechanisms that evolved to reproduce and sustain life are not cognitive. However, even here, there is debate. We know that most living organisms process information selectively and in ways that seem representational or at least intelligent. For example, organisms that navigate complex environments, seek food, and avoid predators must process relevant information to perform the behaviors necessary for survival. Attention evolved to perform these tasks at various levels of sophistication. But plants and bacteria select information in complex and intelligent ways. Are they "paying attention?" Could they even be phenomenally conscious? How to decide this pressing issue without a theory of conscious information?

There is a fundamental distinction between the mechanical (or strictly causal) transmission of information and the meaningful usage of information, which is deeply related to the issue of agency as a hallmark of cognitive processing. We need to separate primitive cognition from machine output without being too anthropocentric. Even bacteria can be drawn towards light and consume other organisms. Unicellular organisms are consumers of information regarding their immediate environment, but it seems very difficult to attribute any kind of mentality or cognition to them. After all, our own cells consume information in a similar fashion, with the rest of our bodies as "environment." Still, bacterial "cognition" could be a kind of agency without full cognition because bacteria are goal-oriented, adapt to situations, and learn from their environment [15]. Bacteria appear to be neither full-blown cognizers nor mere machines.

Different varieties of cognitive agency, from the very minimal to the very complex and highly integrated, can be understood in terms of various types of attention (see [17] for a classification of different kinds of attention; see [11] for a characterization of attention as epistemic agency). Where does consciousness fit in? Attention is clearly informative. In fact, attention is the clearest example of a cognitive capacity that can be computationally and informationally described. Some kinds of attention are best understood in terms of "just noticeable differences" and according to the classic signal-to-noise ratio. Other kinds of attention, such as object and feature-based, are more semantically structured, inferentially specified, and algorithmic [20]. Finally, self-monitoring and meta-monitoring kinds of attention can also be understood in terms of meta-hypotheses or meta-functions that specify information classification and salience at early stages [22]. In fact, attention is defined by its functions and its informational structure. It is highly informative in ways we can model and understand. But by definition, consciousness cannot be understood in any of these ways per the hard problem.

Bacteria seem to be consumers rather than mere transmitters of information, and thus they seem to have a particular type of agency. Still, it is difficult to justify the claim that they are attending to the environment. Attention seems to be more semantically structured, intentional (or directed towards contents), and algorithmic or inferentially structured. Perhaps this kind of information consumption and minimal agency is more prevalent in nature than we assume, and maybe it involves a primitive kind of mentality that is necessary for attention: a primary signal-to-noise information gathering that helps structure basic behavior.

What is the status of information processing in plants? Figdor's "literalism" holds that scientific statements like "plants make decisions" or "bacteria converse" should be interpreted literally [12]. Across human and non-human contexts of evaluation, we mean the same things with these statements or predicates. The central claim here is not that plants in fact think and cognize. Instead, the claim is that they could because the sentences used to describe their cognition or communication capacities should be considered as being either true or false. The emergent field of "plant cognition" confirms this hypothesis [27], and the verifiable statements about plants could, in principle, include phenomenally conscious states (e.g., a plant being in pain or having a certain motivation).

The prospect of developing a theory of plant cognition presents intriguing possibilities. From a scientific perspective, these possibilities must be studied by analyzing information processing, how it is done or measured in plants, and what kind of information processing it involves (is it really cognitive?). To assess whether the statements about plant cognition are true or false, we cannot reason simply by comparison, imagination, or analogy. It is one thing to attribute by analogy or by imaginative projection conscious states to plants or bacteria. It is an entirely different task to understand what their being conscious could mean in terms of information processing.

Attention skills are perfectly suited for this kind of cross-species comparison precisely because they are defined in terms of their informational structure and functional role. Research on different cognitive capacities across species relies on this functional role for fruitful comparisons. The navigational and largely statistically structured capacities of animals that could potentially be found in plants are very well-documented and include magnitudes such as time, rate, space, and number [16, 21]). Attention in these cases is distributed and allocated in terms of likelihoods and predictions about speed and trajectory. But there is a debate about whether objects are the basis for spatial attention's allocation and statistical properties [26, 28]. If objects are at the basis of the structure of perceptual space (perhaps differing in structures and geometry from one modality to the other), then attention would involve not only statistical predictive signal-to-ratio information, but also built-in assumptions about which aspects of space constitute perceptual objects. This can all be measured in terms of function and structure.

One way to approach this difference between statistical information and more precise categorical information about objects is their format. But even if one decides to approach this issue in terms of analog and digital formatting, that is still very much a functional and structural characterization. The essential point is that structurally-dependent attention is relevant for navigation that is fuzzy or approximate rather than precise and algorithmic. Plants and bacteria may "attend" to these statistically salient features in ways that differ radically from animals with a brain and a nervous system. More precise, categorical, and even symbolic-based attention may depend on conceptual or predicate-like classifications of information that depend on specialized brain areas that classify quite specific information in an algorithmic and inferential fashion.

As mentioned above, none of these functions can be interpreted as involving phenomenal consciousness because of the hard problem.

However, they can easily be interpreted in terms of attention routines. In fact, interpreting them as attention routines is perhaps the best scientific approach to understanding these distinctions.

3 Information and scientific theories of consciousness

All the major scientific theories of phenomenal consciousness strongly emphasize information processing, most of them in terms of the structure and function of neural activations or networks. This is not because they willfully ignore the hard problem. Rather, it is because, otherwise, they would not be scientific approaches to consciousness—various definitions of information are at the very basis of our understanding of cognition. The "global neuronal workspace theory" (or GNWT) explicitly explains consciousness in terms of neural activation concerning algorithmic or statistically relevant information. It implicitly assumes that attention is necessary for consciousness. Thus, GNWT is explicitly a functional or structural theory of consciousness. Higher Order and First Order theories of consciousness are undoubtedly compatible with functional and structural interpretations of consciousness, so neither entails a non-functional or "non-informational" approach. We expand upon the Integrated Information Theory (or IIT) because it is the theory of consciousness that most explicitly emphasizes information processing and most vigorously accepts the hard problem.

A significant difficulty to examine in future research is whether a non-informational theory of consciousness is both epistemically and metaphysically impossible (inconceivable and outside the scope of reality). If it is both epistemically and metaphysically impossible, what does that mean for the hard problem? We set aside this thorny issue but want to highlight the importance of getting clear on providing an informational account of consciousness.

Another complication concerns the notion of semantic information. The difficulty here is to address how different theories of semantics (e.g., situation, possible-world, or teleological) inform algorithmic information. But note that on all accounts are susceptible of functional analysis. For higher-order theories, adopting a specific semantic framework may have significant repercussions for the kind of regulatory meta-preferential or meta-monitoring routines involved in higher-order representation.

IIT deserves more discussion because of its phenomenal axioms designed to start with phenomenological insights in order to inform the causal, conceptual, and structural account of integrated information—a seemingly "bite the bullet" approach to the hard problem [25]. If the integrated information approach is correct, then there must be a radical difference between conscious and unconscious information, one that not only entails a difference in information complexity and network activation between consciousness and attention, but a difference grounded on a metric of intrinsic information, called Φ.

However, IIT has a problem [2, 19]. As its name implies, IIT is a theory that relies on the concept of information. As such, it starts with observable processes that can be associated with a joint probability distribution (Oizumi et al.). This joint can thus be used to compute the amount of information, following some schema such as computing Shannon's entropy. Nevertheless, as we pointed out in [2] and [19], joint probability distributions do not exist for contextual processes, exactly the types of processes crucial for understanding conscious experiences (consider, for instance, the case of indexicals). If we consider consciousness as contextual, as many do, how do we measure contextual information? Except for von Neuman's entropy (and perhaps extensions of Shannon's entropy to include extended probability theories [1]), no appropriate measure of information exists that could provide a measure of contextual information. We cannot, for instance, use algorithmic information, since no true contextuality can exist in algorithmic processes. So, how can we have a scientific theory of consciousness that relies on information without a theory of contextual information? We simply cannot. Chaitin [6] examines the information problem regarding phenomenal consciousness, highlighting the need for more precise measures. Our contribution is that contextuality and a clear contrast with attentional information must be at the center of this effort.

The above point makes it evident a distinction between attention and phenomenal consciousness regarding information. Whereas attention is also contextual, its contextuality is purely epistemic, in the sense that more detailed information about the environment and contextual cues could be used to identify the context objectively. Therefore, a theory of information for processes involving attention seems quite feasible. However, phenomenal experiences' contextuality is not directly accessible to a third party: this is the core of the hard problem. Thus, imposing to phenomenal consciousness an information measure implies an epistemic solution and an ontological structure where "things" are actually non-contextual.

4 Conclusions

In this paper, we discussed the connection between information, attention, and phenomenal consciousness. We saw that attention can and is typically described by information measures, such as Shannon's entropy or algorithmic information. This ability to measure attentional information comes from attention being non-uniquely contextual. Phenomenal consciousness, on the other hand, is uniquely contextual or indexically "centered." For instance, consider the concept of "self-location" or indexical sentences such as "I am hungry now." Those are uniquely contextual and "solipsistic," unlike the semantic context that frames the publicly available and "algorithmic" conversational common-ground. Therefore, any scientific description of phenomenal consciousness using information needs to rely on a theory of information that is indexically contextual. Unless we move to quantum information, with all its difficulties (both epistemic and ontological) and its limitations (following quantum logic structure), such a scientific description is still elusive.

References

[1] J Acacio de Barros. "On information, quanta, and context". In: *Unravelling Complexity: The Life and Work of Gregory Chaitin.* Ed. by F.A. Doria and S. Wuppuluri. Singapore: World Scientific Publishing Co. Pte. Ltd., 2020.

[2] J. Acacio de Barros, Carlos Montemayor, and Leonardo P. G. De Assis. "Contextuality in the Integrated Information Theory". In: *Quantum Interaction.* Lecture Notes in Computer Science. Springer, Cham, 2017, pp. 57–70. DOI: 10.1007/978-3-319-52289-0_5.

[3] Ned Block. "Attention and Mental Paint". In: *Philosophical Issues* 20.1 (2010). Publisher: Wiley Online Library, pp. 23–63.

[4] Ned Block. "Consciousness, accessibility, and the mesh between psychology and neuroscience". In: *Behavioral and brain sciences* 30.5-6 (2007). Publisher: Cambridge University Press, p. 481.

[5] Ned Block. "Perceptual consciousness overflows cognitive access". In: *Trends in cognitive sciences* 15.12 (2011). Publisher: Elsevier, pp. 567–575.

[6] Gregory Chaitin. "Consciousness and Information, classical, quantum or algorithmic?" In: *A True Polymath: a Tribute to Francisco Antonio Doria.* Ed. by J Acacio de Barros and Decio Krause. Vol. 2. Brazilian Academy of Philosophy. Rickmansworth, UK: College Publications, 2020, pp. 260–269.

[7] David J Chalmers. "Facing up to the problem of consciousness". In: *Journal of consciousness studies* 2.3 (1995). Publisher: Imprint Academic, pp. 200–219.

[8] David J. Chalmers. *The Conscious Mind: In Search of a Fundamental Theory.* en. Oxford University Press, Mar. 1996. ISBN: 9780199839353.

[9] Frans De Waal. *Mama's Last Hug: Animal Emotions and what They Tell Us about Ourselves.* WW Norton & Company, 2019.

[10] Daniel C Dennett. *Consciousness explained.* Boston, New York, London: Little, Brown and Company, 1991.

[11] Abrol Fairweather and Carlos Montemayor. *Knowledge, dexterity, and attention: A theory of epistemic agency.* Cambridge University Press, 2017.

[12] Carrie Figdor. *Pieces of mind: The proper domain of psychological predicates.* Oxford University Press, 2018.

[13] W Tecumseh Fitch. *The evolution of language.* Cambridge University Press, 2010.

[14] Keith Frankish. "Illusionism as a theory of consciousness". In: *Journal of Consciousness Studies* 23.11-12 (2016). Publisher: Imprint Academic, pp. 11–39.

[15] F. C. Fulda. "Natural agency: the case of bacterial cognition". In: *Journal of the American Philosophical Association* 3.1 (2017). Publisher: Cambridge University Press, p. 69.

[16] Charles R Gallistel. *The organization of learning.* The MIT Press, 1990.

[17] Harry Haroutioun Haladjian and Carlos Montemayor. "On the evolution of conscious attention". In: *Psychonomic Bulletin & Review* 22.3 (2015). Publisher: Springer, pp. 595–613.

[18] Corey J Maley. "Analog and digital, continuous and discrete". In: *Philosophical Studies* 155.1 (2011). Publisher: Springer, pp. 117–131.

[19] C. Montemayor, J. A. de Barros, and L. P. G. De Assis. "Implementation, Formalization, and Representation: Challenges for Integrated Information Theory". en. In: *Journal of Consciousness Studies* 26.1 (2019), pp. 107–132. URL: https://www-ingentaconnect-com.stanford.idm.oclc.org/content/imp/jcs/2019/00000026/f0020001/art00006 (visited on 02/22/2019).

[20] Carlos Montemayor. "Inferential Integrity and Attention". In: *Frontiers in Psychology* 10 (2019). Publisher: Frontiers Media SA.

[21] Carlos Montemayor. *Minding time: A philosophical and theoretical approach to the psychology of time.* Vol. 5. Brill, 2012.

[22] Carlos Montemayor and Harry H Haladjian. "Perception and cognition are largely independent, but still affect each other in systematic ways: arguments from evolution and the consciousness-attention dissociation". In: *Frontiers in psychology* 8 (2017). Publisher: Frontiers, p. 40.

[23] Carlos Montemayor and Harry Haroutioun Haladjian. *Consciousness, Attention, and Conscious Attention.* en. MIT Press, Apr. 2015. ISBN: 978-0-262-02897-4.

[24] Thomas Nagel. "What is it like to be a bat?" In: *The philosophical review* 83.4 (1974). Publisher: JSTOR, pp. 435–450.

[25] Masafumi Oizumi, Larissa Albantakis, and Giulio Tononi. "From the Phenomenology to the Mechanisms of Consciousness: Integrated Information Theory 3.0". In: *PLoS Comput Biol* 10.5 (May 2014), e1003588. DOI: 10.1371/journal.pcbi.1003588. URL: http://dx.doi.org/10.1371/journal.pcbi.1003588 (visited on 12/06/2015).

[26] Zenon W Pylyshyn. *Things and places: How the mind connects with the world.* MIT press, 2007.

[27] Miguel Segundo-Ortin and Paco Calvo. "Are plants cognitive? A reply to Adams". In: *Studies in History and Philosophy of Science Part A* 73 (2019). Publisher: Elsevier, pp. 64–71.

[28] Anne Treisman and Stephen Gormican. "Feature analysis in early vision: evidence from search asymmetries." In: *Psychological review* 95.1 (1988). Publisher: American Psychological Association, p. 15.

t We Can't Know of the World: Science and the Limits of Knowledge[1]

Marcelo Gleiser[2]

Department of Physics and Astronomy
and
Institute for Cross-Disciplinary Engagement
Dartmouth College, Hanover, NH 03755, USA

Abstract

For the past five decades, a large number of high energy physicists have joined the quest to find a unified theory of the fundamental interactions, with the goal of bringing together all forces of Nature under a single description, based on an overarching symmetry principle. I present general epistemological arguments to argue that notions of final unification are unfounded, and lead to much confusion and frustration. This will allow me to explore the conceptual framework of theoretical physics and what kinds of problems we can expect it to solve. I illustrate my argument with a list of scientific questions that are, given our current understanding of the laws of nature, unknowable, including the Big Bang singularity. As my argument unfolds, I will reminisce on my interactions with F A Doria, a transformative mentor during my early career.

1 Introduction

It was my last intro physics exam as a chemical engineering student at the Federal University of Rio de Janeiro (UFRJ). By that time, after much internal struggle and doubt, I had decided to switch to

---

[1] Essay written for Festschrift in honor of Francisco A. Doria, ed. by José Acácio de Barros and Décio Krause.
[2] Marcelo.Gleiser@dartmouth.edu

a major in physics at the Pontific Catholic University of Rio (PUC-RJ), then home of the top physics department in the country. As I hand-delivered the exam to the professor proctoring it, he looked at me, somewhat stunned. "Are you related to Luiz Gleiser?" he asked. "Yes, he's my older brother," I answered. The man's face brightened in a wide smile. "Amazing! I've known your brother for years." That was the beginning of what was to become a transformative friendship in my life. The encounter was pure serendipity. Doria was not my professor at the time. He was just doing a favor to a colleague. Years later, I wrote a book with the title *The Simple Beauty of the Unexpected*, about, among other things, transformative encounters in our lives [1]. Our encounter certainly qualifies as both unexpected and beautiful.

Even though I transferred to PUC-RJ, Doria and I kept in touch. Occasionally, I visited him in his home in nearby Petrṕolis, nested in the spectacular mountain range that runs along the coast of Rio, about 30 or so miles inland. There, he initiated me into what I could only call cross-disciplinary science-philosophy thinking, the merging of ways to consider the nature of physical reality. The trips felt to me like a pilgrimage. We talked about math, physics, fundamental questions about space, time, and matter, and how they are interrelated through symmetry and conservation laws. Those were unforgettable afternoons, sitting at Doria's varanda, surrounded by the singing of bem-te-vis and the tropical beauty of colorful hydrangeas and bougainvilleas.

A couple of years later, in August 1981, major in hand, I was back at UFRJ, this time as Doria's Masters student. I was slated to go to work with eminent theoretician Jorge Andre Swieca, who had been my undergraduate advisor at PUC-RJ and had recently transferred to the University of São Carlos, in São Paulo. Tragically, Swieca took his own life a week after I had come to visit him to arrange the details of my studies. Doria rescued me from my academic orphanhood and took me under his wing, to work on gauge field theories, focusing on the problem of gauge field copies [2], an ambiguity in connecting math and the reality of gauge fields. This was my first encounter with real research, and I was hooked for life. With additional help from another Francisco, Francisco Amaral, I finished my thesis and continued my graduate studies at the University of London. Doria and I kept an active letter exchange over the years, where we would bounce ideas at one another, especially the not so conventional ones. It was largely due to him, and his mentorship, that I would venture out of the box once in a while and take a hard critical look at what myself and colleagues were doing. Contrary to many physicists, I see philosophy as my friend, a guiding light when we step out on a limb into unchartered territories. Doria encouraged this openness to the Humanities, and I am eternally grateful to him for it.

In what follows, I will use simple epistemological arguments to address the limits of knowledge and the quest for final theories of Nature, including questions related to quantum gravity. Many of my arguments are explored at a deeper level in my books *A Tear at the Edge of Creation* [3] and *The Island of Knowledge* [4], although some appear here for the first time.

2 What Reality?

How much can we know of the world? This, of course, is the central question for physics, and has been since the beginning not just of modern science as we know it, but of Western philosophy when, around 650 BCE, Thales first speculated on the basic material fabric of reality. The essential tension here is one of perception. To describe the world we must see it, sense it, and go beyond, measuring it in all its subtle details. The problem is the "all." We humans are necessarily blind to many aspects of physical reality and, those that we do capture are necessarily colored through the lenses of our perception. We can't, not even in principle, begin to understand what "all of reality" means. With our instruments and tools of observation, we capture fragments of it, which we then describe through our mathematical models and theories the best way we can. To quote philosopher Thomas Nagel, there is no "view from nowhere," the God's eye view of reality in all its detail [5]. We may be able to imagine such a thing, but we cannot ever grasp it. There is no such thing as complete knowledge of anything, let alone of the natural world around us.

There are two main reasons for our shortsightedness. One is related to technology. The other to our physical, body-centered connection with Nature. Below, are some thoughts on both.

2.1 Technological Limitations to Knowledge

Let's start with technology. Data gathering depends on the accuracy of tools and instruments. What we "see" is contingent on the tools we use to see with. As a consequence, worldviews shift as our tools of exploration evolve. *There is no perfectly clear snapshot of reality, only a partially obscured view.* An obvious example is that of astronomy before and after the telescope. or that of biology before and after the microscope. Both inventions changed, in irreversible ways, the way we think about the skies and about life. As telescopes and microscopes evolved from their initial 17th-century progenitors and became progressively more accurate, worldviews changed again and again; from a static to an expanding universe dominated by dark energy and dark matter, and from cells to the genetic revolution based on the double-helix DNA model and, now, on DNA-sequencing technologies [6].

There is no question that as our tools of exploration advance, we are able to construct a more detailed picture of physical reality. The question of interest to us, however, is not one of progress but one of limits. Can technology advance to the point of gathering all the meaningful information about Nature, so as to effectively exhaust it? We can never know. Even if our tools become extremely fine, picking up details at what is currently unimaginably small and large scales, we can't determine that we have reached the limit of what is "out there" in the world.

Perhaps a useful analogy is a fishnet. We can make it thinner and thinner, and thus capture more and more species of marine life. As we explore smaller microscopic scales, we can't be sure that there is no limit to how thin the net needs to be to capture all there is to be captured. Life will certainly end below a certain scale, but there will be molecules and atoms and subatomic particles too, some of them requiring more sophisticated approaches to be "captured" and identified. Unless one can prove a definite theorem demonstrating that subatomic particles reach a final lower level of irreducibility, or that there is an ontological shift in the fundamental nature of reality, as in string theory, there is no way of knowing where to stop the search. And since the searching strategies depend on our current models, we reach a strange sort of loop, where we can only look for what we expect to see, like the fabled Ouroboros closed on itself, while the rest of reality is "out there," unreachable.

Of course, we are not driving completely blind here. Fundamental physics does provide some guidance as to where the limits for material structure may be. The most essential of these limits is based on the Planck units, taken as the absolutes for the breakdown of our current descriptions of physical reality based on a differentiable spacetime and quantum fields. Thus, we have the Planck length of 1.62×10^{-35}m, the Planck time of 5.39×10^{-44}s, and the Planck energy of 1.22×10^{19}GeV. Even if we can't establish for sure that the Planck units do represent an ultimate limit to physical description, assuming that they do we quickly realize that our current experimental knowledge of particle physics is about 15 orders of magnitude away.

This so-called "desert" is troublesome. Given all that happens between the energy scales of 10 TeV down to 0.5 MeV, covering essentially all of the Standard Model (without neutrinos), it's hard to image a lack of heavier particles from 10 TeV to the Planck scale. Even if we assume that there is such a thing as a Grand Unified theory (GUT), with massive excitations of order 10^{15} GeV or so, the desert is still vast, about 11-12 orders of magnitude.

This situation is in sharp contrast with what was going on when I was a Masters student with Doria in the early 1980s. At the time, and in the subsequent years of my PhD, enthusiasm about the idea of a GUT, possibly extended by supersymmetry, reigned supreme. A young theorist interested in high-energy had to work on something related. There was little doubt that supersymmetry was the way to go and that it would solve the cosmological constant problem by setting it to zero, once virtual fermions and bosons on the quantum vacuum were properly accounted for. When Green and Schwartz proved in 1984 [7] that anomalies in type I string theory cancel for the gauge group SO(32), there was explosive enthusiasm that strings may indeed lead to a final unified theory of the fundamental interactions, even if it asked for the existence of supersymmetry and extra spatial dimensions.

This is now history, and it remains unclear whether superstrings have anything to say about the fundamental nature of physical reality; no traces of supersymmetry or extra dimensions have been found, despite extensive search for the lightest supersymmetric particles at high-energy colliders such as the LHC and dark matter detectors.

We remain uncertain as to the ontological structure of Nature, whether it is described by fundamental quantum fields and their related particle excitations, vibrating strings, or something completely different and unexpected. We also remain uncertain as to whether spacetime should or not be quantized, even if the consensus is that it must. The usual arguments for spacetime quantization are that we can't have theories with singularities, as is the case of the big bang model of cosmology and black holes. Something must interfere down below to stop this nonsense, and this something must be a different layer of description for the nature of physical reality in terms of quantum spacetime.

There is an implicit extrapolation here, one that is not often disclosed. Quantization of fields have been extremely successful as a strategy, as we know from the success of the Standard Model and renormalization techniques in quantum field theory: every known particle of matter or of force is an excitation of an underlying quantum field. The ontology here is fields first, at least when it comes to "stuff." Spacetime remains the background where all this takes place, even if, beyond Newton, its properties are influenced by the matter content. The extrapolation happens by assuming that since general relativity is a classical field theory, it must, just like the other field theories for the electromagnetic and strong and weak nuclear forces, have a quantum version. That is, spacetime is also "stuff." But how do we know this? Attempts to use the same techniques that were successful to quantize matter and gauge fields fail spectacularly when applied to gravity. String theory and loop quantum gravity are the two current approaches to go beyond the impasse, both assuming that gravity, like everything else, must have a quantum version.

This amounts to an objectification or "matter-ification" of spacetime, the scaffolding originally used to construct our field-based description of physical processes between material entities. General relativity has indeed established a deep relationship between energy and spacetime, but comes short of stating that spacetime is energy. Indeed, we don't know how to ascribe energy to spacetime itself, as the energy-momentum tensor appearing in Einstein's equations comes from matter and force fields and not from gravity itself. When we write $G_{\mu\nu} = 8\pi G T_{\mu\nu}$, we are distinguishing spacetime geometry from the rest of material contributions: they have a separate existence and affect one another in ways we determine by solving Einstein's equations. In John Wheeler's famous aphorism, "Spacetime tells matter how to move; matter tells spacetime how to curve" [8]. If spacetime is just stuff, Wheeler's aphorism would be something like "see spacetime and matter dissolve into the primal quantum foam, the fundamental substrate of reality." Unfortunately, despite progress with quantum loop theory [9], we are far from a prescription like this, remaining as ignorant of the fundamental substrate of reality as we have ever been.

Attempts to capture the "quantumness" of spacetime by measuring arrival-times for photons from gamma-ray bursts have so far yielded negative results [10], tightening the constraints on the lumpiness of spacetime to 1/525 of Planck's length. It is possible that the photons in this observation didn't leave the source at the exact same time, and the timing error of 1% is larger than needed for a definite proof. Still, the technique points to a way of possibly measuring an effect related to the quantum properties of spacetime, although other potential explanations may also be relevant.

Another approach attempts to rule out the existence of spacetime singularities through bounce mechanisms [11]. In this case, spacetime stops contracting at a certain length-scale, possibly larger than where quantum effects become relevant. In this case, the Universe would never go into a quantum gravity stage near the big bang singularity, and general relativity can remain a classical theory. Presumably, the same would happen inside black holes, where similar extreme physics would become relevant as stellar matter collapses into the central singularity. Yet another, considers spacetime as an emergent theory, as a mean field approximation of underlying microscopic degrees of freedom, similar to the fluid mechanics approximation of Bose?Einstein condensates, an idea that dates back to Andrei Sakharov [12]

All of these initiatives face the essential difficulty that we don't have a fundamental theory of matter and its interactions valid up to energies where such questions become relevant. As mentioned above, the "desert" is wide-ranging and, so far at least, profoundly silent with respect to any possible new physics. We must keep pushing the boundaries of the known, but also remain humble to the vastness of our current ignorance.

2.2 Human Limitations to Knowledge

The British philosopher and historian of ideas Isaiah Berlin had much to say about statements related to absolute declarations: "A sentence of the form 'Everything consists of' or 'Everything is' or 'Nothing is' unless it is empirical states nothing, since a proposition which cannot be significantly denied or doubted can offer us no information" [13]. In other words, all-encompassing statements based on subjective authority, which cannot be comparatively contrasted or measured, carry no information: they are articles of faith, not reason. We may have tastes and intuitions that may guide our thoughts toward certain ideas related to the totality of knowledge, but they are just that, tastes and intuitions. To make sense of the One, we need the

Other.

The conceptual complications we have encountered at the extremes of physics, near the big bang or black hole singularities – let's call them essential singularities – point to a deeper philosophical problem, one that should not be taken lightly. This is the problem of the First Cause, what Aristotle deemed his "Unmoved Mover," and the Judeo-Christian-Islamic monotheistic faiths attribute to God, the uncaused all-powerful entity that is ultimately responsible for the causes that set the cosmos into motion. There is no known physics for the essential singularities, as we noted above. The essential question, of course, is whether there can be. Can we use the current framework of quantum and classical field theory to address the hardest of all questions, that of the origin of all things? Or, perhaps, there are ways to avoid spacetime singularities altogether?

Physics works under a very clear framework. In order to determine the time evolution of a system, we need to state its initial conditions, the state of the system at time zero. This implies knowledge of the system at the beginning, something we obtain through measurement or as we prepare the system in a certain initial state. For example, knowing the position and velocity of a planet at a certain time, we can predict its future motion. Or we prepare a gas at a certain temperature and pressure, and observe its properties as we change these parameters. In cosmology, this approach is impossible. How can we get out of the box if the box is all there is? We may restrict the initial conditions and the range of values of the fundamental constants given what we know about the universe today, but we can't be sure that our conclusions are in any way final. The best that we can do is aim for consistency, so that our models of the early Universe evolve into the Universe we measure today. The clues we gather about the Universe's distant past can only give us a fragmented picture of what happened close to the beginning. And even then, we only have a picture within our cosmic horizon, the distance light has traveled since the Big Bang, about 46 billion light-years. We cannot get information from what is outside of it. This means that whatever parts of the Universe lie outside our cosmic horizon are unknowable to us. We must come to terms with this unknowability and what it means to the scientific enterprise. Let's explore this a bit.

Consider the metaphor of the Island of Knowledge [4, 14]: if what we know fits in an ever-growing island, this island is surrounded by the ocean of the unknown, what we don't know about the physical universe. (You can add here other unknowns pertaining to he Humanities, but let's stick with science.) However, the acquisition of knowledge, by its very nature, is an endless pursuit. As our Island of Knowledge grows, so do the shores of our ignorance, the boundaries between the known and the unknown. The more we know, the more we discover we don't know. As the history of science teaches us over and over, new discoveries, new tools, have the power to change entire worldviews.

A key ingredient of the Island of Knowledge metaphor that has not been explored before Ref. [4] is that we are not only surrounded by unknowns, but that some of these unknowns are unknowable: there are well-posed questions that science cannot address. There are natural phenomena that we can't explain, possibly ever. This declaration may fly in the face of those who believe in a sort of scientific triumphalism, that science can conquer it all. Well, it can't. We scientists must have the integrity and clarity of vision to know where to draw the line, acknowledging what we can't handle, at least through the scientific method as we now understand it. Here are a few current examples of unknowables, from the cosmic to the cognitive: (More details on Refs. [1, 4])

- We can't know what's beyond our cosmic horizon, the bubble of information defined by the distance light has traveled since the Big Bang, around 46 billion light-years;

- We can't explain in deterministic fashion the essential randomness that takes over at the quantum level, having to accept that the possible outcomes of a measure are probabilistic;

• We can't construct a self-referential logical system that is closed, in the sense that every statement within that system can be proved, as the Austrian mathematician Kurt Gödel showed with his Incompleteness theorems (this is something very dear to Doria [15]);

• A computer cannot include itself in a simulation; it is thus impossible in principle to simulate the Universe as a whole since the simulation would by necessity need to include itself.

• Humans may be cognitively-impaired to understand their own consciousness, a problem known in the cognitive neurosciences and related philosophical discussions as the "hard problem of consciousness" [16].

Of course, the notion of an unknowable must be considered with care, given that what may seem unknowable today may become knowable tomorrow. However, for the examples above, this unknowable-to-knowable transition would require very fundamental revolutions in our understanding of physical reality. In the same order as above:

• Faster-than-light travel needs to be possible; or inter-universe-connecting traversable wormholes need to exist and be stable;

• A whole new way of thinking about quantum physics needs to be developed, one that explains the apparent randomness of results from measurements. Previous efforts based on local "hidden-variable" theories don't work. [17]

• A new, self-contained logical structure for mathematics would have to be created;

• New concepts in computing where a machine can simulate itself would need to emerge, an apparent impossibility;

- The human subjective sense of self that baffles the cognitive neurosciences would have to be explained as an emergent property of complex neuronal activity.

In one way or another, all of these unknowables share a common characteristic: they require achieving some sort of total knowledge, an all-encompassing explanation of a given physical and/or biological system from within. Again, in the same order as above:

- To know the Universe as a whole, without being able to step outside of it;

- To explain all possible results in a quantum system deterministically, including their relative probabilistic weight;

- To have a complete mathematics;

- To have a complete simulation, one that includes itself;

- To have the brain explain itself in its entirety.

The fact that our knowledge of the world is incomplete should not be seen as an intellectual weakness or a defeat of our reasoning powers. Instead, it should be seen as liberating: the incompleteness of knowledge frees us to explore the ocean of the unknown without the burden of finding some kind of final truth. Every meaningful discovery leads to new questions. The expectation that we can grasp some kind of final truth transforms the scientific pursuit of knowledge into a religious pursuit, given that only in religion such notions of finality are acceptable. The ultimate nature of reality is a notion that mirrors the presumed reality of the infinite, thinkable but not provable. We can't embrace it in its totality, only consider it as an idea and use it as a conceptual tool.

Science is an ongoing construction, a narrative we devise to make sense of the natural world. To set as its loftier goal the unveiling of some kind of final truth takes away from its exploratory excitement, overburdening it with a sense of religious quest. The vastness of Nature's wilderness should not be limited to climbing a single peak, as if it were a pilgrimage to a fixed destination. There are a multitude of peaks out there, and if we look carefully there will always be new ones to be discovered.

3 Final Thoughts: Mapping the Territory

This being a tribute to Doria, I couldn't end without bringing up the Argentinian writer Jorge Luis Borges. Few, if any, writers have explored the notion of the infinity and the limits of human knowledge as brilliantly. In his one-paragraph short story, *On Exactitude in Science* [18], Borges explores the futility of believing that science can provide a perfectly accurate description of physical reality:

> ...In that Empire, the Art of Cartography attained such Perfection that the map of a single Province occupied the entirety of a City, and the map of the Empire, the entirety of a Province. In time, those Unconscionable Maps no longer satisfied, and the Cartographers Guilds struck a Map of the Empire whose size was that of the Empire, and which coincided point for point with it. The following Generations, who were not so fond of the Study of Cartography as their Forebears had been, saw that that vast Map was Useless, and not without some Pitilessness was it, that they delivered it up to the Inclemencies of Sun and Winters. In the Deserts of the West, still today, there are Tattered Ruins of that Map, inhabited by Animals and Beggars; in all the Land there is no other Relic of the Disciplines of Geography.
>
> – Suarez Miranda,Viajes de varones prudentes, Libro IV,Cap. XLV, Lerida, 1658.

The only perfect map is one that reproduces precisely all the details of whatever it is mapping, or the "territory." Hence the paradox: if the map is as large as the territory, it has no value whatsoever. A perfect map is "useless" [19].

From the title of the story, it is clear that Borges is poking fun at scientists who believe, quite naively, that what they do is actually producing a perfect map of reality. For science can be understood as a map, a representation of what we see of the world, while Nature, then, is seen as the territory. The analogy is extremely apt, as it captures both the goals and the frustrations of science: we want to know as much as possible about the world and turn it into a description that we can share (the map). The more we know, the better the map is. There is tension between our curiosity to always know more and our myopic gaze, the impossibility to see all. This tension is a good thing, one that inspires our creativity and inventiveness.

The danger, as Borges so cleverly admonishes us with this short story, is that our ambitions can lead us astray. Unchecked, the urge to produce better and better maps of reality, with the intention of reaching that final, perfect description of the territory, is a form of blindness. Borges, who suffered from cataracts that eventually made him blind, understood this better than most. Even in full sight, there is a lot of the world that remains unseen.

By its very nature, science can never reach a final state of knowledge in any discipline. Just as it is pretty much impossible to catalogue all species of insects or of fungi in the world, we can't ever be sure that what we think is the final understanding of subatomic particles and their interactions is truly final.

In the case of bugs and fungi, it's not just practically impossible to locate them all in the vastness of Earth's surface, but some species would become extinct during the survey, while others may be undergoing mutations and changes. There is an elusiveness to the whole enterprise, one that should both motivate and inspire a healthy dose of humility.

In the case of elementary particles, there is always the possibility that some of them escaped our detectors and search algorithms, our "fishing nets." As we argued earlier, we can't ever be sure that we have a net fine enough to capture all there is to capture of the subatomic world for the simple reason that we can't ever know all there is to capture!

A map is a device that serves a very precise purpose: to guide you from point A to point B. An efficient map is one that does its job by stripping out all the unnecessary details while still fulfilling its function. That's what scientific models do, they represent the essential aspects of reality we want to study, leaving out what's not needed. Our models are our maps, but they are not the territory.

Thinking of science as the map and Nature as the territory, Borges teaches us an important lesson, one that I think Doria would agree with. We should be proud of the maps we can make of the world, and strive to perfect them. But we should also be aware that all maps are limited and provide only limited information of the territory. We see the world through very human eyes, and our imperfect maps reflect this.

Acknowledgements The author thanks José Acacio de Barros and Décio Krause for their invitation to contribute to this volume. He acknowledges partial support from the Institute for Cross-Disciplinary Engagement at Dartmouth through a grant from the John Templeton Foundation.

References

[1] Gleiser, M., *The Simple Beauty of the Unexpected: A Natural Philosopher's Quest for Trout and the Meaning of Everything*, (ForeEdge, University Press of New England, Lebanon, USA, 2016).

[2] Doria, F. A., *The geometry of gauge field copies*, Comm. Math. Phys . **79** (1981), no. 3, 435–456.

[3] Gleiser, M., *A Tear at the Edge of Creation: A Radical New Vision for Life in an Imperfect Universe*, (Simon & Schuster, New York, NY, 2010).

[4] Gleiser, M., *The Island of Knowledge: The Limits of Science and the Search for Meaning*, (Basic Books, New York, NY, 2014).

[5] Nagel, T., *The View from Nowhere*, (Oxford University Press, Oxford, Uk, 1986).

[6] Adams, J. U., *DNA Sequencing Technologies*, Nature Education, **1** (1), 193 (2008).

[7] Green, M. B.; Schwarz, J. H., "Anomaly cancellations in supersymmetric D = 10 gauge theory and superstring theory". Physics Letters B **149**, 117 (1984).

[8] Wheeler, J. A., in *Geons, Black Holes, and Quantum Foam: A Life in Physics*, pg. 235 (W. W. Norton, New York, NY, 1998).

[9] Smolin, L, *Three Roads to Quantum Gravity*, (Basic Books, New York, NY, 2001).

[10] Robert J. Nemiroff, Ryan Connolly, Justin Holmes, and Alexander B. Kostinski Phys. Rev. Lett. **108**, 231103 (2012).

[11] Battefeld, D. and Peter, P., *A Critical Review of Classical Bouncing Cosmologies*, Phys. Rep. **571**, pgs. 1-66 (2015). arXiv:1406.2790v4. For a nontechnical description, see https://www.quantamagazine.org/big-bounce-models-reignite-big-bang-debate-20180131/.

[12] Sakharov, A. D., *Vacuum Quantum Fluctuations in Curved Space and the Theory of Gravitation*, Sov.Phys.Dokl. **12**,1040-1041 (1968); for a modern perspective see, Visser, M., Mod.Phys.Lett. A**17**, 977, (2002).

[13] Berlin, Isaiah, *Concepts and Categories*, ed. by Henry Hardy. Viking, New York, 1979.

[14] I note that I'm not the first to come up with such a metaphor, or something related to it. Here are some similar ideas I was able to find *a posteriori*: (i) Austrian physicist Victor Weisskopf: "Our knowledge is an island in the infinite ocean of the unknown, and the larger this island grows, the more extended are its boundaries toward the unknown," Victor Weisskopf, *Knowledge and Wonder: The Natural World as Man Knows It* (Garden City, NY: Doubleday, 1962). (ii) The following statement is atributted to John A. Wheeler: "We live on an island surrounded by a sea of ignorance. As our island of knowledge grows, so does the shore of our ignorance," strikingly close to my own. The only source for this I could find is in John Horgan's *The End of Science: Facing the Limits of Knowledge in the Twilight of the Scientific Age* (New York: Broadway Books, 1996), pg. 83. (iii) Sir William Cecil Dampier, in his *A History of Science and Its Relations with Philosophy and Religion*, 4th ed. (Cambridge University Press, Cambridge, UK, 1961), wrote in pg. 500: "There seems to be no limit to research, for as been truly said, the more the sphere of knowledge grows, the larger becomes the surface of contact with the unknown". (iv) The metaphor appears, perhaps for the first time, in Friedrich Nietzsche's *The Birth of Tragedy*: "For the periphery of the circle of science has an infinite number of points; and while there is no telling how this circle could ever be surveyed completely, noble and gifted men nevertheless reach, e'er half their time, and inevitably, such boundary points on the periphery from which one gazes into what defies illumination." In *Basic Writings of Nietzsche*, trans. Walter Kaufmann, (Modern Library, New York, NY, 2000), pg. 97.

[15] Chaitlin, G., da Costa, N., and Doria, F. A., *Gödel's Way: Exploits into an Undecidable World*, (CRC Press, London, UK, 2012).

[16] Chalmers, D., *Facing Up to the Problem of Consciousness*, Journal of Consciousness Studies **2** no. 3, pgs. 200?219 (1995).

[17] Handsteiner, J. et al. *Cosmic Bell Test: Measurement Settings from Milky Way Stars*, Phys. Rev. Lett. **118**, 060401 (2017).

[18] Borges, J. L., *On Exactitude in Science*, Collected Fictions, trans. Andrew Hurley (Viking, New York, NY, 1998). https://kwarc.info/teaching/TDM/Borges.pdf

[19] Adapted from Gleiser, M., *There is No Perfect Map*, Orbiter Magazine, Aug. 9, 2018, https://orbitermag.com/there-is-no-perfect-map/

A Question of Identity

Muniz Sodré

Are the behaviors and properties of matter truly linked to the structure of space and time? Modern physics research seems to have no doubt about this question, which can be described as a problem of identity.

Physics, as we all know, is the study of the material world and of the laws which describe its inner workings, where the word "identity" has a scope that is at the same time limited and enormously complex. Limited because it is a subject for brilliant or specialized minds who try to respond mathematically to why the universe is the way it is; complex because the magnitude of the questions is such that even mathematics, shrinking in the face of infinite numbers, may not be able to explain anything about the point of the universe called "singularity," where the spacetime curvature becomes infinite.

But this word "identity," fished from the ancient, cloudy waters of metaphysics, slips in an equally problematic way to questions formulated by the sciences of the "historic continent" in various gradations, from psychology to anthropology. It certainly does not have the mathematic complexity of physics, but it does run into walls which are difficult to climb over or knock down.

This cursory approximation between the two fields, incidentally, brings to mind the figure of professor and researcher Francisco Doria, the only intellectual within my circle of acquaintances to adroitly navigate the problem of identity, whether in physics, or in the sciences of man.

In the latter case, of which identity are we speaking?

Identity is the word that appears when designating the organized group of conditions which govern and classify individual action or even that of a group in an interactive situation, allowing it to behave as a social actor.[1] Traditionally, it asserts itself as something which predicates a subject as a property or an attribute of being. It explains, for example, that each being or each man is one in himself, each one is himself. For this, identity is something implicit in whatever representation we make of ourselves, that which reminds us of us.

The word *idem* (Latin version of the Greek word *auto*, "the same") refers to the stability of representations, made possible by symbolic order and by language, and also to the unity of the subject in itself. In scholastic Latin, the term comes from *identitas*, that is, the permanence of the object, unique and identical to itself despite internal and external transformative pressures. Identity - or conformity by similarity or equality among diverse things - is, thus, the character which is called "one," although it may be "two" or "other," by form and effect. It is what Plato has the "Stranger" say in the Sophist dialog: "Each of them is, then, other than the remaining two, but the same itself".

In reality, various ideas are grouped together under the term "identity," as observed by Green[2]. In first place is the notion of permanence, of maintaining constants; in second, delimitation, which allows for making distinctions and circumscribing the unit; and, finally, the idea of a relationship of similarity among elements, which allows for the recognition of something like it.

As far as *identification*, it contemporarily designates the introjective process of forms of a structured identity, not as a pure and simple, realist and individual assimilation of the other and the individual, but in the universal, that is, at the logical level (however, in the structure of conscience, without content or meaning) of the reasoning implied by this assimilation. Conscience, whilst a symbolically determined form, is the place of identity, although principally

---

[1] Cf. Sodré, Muniz. Claros e Escuros — identidade, povo, mídia e cotas no Brasil. Vozes, 2016. This book continues the discussion of identity issues.

[2] Cf. Green, A. Átomo de parentesco y relaciones edipicas. In: Lévi-Strauss, C. (org.). *Seminario La identidad*. Petrel, 1981, p. 88.

of difference. Conscience oscillates between closing and opening, between the fullness of determination and the free flow, empty or undetermined by the events of the world. This oscillation, the game of alternating between full and empty forms, is what can be called the experience of difference.

But being, being one, and recognizing one are solidary characteristics, postulates of the philosophical conscience. Since its first formulation by Aristotle, the principle of identity ("A is A and is not not A") - pillar of representative thought, which tries to model relationships between men, as well as between them and the world - is presented as a "unit," that is, the possibility of repeatedly talking about a thing as itself, guaranteed by representation (the law, the word). The *paideia*, the Western culture of science, skill, and spirit, has as a presupposition the identity/unit of things, in which rational knowledge of the world comes from a subject distinct from the object, in its turn made an "other" or made into a negative form titled "difference." For this reason, many thinkers proclaimed this principle as a fundamental law of spirit, as the "principle of principles".

To say *human identity* is to designate a complex binding which links the subject to a continuous framework of references, composed of the intersection of one's individual history with that of the group in which he lives. Historically, "identity has known sources that are also criteria for community belonging. The archaic sources: ethnic and religious, the most enduring, are the most deeply rooted, the less 'chosen'. Modern sources: *national identities*, which inherit from their precedents, but are inspired not just in a common past, but also in a shared project for the future; *social identities*, corporate or ideological, steered by philosophies of history".[3] From there proceed the formative principles of individual subjectivity noted by psychoanalytic theory as the *ego ideal* and the *superego*.

In other words, every singular subject is part of a historical-social continuity, affected by the integration into a global context of needs (natural, psychosocial) and of relationships with other individuals,

---

[3] Soriano, Paul. Op. Cit., p. 68.

filled with lapses and intervals, which assure this integration. The identity of someone, a "one in itself," is always given by the recognition of the "other," or rather, by the representation which classifies it as *socius*. Empirical subjectivity (the concrete singularity) as well as the person (morally and juridically oriented) are initially composed of a communicative *factum*, in which the *I* is continuously reconstructed by the *other* with whom it binds and by the full private and institutional repertoire of facts and illusions present in anterior communications, that is, of communication with itself and with the world.

No matter how illusory the construction of a subjective identity is, its effects are important for constructing the core of a sense of innerness and irreducibility in a person - "More intimate to me than I am to myself" in the words of St. Augustine. On personal identity, the subjective structure which engenders the representation of I, Tarde says that "it is the permanence of a person, the personality as seen from the point of view of its duration".[4] To he who finds the individual in the reason for his own actions, identity is based in the memory and in the "bundle of habits, prejudices, talents, [and] knowledge compatible with the slowly variable character".

The sociologist Tarde ascribes the notion of identity in duration (finite dimension of time). He does not conceive of it, however, as a closed cosmos, but as an aesthetic order of regulation of the subject, something like the bed through which a river runs; in appearance, fixed and predetermined, the bed transforms imperceptibly. Individual personality suffers changes (an effect of its continuous "dialog" with society or the external world) or, before, variation over a "more or less identical" foundation.

The sociological argument was deepened philosophically by Heidegger when he criticized the conventional understanding of the principle of identity.[5] To begin, it is important to establish a dis-

---

[4]Cf. Tarde, G. Les lois de l'imitation. Slatkine, 1919. A propos, cf. tb. La philosophie pénale

[5]Heidegger, M. Identidade e diferença– O princípio de identidade e constituição onto-te-lógica da metafísica. Editora Abril [Col. Os Pensadores], 1973, pp. 377-400.

tinction between the terms idem and *ipse*. Idem means "the same," however in a relational notion, that is, of equality or similarity between to terms of comparison. *Ipse*, on the other hand, is "the same," but not related to the other; it is not "equal," but "itself." *Ipse* accommodates the equality of all differences: it is not a path ready to walk, but the path that arises together with and upon walking.

Heidegger moves on by saying that the general formula of the principle of identity ($A = A$) designates similarity or equality between two elements of an equation (one A is similar to another); thus, it has to do with the meaning of *idem*. However, to be the same, it is enough to be "one" and not "two" (in other words, each element is itself), whereas the unit with itself - identity question, par excellence - is in fact represented by the word *ipse* (A is A).

Nevertheless, there resides in each identity (A is A) a relationship, mediated through language, with the unit. This indicates that the unit is not a "uniformity without vigor", in that is presupposes the relationship "of the same with itself".[6] Carneiro Leão states: "Identity is neither static nor is it ready and finished in one go. Identity is a dynamic continuum of dialectical realization of equality and difference. We are equal to any other being, but with a profile and way of being that is different.".[7]

Synthetic mediation, which forms a fundamental trait between the *Being* (here understood as a primordial experience forgotten or concealed by metaphysics) and the *entity* (the man) would be, to Western thought, the unit of identity that, to Heidegger, is not an attribute of the Being. The philosopher agrees with Parmenides, who, since before the classical tradition of philosophy, affirms that to be and to think are attributes of identity (and not the contrary, as sustained by metaphysics). Dependent on a unit called "the same", to be and to think complement each other, are co-belonging and creators of a *common-belonging*, whose strong accent is on "belonging": although complementary, to be and to think maintain

---
[6]Carneiro Leão, Emmanuel. Op. Cit. p. 100.
[7]Carneiro Leão, Emmanuel. Op. Cit. p. 100.

their own prerogatives. Thus, rather than thinking of identity as a property of being or a predication of the subject, the philosopher places it in a game of reciprocal appropriation or "transpropriation" between the man (possibility of thinking) and the Being: The man receives from the Being the very possibility of identifying with him, who in his turn is only revealed through interpellation (something like a judge, who only speaks when prompted).

In this movement the difference between Being and entity appears. There is no relationship of equality or pure uniformity between the Being and the man, but there is a relationship of difference, historically moved by happening. Heidegger points to liberty (the undetermined) as something more primordial than the attraction of the spirit by eternal structures: liberty is what makes structures happen, and man's identifications come out in the free choice of common-belonging. In this way, it overtakes the principle which says "A is A" and proclaims liberty as the essence of truth. There is no dialectics of the abstract, for dialectics is the creation of liberty, the space of happening.

Thus, the questioning of the principle of identity is done in various theoretical fields and in different terminologies - in sociology, philosophy, and psychoanalysis: sociology applies it to the "individual"; philosophy alternates "man" with "presence"; psychoanalysis centers on the "subject." Redefining the subject as an effect of language, psychoanalyst Jacques Lacan distinguishes it from the "biological individual" and the "ego" as a psychic instance, presenting it as a trait different among pure forms (the signifiers) which combine in language.[8]

The subject is composed of relationships among differences, diverse marks arising from others, with which he progressively identifies. This means that the man makes a more or less objective representation of himself, added from the representation of images which he supposes are present in the conscience of the other. This representation supports itself in a symbolic matrix, in a primordial "I", which maintains itself, but also "prefigures the transferor

---

[8]Cf. Lacan, J.1961: Seminário do 6/12 [s.d].

destination; it is full of correspondence which unites the *I* to the statue in which the man projects himself like the phantoms which dominate him, finally to the *automaton* in which, in an ambiguous relationship, he tends to finish the world of his making".[9]

In other words, the subject does not support himself in any solid representation, in any ontological base, because he would be pure, differential relationship among forms and not something-in-itself or itself. Conceived as a fundamental lack-of-being, there is no way to see any reality - the weight, the gravity of the already given, already prepared and finished - in his identity. The stable unit, however, is nonexistent. We believe we are something we are not, and what we are originates from a differential mark, produced by a structure (language) which is beyond the individuals. In this way, there is no identity, but there are identifications, understood as the subject's occupation of different positions. Rather than a fixed and substantial order of constitution of the subject, what exists is a movement of internalization of behaviors, attitudes, and customs originating from significant patterns in the familial and social environments.

Despite its cohesion, the unit is always incomplete, and, at most, illusory, thus the search for other marks by means of identifications, in an attempt at wholeness. Identification is the dynamic factor of individual integration with the group and mobilization of its affects and choices. One may study it in the individual, in groups, and also in the national people. This study interested Francisco Doria throughout his academic life, in which he gathered his concerns related to physics and mathematical logic together with a historiographical line oriented toward roots of identity - mythical and real - of the Brazilian elite.

Reality and myth are updated in a method, which is diverse in its variation of national realities. Brazilians live in the core of patrimonialism, which is the assumed form of power in Brazil, as much in the organization of the State as in its social relations. It is the form we inherited from the transplantation of the Iberian State

---

[9]Lacan, J. Écrits. Seuil, 1966, p. 95.

apparatus. A text unique for its comprehension of this patrimonialism is *Livro da Virtuosa Benfeitoria*, written in the first half of the 15th century by the once infant and later Regent Dom Pedro (1392-1449), who promoted discoveries on the West African coast and in the Atlantic Ocean.

In *Livro*, Dom Pedro advocates a kind of society in which "betterments (donations, benefits, favors) from the ruler are exchanged for loyalty from and permanent submission of the subjects." The betterment implies a true doctrine of favor. The ruler is more naturally gifted for such a doctrine than other men, "who, although they know and wish to put it into practice, for worry of scarcity, cease to do that which they desire". To the ruler, therefore, falls the duty of the favor, of the mercy, an "honorable project".

What is notable in this whole text is its clarity as a source of ideology for the nascent, patrimonial Portuguese State, stating, bluntly and with philosophical tones, the category of the privileged favor, persistent until today as a social form in Brazilian life. Through cooption, others may begin to integrate such a community or Estate, united by the business interests of a ruler. This business society, administrated by a ruler with the help of a bureaucracy composed of bourgeois and literate soldiers, merges with the patrimonial State. Morality, laws, and religion converge in a social form of a patrimonial nature, that is, a political-social "way" in which the circulation of wealth obeys familial, clan-, or group-based criteria.

The thesis on the transplantation to Brazil of this patrimonial social order characteristic of the House of Aviz is exceptionally well demonstrated by Faoro.[10] It is the form of power - in which employees and graduates use the laws and the bureaucracy as instruments to oppress inferior classes, as a filter for the "*surplus-jouissance*" of the ruling group - which is transplanted to Brazil: "The reproduction of power, the recreation in the colony of the power structures which existed in Portugal is linked to these people without much money, but with licenses in their pocket, *sesmarias* land concessions,

---

[10]FAORO, R. Os donos do poder– A formação do patronato político brasileiro. Globo/USP, 1973.

and names which intimidate and demand respect there in the Court and here in the colony".[11]

From early on, colonization obeyed the logic of a capitalism politically oriented towards founding an extension of the Portuguese State in South America. Important figures from the ruling Estate were transferred to Brazil, which is in fact a "king's business," integrated into the patrimonial structure financed by bankers and major European merchants, notably the Genovese and crypto-Jews. The destitute were approached with the possibility of a good life in the New World, where they would participate as a member of the ruling elite. Work would be carried out by the increasingly insufficient number of slaves in order to fulfill the Crown's tax requirements.

The spirit of centralized patrimonialism, which the ruler only enforced on the coast, expanded to the Brazilian Sertão with the opening of the mines. Noble landowners (rural lords), New Christians, peddlers, and administrators composed the ruling oligarchy. Employees valued their positions as though they were titles, although they were privatized, arbitrary, and inefficient; this is why the bureaucratic-patrimonial State, independent from society, is perceived as something apart, a "monster with no soul,""the licensor of violence."

Brazil began the 19th century with a despotic, obscurantist State given over to the domain of the kingdom's military, while rural lords found themselves isolated, confined to their regions. What remains unchanged is the political foundation: the kingdom should serve at the pleasure and enjoyment of the ruling Estate. Thus, the empire in Brazil established the Nation-State. Its independence maintained the bureaucracy of Dom João VI's Court, and the bureaucratic-patrimonial State continued, as during the colonial period, the intermediary of all economic activities, orienting them in favor of its clients and sponsors, the "business" of the nation.

Who are these clients and sponsors? First, England; after, on the domestic level, the Brazilian elite or aristocracy, descendants of the

---

[11]DÓRIA, F.A. Os herdeiros do poder. Revan, 1994, p. 44.

land's first captain-donees, general governors and high functionaries of the colonial government. In general, the families were originally linked to the dominant Portuguese class - whether the old nobility from the time of the House of Burgundy, or the bourgeois lifted to nobility under the Aviz dynasty.

The oligarchical power of the rural lord, in turn, assumed the form of an extensive family, whose contours appear more clearly in the Northeast (18th and 19th centuries), with the classical model of the family of the plantation owner, the typical lord of the sugar plantations. This is the head of a family which includes children and grandchildren, brothers without land, nephews, distant relatives and bastards. In the *senzala*, under the shadows of the *casa-grande*, lives the slave, the lowest rung on the ladder.

All of patrimonialism implies a complex set of relations maintained by family, clan, or associated group, which looks to preserve the arrangement through an internal distribution of goods. The economic factor is important, but the essential reasoning of patrimonialism is political-cultural: the maintenance of that specific group and its intrinsic familiarity. The reinterpretation of this Iberian social form conveniently speaks to the discourse and narratives of the invention of a national identity. During a full century, starting with Brazilian Independence, the establishment of a national identity, the definition of "Brazilian-ness," indeed held great political importance to a ruling class destined to perpetuate the nation as a "business," no longer that of the king of Portugal, but that of the rural oligarchy in coalition with the bureaucrats who administered the agro-exporting State.

For the ruling elite, it was necessary to have an identifying profile with some kind of value against Europe and, at the same time, maintain the social space dominated by blacks and natives, which effectively constituted the concrete possibility of people in the sense of native forms of subjectivation. This is why the identity question has been so important for the intelligentsia since independence - Machado de Assis, Modernism, Cinema Novo. In this context, it is relevant to the research of professor Francisco Antonio Dória.

Gödelian Rationality

Sami Al-Suwailem

Abstract

Incompleteness of mathematics is one of the greatest discoveries of the 20th century. How does this phenomenon reflect on economic phenomena? If the economy is modeled as an axiomatic formal system, then externalities (e.g. social dilemmas and tragedy of the commons) can be viewed as the economic counterpart of formal incompleteness. In such an environment, if agents believe they are rational, they become irrational. And if they believe the market is efficient, it becomes inefficient. Gödel's theorem, therefore, calls for a fresh review of the nature and consequences of rationality.

1 Introduction

In the Preface to his book, *Forever Undecided*, Raymond Smullyan (1987, p. xi) writes:

> Is it possible for a rational human being to be in a position in which he cannot believe that he is consistent without losing his consistency in the process? That is one of the main themes of this book. It is modelled on the famous discovery of Kurt Gödel (the so-called Second Incompleteness Theorem) that any consistent mathematical system with enough power to do what is known as elementary arithmetic must suffer from the surprising limitation that it can never prove its own consistency!

This is a quite fascinating result. It redefines rationality at a foundational level. Few economic studies discuss the implications of this result to the characterization of *homo economicus*, and how this reflects onto the behavior of markets.

In his First Incompleteness Theorem, Gödel shows that, in a formal system rich in arithmetic, a sentence can be constructed that roughly states the following: "There exists a proposition p that is true if, and only if, it is not provable in the system." A direct consequence of this result is that such a formal system cannot prove its own consistency. If it does, it ceases to be consistent. The latter result is known as the Second Incompleteness Theorem.

If a competitive market is modeled as a formal system, what would be the economic interpretation of Gödel's sentence? This chapter argues that Gödel's sentence corresponds to the following statement: "In a market system, there exist Pareto-improving opportunities that can be realized only through non-competitive mechanisms." Social dilemmas (e.g. Prisoner's Dilemma and Tragedy of the Commons) are prominent examples of situations where such opportunities arise. Based on the Second Theorem, it follows that, in such an environment, if rational agents believe they are rational, they will become irrational, i.e. inconsistent.

While Gödelian agents are perfectly rational in the neoclassical sense (they are supercomputers), they face unbreakable limits not acknowledged by the neoclassical theory. Gödel's theorems provide a deep insight into the limits of rationality and how they impact market behavior.

The chapter is organized as follows: Section 2 presents a quick overview of Gödel's two theorems. Section 3 sets the stage for exploring the economic content of Gödel's theorems. It proposes that externalities (including social dilemmas) are economic instances of Gödel's First Incompleteness Theorem. Section 4 presents formal arguments for the Second Incompleteness Theorem; namely, if rational agents believe they are consistent, they will become inconsistent. Section 5 discusses rationality in economic theory in the light of the Second Theorem. The conclusion is given in Section 6.

2 Gödel—In a Nutshell

There are many interesting and accessible expositions of Gödel's theorems. Here is a sketch of the main idea, following the exposition of Smullyan (1987). Advanced treatment can be found in Smullyan (1992) and Smith (2013), among others.

Gödel showed that in a formal system rich in arithmetic, S, a theorem P can be shown to hold:[1]

$$(P) \qquad p \leftrightarrow \neg Bp$$

The symbol $\leftrightarrow$ denotes "if and only if;" B means "there exists a proof in S of proposition" p; while $\neg$ denotes negation. $\neg Bp$ means there is no proof of p in S. Theorem P says that a proposition p is true if, and only if, it is not provable in S. Alternatively, P says that the two sentences, "p is true" and "is not provable" are equivalent.

Note that, while P is provable in S, this does not imply that (lower case) p is provable in S. All we can prove is that, *if* the sentence p is true then it is not provable in S, and *if* it is not provable in S then it is true. But S does not tell us whether p is actually true or not.

To find out if p is true or not, we revert to a higher system, S', of which S is a subsystem. That is, we look at S from the "outside." In S', we can see (or prove) that p is actually true in S. As Martin Davis points out (2000, p. 100), "Gödel found that there are propositions that viewed from the outside of such systems could be seen to be true, yet could not be proved inside of them."[2]

Hence, we have the following two results:

1. "p is true if, and only if, it is not provable in S."
   We can prove this statement within S.

2. "p is actually true."
   We can prove this statement only outside S.

---

[1] The derivation will be presented in section 4.2
[2] See also Tieszen (2017), loc. 915.

From these two results, it follows that p is actually not provable in S. This essentially is Gödel's First Incompleteness Theorem.

2.1 Inexhaustibility

But that is not the end of the story. In the expanded system S', we will find another theorem, P', concerning a sentence p', which is provable to be true in system S'', in which there will be another theorem P'', *ad infinitum*. In any formal system rich in arithmetics, there will always be an unprovable sentence. In fact, there will be infinitely many sentences of this sort, as Gödel (1947, p. 272) points out. If there were only a finite number of such sentences, they could be simply added as axioms to the system making it complete (Stillwell, 2016, p. 101).

The basic message of Gödel's theorem, therefore, is that truth extends beyond provability. As Smullyan (1992, p. vii) remarks, "provability is arithmetic; truth is not, hence the two do not coincide." Formal reasoning will never exhaust true statements.

2.2 "I am Provable Not"

So far we have no idea what the sentence p says. Since p is equivalent to $\neg Bp$, we can formulate it as follows:

(p) p is not provable in S.

The sentence p above says that it is not provable in S. If p is, in fact, true, then it is not provable, which is what it states, and thus no contradiction arises.

If, on the other hand, p is false, then it must be provable in S. How could it be possible that proving something makes it false? If we believe that S proves only true statements, then obviously p cannot be false.[3] However, this belief about system S cannot be

---

[3]See: Hofstadter (1979, p. 448), Smullyan (1987, pp. 183-186), and Davis (2000, pp. 118-120).

proved within S itself. If it can, then p will become provable, which contradicts the statement, as will be discussed shortly.

Another way to look at it is that p states that it is not provable. So if p were false, then p will be provable. But p states that it is not provable! So we end up with a sentence that says: "The sentence that says it is not provable in S is provable in S". This is just like having a sentence q that says: "q is invisible." If q were false, then q must be visible. But what is q? It is the sentence that says it is invisible! So we end up with: "The sentence that says it is invisible is visible."

Hence, if p were provable we end up with a contradiction.[4] If S is consistent, i.e. involves no contradictions, then p cannot be provable in S.

2.3 Consistency

But how do we know that S is consistent? In other words, can we prove the consistency of S within S?

Suppose we can. Then, we can prove within S that p is true. The above reasoning ("if S is consistent, then p cannot be false. Therefore it must be true"), will not be from outside S anymore. It becomes within S. And if this is the case, then we are able to prove p within S. But this contradicts what p states, namely that it is not provable in S.

So now we can see how S differs from S'. The latter is simply S with the assertion (or proof) that S is consistent. Consistency of S is asserted within S' but outside S. If it were an integral part of S itself, i.e. if S has an axiom (or a theorem) stating that S is consistent, then we end up with a contradiction. By the same line of reasoning, we can see why S' cannot assert its own consistency. In other words, if S, or S', is able to prove, or simply assert, its own consistency, it becomes inconsistent! It can do so only if it is inconsistent because then it can prove anything (if a system can

---

[4]See: Nagel and Newman (2001), pp. 99-100, and Smullyan (1992) p. 108. The detailed derivation will be presented in section 4.2 below.

prove that 1 = 0, then it can prove any statement).

Based on the above discussion, we can replace the statement p above with the following statement:

(k)   System S is consistent.

We may accordingly rewrite theorem P to become:

(K) $k \leftrightarrow \neg Bk$

See Boolos (1979, p. 123).

2.4 To Prove or Not to Prove ...

One important consequence of Gödel's Second Theorem is that we cannot prove that we cannot prove a contradictory statement. For example, we cannot prove that we will never prove "two plus two equals five"! George Boolos (1994) puts it as follows:

- We can prove that $2 + 2 = 4$.

- We can prove that we can prove that $2 + 2 = 4$.

- We can prove that $2 + 2 \neq 5$, and we can prove that we can prove it.

- Now, can we prove that we will *never* prove that $2 + 2 = 5$?

Surprisingly, Gödel's Theorem says we cannot. In Boolos's (1994, p. 1) words:

> So, we now want to ask, can it be *proved* that it can't be proved that two plus two is five? Here's the shock: no, it can't. Or to hedge a bit: *if* it can be proved that it can't be proved that two plus two is five, *then* it can be proved as well that two plus two is five, and math is a lot of bunk. In fact, if math is not a lot of bunk, then no claim of the form "claim X can't be proved" can be proved.

So, if math is not a lot of bunk, then, though it can't be proved that two plus two is five, it can't be proved *that* it can't be proved that two plus two is five.

By the way, in case you'd like to know: yes, it can be proved that if it can be proved that it can't be proved that two plus two is five, then it can be proved that two plus two is five.

Simply put, a formal system cannot identify its own boundaries. It can prove an infinite number of theorems, but it cannot prove that "proposition X is not provable in the system," whatever X is. If the system happens to prove such a statement, then it becomes inconsistent, and thus will be able to prove any proposition, including X.[5] This, in essence, is Gödel's Second Incompleteness Theorem. It is a remarkable result. It will have substantial implications for the nature of rationality of economic agents.

3   The Market as a Rich Formal System

After World War II, a general trend emerged, particularly in the US, towards formalizing economic theory (see: Weintraub, 2002). The market is presented as an axiomatic formal system, where choices and prices are derived through theorems proved in the system. Axiomatic systems have their merits, but they also have their deficiencies. There is a "price" for a consistent formal system: incompleteness. That is, there will be (many, many) undecidable statements in the system that are true but not provable, as we have just seen.

---

[5] A person once asked Bertrand Russell, "You say that a false proposition implies any proposition. For example, from the statement $2 + 2 = 5$, could you prove that you are the Pope?" Russell replied, "Yes," and gave the following proof. "Suppose $2 + 2 = 5$. We also know that $2 + 2 = 4$, from which it follows that $5 = 4$. Subtracting 3 from both sides of the equation, it follows that $2 = 1$. Now, the Pope and I are two. Since two equals one, then the Pope and I are one! Therefore, I am the Pope". Smullyan (2014, p. 9).

3.1 Epistemology vs. Ontology of Gödel's Theorems

What is the economic significance of incompleteness in economic models? There are two perspectives for the answer to this question.

One is epistemological. From this perspective, Gödel's theorems put a limit on how much we are able to know. This means that there is an inherent uncertainty in economic modeling that cannot be avoided. For example, Al-Suwailem, Doria, and Kamel (2018) discuss the implications of such uncertainty. The main conclusion is that formal modeling, while helpful, is by no means a substitute for good judgment and qualitative analysis. Ignoring the inherent limitations of formal models might actually compound uncertainty and make the market more risky and less stable.

The other perspective is ontological. When modeling social or natural systems, mathematical concepts will have a corresponding counterpart in the real world. For example, the mathematical concept of fixed points corresponds, in a model of the economy, to the market-clearing price vector. Gerald Debreu (2008, p. 455) points to "the perfect fit between fixed points and the social science concept of equilibrium." Debreu presents other examples of the "perfect fit" between mathematical concepts and economic phenomena.

In this context, we may ask: what is the economic phenomenon that corresponds to the incompleteness phenomenon? Since incompleteness is a mathematical phenomenon, it applies to both natural and social systems. In general, one might argue that incompleteness corresponds to *emergence*: Properties of the system as a whole, not of parts *within* the system (Anderson, 2011, pp. 134-139). For natural systems, emergence is integral to natural processes. But for social systems, it might lead to unexpected consequences. Unlike atoms or molecules, humans are equipped with imagination and free will. By their imagination, they are able to "see" the emergent properties despite being inside the system. By their free will, they are able to take advantage of them. This results in a divergence between the group's rationality and the individual's rationality. This divergence lies behind many of the most important events in economic history, including bubbles and crashes, as well as environmental crises. These phenomena are usually described in textbook economics as "externalities."

There are good reasons to believe that externalities can be seen as an economic counterpart to formal incompleteness. If so, then externalities cannot be considered as exceptions to the standard general equilibrium model. They are an inseparable part that arises inevitably from the mathematical formalism used to model the economic system.

3.2 Self-reference in Economic Models

Although general equilibrium theory assumes agents to behave individualistically and competitively, formal modeling will necessarily generate interdependence and self-reference in the manner that Gödel had shown. To see how self-reference arises from seemingly innocent parts, consider the following two sentences:

$p(1)$ The following sentence is false

$p(2)$ The preceding sentence is true

Collectively, we get a paradox equivalent to the famous Liar's paradox: "This sentence is false." As Douglas Hofstadter (1979, p. 21) points out, each sentence separately is harmless and even potentially useful. The "blame" for this paradoxical self-reference can't be pinned on either sentence–only on the way they relate to each other.

Even more puzzling is Yablo's paradox (Yanofsky, 2013, pp. 24-26). In this paradox, we have an infinite sequence of statements:

$p(1)$ Statement $p(i)$ is false for all $i > 1$

$p(2)$ Statement $p(i)$ is false for all $i > 2$

...

$p(n)$ Statement $p(i)$ is false for all $i > n$

Again, although no explicit circularity exists in this structure, it is paradoxical nonetheless.

Bertrand Russell and Alfred Whitehead (1910-1913) were very careful in their monumental *Principia Mathematica* to eliminate any kind of circular reference to avoid paradoxes. The *Principia* was written with utmost rigor and objectivity. As John Stillwell (2010, p. 69) explains, the weakness of the *Principia* (and all similar systems) is its very rigor and objectivity. Since it is able to describe mathematics with complete precision, *Principia* is itself a mathematical structure, which can be reasoned about. Despite the great efforts by the authors to avoid paradoxes, Gödel in 1931 presented an ingenious argument of how self-reference may still arise such that, if the *Principia* were consistent, it must be mathematically incomplete.

The mathematization of economic theory has more or less the same objectives: rigor and completeness. The theorists of general equilibrium models, just like the authors of *Principia*, eliminated by construction any form of explicit interdependence among agents. Yet, self-reference emerges collectively even if agents were assumed purely individualistic and independent of each other. The resulting interdependence and indirect self-reference reflect fundamental nonconvexities in consumption and production sets, which are considered as a root cause behind externalities (Cornes and Sandler, 1996, pp. 44-49). Nonconvexity is a major point of departure of complexity economics from neoclassical economics (Al-Suwailem, 2010).

3.3 Undecidable Dilemmas

The neoclassical theory holds that the market mechanism guarantees to solve the problem of efficient allocation of resources. That is, general competitive equilibrium is able to achieve Pareto-optimal allocation of resources, in accordance to the First Fundamental Welfare Theorem.

The Incompleteness theorem, if interpreted in an economic context, implies the divergence of the mechanism and the solution, similar to the divergence of provability and truth in axiomatic formal systems. Accordingly, there will be economic situations whereby the market mechanism fails to produce an efficient allocation, and the efficient allocation can be achieved only through non-market mechanisms. If the market is supposed to produce only efficient allocations, then these situations will be undecidable. We argue that externalities, most notably Social Dilemmas, are prominent examples of such undecidable situations.

This can be clarified by looking at a special case of Social Dilemmas, a 2-person one-shot Prisoner's Dilemma game. As is well known, this is a situation whereby each player may choose either to cooperate or to defect against the other player. So we have something like that in Table 1.

Table 1: Prisoner's Dilemma

		Player 2	
		Cooperate	Defect
Player 1	Cooperate	(a, a)	(c, b)
	Defect	(b, c)	(d, d)

$$b > a > d > c$$

The competitive solution, i.e. the dominant strategy for each player, is *Defect*. Each player is better off unilaterally to *Defect* regardless of the choice of the other player. So the Nash equilibrium is (d, d) whereby no player is better off to deviate unilaterally. But, obviously, this outcome is Pareto-inferior to (a, a) resulting from the dominated strategy *Cooperate* (see: Binmore, 1994, pp. 102-104).

The dilemma arises from the divergence between dominance and efficiency. A dominant strategy is supposed to produce a dominant payoff, but the result of the *PD* game is the opposite. Dominance becomes a self-defeating criterion.

On the other hand, there is no mechanism in a competitive market to achieve the Pareto-superior outcome (a, a). The dominant strategy produces an inefficient outcome, while the efficient outcome cannot be achieved except by a dominated strategy. As R. Duncan Luce and Howard Raiffa (1957, p. 107) point out, the *PD* game "has no jointly admissible equilibrium pair". Nash equilibrium is defined unilaterally, and it is true that no player can unilaterally deviate from (d,d). But this shows that the Pareto-superior outcome (a,a) is not achievable via a competitive mechanism. Thus, Pareto-optimality is not solvable in a competitive market system.[6] A game is solvable when each player has a dominant strategy (see Binmore, 1992, p. 13). But in a *PD* game, the solvable outcome is not admissible, while the admissible outcome is not solvable. Following Douglas Hofstadter (1985, p. 764), the one-shot *PD* game is arguably undecidable within a competitive system.[7] The *dilemma* is real.

The Prisoner's Dilemma game shows how (indirect) self-reference thwarts the solvability of the game. The choice of one player is reflected back onto his payoffs through the choice of the other player, causing the failure of the dominant strategy to achieve a Pareto-superior outcome. This kind of self-reference lies at the heart of the weird nature of Gödel's sentence: $p \leftrightarrow \neg Bp$.

Gödel's theorem shows that truth extends beyond provability. Translating into *PD* game's parlance, efficiency extends beyond competitive strategy. Value in these cases can be achieved only through non-competitive arrangements, as Nobel laureate Elinor Ostrom (1990) shows for the Tragedy of the Commons. This is an active area of research in experimental economics (e.g. Croson, 2008, and Chen and Ledyard, 2008).

It is well known that, with externalities, competitive general equilibrium fails to be Pareto-efficient (e.g. Geanakoplos and Polemarchakis, 2008, p. 682; Ledyard, 2008; and Starr, 2011, p. 207). What is not so well known, perhaps, is that such externalities are

---

[6]One way to show that the game is not solvable is to write a computer program to find the Pareto-efficient outcome of the game. The program will never come to a halt. If the program starts at (a,a), it will move to (b,c) or (c,b), then to (d,d), back to (a,a), *ad infinitum*.

[7]Tsuji, da Costa, and Doria (1998) show that, in general, Nash equilibria in non-cooperative games are undecidable.

logically inevitable if the economy is sufficiently complex. They do not arise because of missing or "incomplete markets" of the Arrow-Debreu model. No matter how many markets do we have, *logical* incompleteness will always emerge due to the structure of the system. As we have seen earlier, formal incompleteness is inexhaustible.

Social Dilemmas are ubiquitous in social systems, and they infiltrate almost every aspect of economic activities. Thanks to Gödel, we now know that such phenomena are logically unavoidable. This coincides with the point made earlier, that in theory, there will be an infinite number of instances of Gödel's sentence. Axiomatic rigor was supposed to make economic theory traceable and predictable. However, Gödel's theorems show how rigor can give rise to unexpected results.

3.4 The Impossibility of a Paretian Liberal

In 1970, Amartya Sen, who won the Nobel Prize in 1998, published a paper with the above title (Sen, 1970). The basic argument is simple but profound: Under fairly general conditions, it is impossible to satisfy individual liberty and Pareto optimality at the same time. More specifically, if agents are allowed to have interdependent preferences, then it is generally impossible to satisfy individual liberty and the Pareto principle. The interesting part is that the formal structure of Sen's impossibility theorem is identical to that of the Prisoner's Dilemma game. Hence, the undecidability result above is consistent with the Sen's impossibility theorem. Below is a sketch of the theorem.

Suppose we have a society of n individuals. Each individual has a preference relation R_i which is the ordering of individual i of the set X of all possible social states, where social states include every individual's ordering. Let R be the social preference relation. R should generate a choice function that determines the best alternatives in every subset of X. The resulting function is called Social Decision Function (SDF). Sen then specifies the following conditions for the collective choice:

- Unrestricted Domain (U): The domain of the SDF includes every logically possible set of R_i.

- Pareto Principle (P): If every individual prefers x to y, then the society must prefer x to y.

- Liberalism (L): For each individual i, there is at least one pair, (x,y), such that, if i prefers x to y, then the society must prefer x to y (in this case we say that person i is "decisive" over the pair x,y). The same applies if i prefers y to x. Sen defines a less demanding condition, Minimal Liberalism, or L*, were the condition L applies to at least two individuals (since if it were to apply to only one individual it will be a kind of dictatorship).

Sen then proves the following Theorem:

There is no Social Decision Function that satisfies conditions U, P, and L.*

See Sen (1970, 2017 ch. 6*).

Following Sen (1970), here is a simple example that shows the impossibility of a collective choice rule: Suppose we have two persons 1 and 2 within an n-person community. Suppose person 1 prefers $w \succ_1 x$, while person 2 prefers $z \succ_2 w$. Suppose everyone, including 1 and 2, prefers x to z. Then by Condition P the community should prefer $x \succ z$. Person 1 should be decisive regarding the pair (w,x), and thus the society, following 1, should prefer w to x. Similarly, person 2 is decisive regarding the pair (w,z). Hence, the society likewise should prefer z to w. But this means that the collective choice rule will generate a cyclic preference: $w \succ x \succ z \succ w$. Accordingly, there is no best element in the set (w,x,z) in terms of social preference, as Sen (1970, p. 154) notes, and every alternative is worse than some other. Therefore, a collective choice does not exist.[8]

---

[8]Sen (1970, p. 153) argues that the IPL theorem goes beyond that of Arrow. Arrow's impossibility theorem applies to social welfare functions (SWF), but

Sen (2017, ch. A5*) notes that the formal structure of the Impossibility of a Paretian Liberal (IPL) result is isomorphic to that of the Prisoner's Dilemma, although the interpretation is different. To see how the two are identical, consider the following table:

Table 2: *PD & IPL*

	Player 2	
	A'	B'
Player 1 A	x	y
B	w	z

If we define $x = (a,a)$, $w = (b,c)$, $z = (d,d)$, then we end up with the *PD* game. Player 1 prefers w to x. Player 2, given the preference of player 1, prefers z to w. And everyone prefers x to z. This will result in the cycle $w \succ x \succ z \succ w$. We could have defined $y = (c,b)$, in which case player 2 prefers y to x. Player 1, given the preference of player 2, would prefer z to y. Everyone prefers x to z, as before. We thus end up with a similar cycle for the social choice: $y \succ x \succ z \succ y$.[9]

The IPL theorem confirms our argument above that a *PD* game is undecidable. The point is that there is an inherent tension between individualistic and collective properties. In Gödel's theorem, the collective property is consistency. It is a property of the system as a whole. In the *PD* game and the IPL, the collective property is Pareto optimality. The individualistic property, on the other hand, is provability in formal systems, and selfishness or "liberal"

---

does not extend to social decision functions (SDFs). The range of an SWF is restricted to orderings, while that of an SDF is preference relations that generate a choice function.

[9]See also Aldrich (1977) and Muller (2003) ch. 27.

preferences in *PD* and IPL, respectively. The key ingredient in both is interdependence.

4 Rational Reasoners

In his Preface, Smullyan (1987, p. xi) writes:

> There are several reasons why I have transferred Gödel's argument from the formal domain of mathematical systems and the propositions provable in them to the realm of human beings and the propositions believed by them. For one thing, human beings and their beliefs are far more familiar to the nonspecialist than abstract mathematical systems, and so I can thus explain the essentials of Gödel's ideas in a language that everybody can understand. Also, putting these matters in human terms has an enormous psychological appeal and turns out to be highly relevant to the ever-growing field of artificial intelligence.

Smullyan presents formal systems as "logicians who reason about themselves." As he points out in the Preface, not only this presentation is quite appealing, it is also relevant to the field of Artificial Intelligence. In the first half of the 21st century, this is even more relevant. We can simply think of such logicians as robots equipped to answer the kind of questions we are interested in.

Interestingly, Kenneth Binmore (1987) argues that rational agents can be represented as Turing Machines. Turing machines can be viewed as a kind of robots that Smullyan alludes to. Thus, the rational reasoner is comparable to the economic man or *homo economicus*. From this perspective, the approach of Smullyan is very relevant to economic theory.

Smullyan provides a hierarchy of different types of reasoners with increasing degrees of sophistication. We shall focus on the most advanced type, he calls it Type G Reasoners. We call them

Rational Reasoners. (From now on, we refer to Smullyan (1987) as
FU.)

4.1 Characters of Rational Reasoners

Rational reasoners will believe propositions only if these propositions
are provable. To say that a reasoner "believes a proposition p,"
denoted as Bp, is equivalent to saying that p is provable within the
analogous formal system. As Smullyan points out (*FU*, p. 108),
the symbol B was used by Gödel to indicate "provability" in formal
systems; it stands for the German word *beweisbar*. The two kinds
of use (for reasoners and for formal systems) are equivalent.

The following properties characterize our rational reasoner (*FU*,
ch. 11):

1. The reasoner thoroughly understands propositional logic. In
   particular, he believes all tautologies. Of course, this assumption is highly idealized, since there are infinitely many
   tautologies. However, we may simply imagine the reasoner
   so programmed that (i) sooner or later he will believe every
   tautology; and (ii) if he ever believes p and ever believes $p \to q$,
   then sooner or later he will believe q (*FU*, p. 69).

2. The reasoner is "aware" of property 1 above. That is, for any
   propositions p and q, the reasoner believes: "If I should ever
   believe p and believe that $p \to q$, then I will also believe q."
   Thus, the reasoner believes the proposition: $(Bp \,\&\, B(p \to q)) \to Bq$. This is equivalent to the proposition: $B(p \to q) \to (Bp \to Bq)$.

3. For any proposition p, if the reasoner believes p, then he
   believes that he believes p, i.e. he believes: $p \to Bp$.

4. The reasoner is aware of property 3 above. That is, for every
   proposition p, the reasoner believes: "If I should ever believe
   p, then I will believe that I believe p." Thus, the reasoner
   believes the proposition: $Bp \to BBp$.

A reasoner with the above properties knows that, if he ever believes a proposition p and its negation $\neg p$, then he will be inconsistent (*FU*, p. 101).

5. The reasoner is reflexive: For any proposition q, there is at least one proposition p such that the reasoner will believe $p \leftrightarrow (Bp \to q)$. The proposition p is called a "fixed point" of the belief system of the reasoner. The system is therefore diagonalizable (Smullyan, 1992, p. 110). All systems investigated by Gödel, including *Principia Mathematica*, are diagonalizable or reflexive (*FU*, p. 147).

   This property implies that the reasoner believes the proposition $(Bp \to p) \to p$. This is called Löb's theorem. Smullyan calls the last result "modesty." The reasoner is modest if it is the case that he believes $(Bp \to p)$ only if he believes p.

With these properties, the reasoner achieves maximum logical power and self-awareness (*FU*, pp. 94-95). Now comes the interesting part.

4.2 Theorem G (After Smullyan)

1. If a consistent reasoner with properties (1–4) above believes a proposition of the form $p \leftrightarrow \neg Bp$, then he can never know that he is consistent. Stated otherwise, if the reasoner believes $p \leftrightarrow \neg Bp$ and believes that he is consistent, then he will become inconsistent.

2. A reasoner with properties (1–5) above *must* confront a proposition of the form $p \leftrightarrow \neg Bp$. Accordingly, if he is consistent, he will never be able to believe that he is consistent.

Proof

1. (i) Since the reasoner believes $p \leftrightarrow \neg Bp$, this implies that he believes $p \to \neg Bp$. From property (2), he must then believe $Bp \to B\neg Bp$. But, from property (4), $Bp \to BBp$. Thus, Bp implies believing Bp and its negation $\neg Bp$. By property (4),

it follows that, *if* the reasoner believes Bp, he will believe he is inconsistent. Hence, $Bp \to B\bot$, where $\bot$ indicates logical falsehood (i.e. $(p \,\&\, \neg p)$ or $1 = 0$). Note that, so far, the reasoner does not believe he is inconsistent; only that *if* he believes Bp then he will believe he is inconsistent.

(ii) Now, if the reasoner believes he is consistent, then he believes $\neg B\bot$, i.e. he believes he will never believe or prove falsehood. Since $Bp \to B\bot$, then, by negation, $\neg B\bot \to \neg Bp$. By property (2), this implies that he believes $\neg Bp$. Since he believes the proposition $p \leftrightarrow \neg B$ by assumption, then he will believe p and thus Bp. But he already believes $\neg Bp$. So the reasoner ends up believing Bp and $\neg Bp$, and thus becomes inconsistent (*FU*, pp. 102-103).

2. We need first to show that a rational reasoner cannot consistently believe $\neg Bp$, for any p.[10]

    (i) It is a tautology that, for any p, $\bot \to p$ (anything follows from a false premise). By property (2), we get $B\bot \to Bp$. By negation, $\neg Bp \to \neg B\bot$. $\neg B\bot$ implies $(B\bot \to \bot)$ (this is a tautology). It follows that $\neg Bp \to (B\bot \to \bot)$. If the reasoner believes $\neg Bp$, then he believes $(B\bot \to \bot)$. By Löb's theorem, $\neg Bp \to \bot$ (*FU*, pp. 163, 167-168).

    This result says that, for any proposition p, a rational reasoner cannot consistently believe that he will never believe p (i.e. he will not be able to prove that a particular proposition is not provable without being inconsistent). A consistent reasoner is unable to decide his own boundaries.

    (ii) Next, since $\neg Bp \to \bot$, then by property (2), it follows that $B\neg Bp \to B\bot$. Since $\bot \to \neg Bp$ is a tautology, then

---

[10]Following Boolos (section 2.4), p can be the proposition "$2 + 2 = 5$." Thus, we cannot consistently prove that we will never prove $2 + 2 = 5$.

$B\bot \to B\neg Bp$. From these two propositions, it follows that $B\neg Bp \leftrightarrow B\bot$. Replacing p with $\bot$, we have $B\neg B\bot \leftrightarrow B\bot$. By negation, the reasoner believes $\neg B\neg B\bot \leftrightarrow \neg B\bot$. Now, denote $\neg B\bot$ with p. Then we obtain Gödel's sentence: $\neg Bp \leftrightarrow p$ (*FU*, pp. 168-169). Thus, a rational reasoner will inevitably confront Gödel's sentence.

(iii) It is straightforward from Löb's theorem to show that, if a rational reasoner believes he is consistent, he will become inconsistent. From (i) above, we have $\neg Bp \to (B\bot \to \bot)$. Replace p with $\bot$, we get: $\neg B\bot \to (B\bot \to \bot)$. A reasoner who believes he is consistent believes $\neg B\bot$. Thus he believes $B\bot \to \bot$. By Löb's theorem, this implies $\bot$. Believing in his consistency makes the reasoner inconsistent (*FU*, pp. 153-154).

Q.E.D.

4.3 Remarks

- It is important to note that the above result doesn't mean that the reasoner is necessarily inconsistent; he might very well be consistent. But, if he is consistent, then he cannot believe (or prove) that he is (*FU*, p. 154). We, as outsiders, may be able to tell (or prove) that the reasoner is consistent. But the reasoner himself, if consistent, cannot prove it.

- The reasoner is aware that he cannot prove his own consistency without becoming inconsistent (*FU*, pp. 102-103, 168). In Boolos' remarks cited above (section 2.4), "it can be proved that if it can be proved that it can't be proved that two plus two is five, then it can be proved that two plus two is five." In other words, a consistent reasoner is aware of his limitation and he can prove it. This limitation is in the form of a biconditional: If the reasoner proves (or asserts) his consistency he will become inconsistent, and vice versa. However, he can prove neither his consistency nor his inconsistency. If he is consistent, he will be forever undecided!

- From a practical point of view, it does make sense that a rational person cannot confirm his own infallibility. As Smullyan points out, to trust the consistency of a reasoner on the grounds that he can prove his own consistency would be as foolish as to trust the veracity of a person on the grounds that he claims to be always truthful (*FU*, p. 111).

- The inability of the reasoner to prove or assert his own consistency does not mean that he has no faith in any of the propositions he is able to prove. In his Gibbs Lecture of 1951, Gödel (1995, p. 309) explains that the reasoner "could perceive to be true *only one proposition after the other*" (emphasis added). The problem, however, is that he cannot assert that *all* the propositions he is able to prove will be true. George Boolos, in his introductory remarks to Gödel's lecture (pp. 292-293), notes "we could successively come to recognize, of each proposition ..., that the proposition is correct; but we could not know the general proposition that they are *all* true."[11] The whole is clearly different from the parts.

- Although the reasoner cannot deductively prove that all his propositions are true, he can nonetheless assert the "empirical" or "inductive" consistency of such propositions (Tieszen, 2013, loc. 1129). According to Gödel (1995, p. 309):

"... we could perceive to be true only one proposition after the other, for any finite number of them. The assertion, however, that they are all true could at most be known with empirical certainty, on the basis of a sufficient number of instances or by other inductive inferences."

---

[11] Peter Smith (2013, pp. 236-237) notes that Peano Arithmetic (PA) can prove the consistency of any finitely axiomatized sub-theory of PA. "It just can't go the extra step of proving consistent the result of putting those sub-theories together into one big theory."

Obviously, empirical consistency can never achieve the certainty of deductive consistency. It does provide, however, a reasonable degree of confidence but with a healthy dose of skepticism to keep the process of creativity and discovery alive. In the words of the Paul Rosenbloom, "Man can never eliminate the necessity of using his own intelligence, regardless of how cleverly he tries!" (in Smullyan, 2013, p. v).

- Gödel (1995, p. 309) also explains how consistency of a well-defined system transcends the system:

"For it makes it impossible that someone should set up a certain well-defined system of axioms and rules and consistently make the following assertion: All of these axioms and rules I perceive (with mathematical certitude) to be correct, and moreover I believe that they contain all of mathematics. If someone makes such a statement he contradicts himself. For if he perceives the axioms under consideration to be correct, he also perceives (with the same certainty) that they are consistent. *Hence he has a mathematical insight not derivable from his axioms.*" (Italics added.)

In other words, consistency is a collective or *emergent* property arising from the joint co-existence of the axioms. This shows why consistency cannot be proved from these axioms.[12]

5 Does *Homo Economicus* Know He is Rational?

Despite the importance of this question, it receives negligible attention in the vast literature on rationality. It seems natural that a rational person would think of himself as rational. Why would self-awareness contravene rationality? We, therefore, need to be clear about what do we mean by "rationality."

---
[12]See also Stillwell (2004) p. 13.

5.1 Shades of Rationality

Gödel's theorems do not exclude all kinds of rational self-awareness. From the discussion in the previous section, we may identify more than one meaning of rationality:

1. Adherence to rules of logic. If the agent believes that he or she adheres to the rules of logic, this will not lead to a contradiction. We have already seen that rational reasoners are fully aware of their logical powers, without this awareness leading to inconsistency.

2. Assertion of the validity of a particular proposition derived from the given set o axioms. A rational agent can believe the correctness of a specific proposition without falling into contradiction, as highlighted by Gödel.

3. Consistency. If rationality means that the reasoner is always consistent, i.e. can never commit a contradiction, then believing in his own rationality is self-defeating. Unfortunately, neoclassical theory makes consistency the most important aspect of rationality (Binmore, 2015, p. 21).

The tension between consistency and self-awareness arises only for agents with sufficiently rich logical powers. As explained above (section 4.1), a reasoner with only properties 1–4 (without the fifth property of reflexivity), will have no problem in believing in his own consistency. This kind of agents, therefore, corresponds to the neoclassical *homo economicus*. Such reasoners will not confront Gödel's sentence, and thus do not face dilemmas like the *PD* game.

So the question now becomes: Do rational agents live in an environment where Gödel's sentence naturally arise? Are they reflexive? As pointed out above, rich systems, like Peano Arithmetic and *Principia Mathematica*, are reflexive. Since the economy, mathematically, is no less rich than such systems, it follows that it must also be also reflexive. Accordingly, rational economic agents possess the properties 1-5 above, and thus they must face situations that correspond to Gödel's sentence. In such an environment, believing in one's own rationality becomes self-defeating.

5.2 The Rationality Dilemma

To see how Gödel's theorem applies to an economic environment, let us go back to the one-shot Prisoner's Dilemma game. As already discussed, *Defect* is the dominant strategy for each player. A dominant strategy by definition will *never* produce an inferior outcome. This amounts to believing $\neg Bp$, where p is the proposition that a dominant strategy leads to an inefficient outcome. If this is the case, then each player will be inconsistent when they *jointly* choose *Defect* in the one-shot *PD* game.

Had each player instead suspended his belief in the infallibility of a dominant strategy, each would have become undecided and thus abstained from making a choice in a *PD* setting. This allows the two parties to negotiate a non-competitive solution to achieve the Pareto-superior outcome. Believing in one's infallibility gives a false impression that the player is able to rationally decide *every* problem. But the indirect self-reference in the *PD* game prevents the players from rationally achieving the efficient outcome.

Gödel's remarks above may shed some light here. Recall that, in a formal system, a provable proposition can be perceived to be correct only one at a time. However, the same cannot be true for *all* propositions. Similarly, a player in the *PD* game may perceive his choice of the dominant strategy to be efficient only "one at a time," i.e. only if the choice is made unilaterally. However, efficiency does not obtain when *all* players choose the same strategy simultaneously. The divergence between individual and collective choice is one of the major lessons from Gödel's theorem. It shows how the "fallacy of composition" may lead to self-defeating decisions. Accordingly, rational players cannot presume that a dominant strategy will *always* produce an efficient outcome for all players. If they do, they fall into contradiction and end up with an inefficient outcome.

6 Economic Modeling

Gödel's theorem has critical implications for modeling economic behavior. These implications, unfortunately, are predominantly neglected in the economic literature. We briefly survey here some important areas whereby the modeling approach violates Gödel's theorem.

6.1 Rational Expectations

The Rational Expectations Hypothesis (REH) presumes that market participants' forecasts are "essentially the same as the predictions of the relevant economic theory" (Muth, 1961, p. 315). Thomas Sargent (2008) explains that REH imposes a "communism of models and expectations." A rational expectations equilibrium asserts that the same model is shared by (1) all of the agents within the model, (2) the econometrician estimating the model, and (3) nature, or the data generating mechanism. The economist and the agents modeled by the economist, are usually placed on equal footing: the agents in the model should be able to forecast and profit-maximize and utility-maximize as well as the econometrician who constructed the model (Sargent, 1993, p. 21; Frydman and Goldberg, 2011, pp. 58-59).

But based on Gödel's theorem, if economic theory is consistent, then the agents modeled by the theory should not be able to know (prove) the consistency of *their* models. To assume that the view from the *inside* the is the same as that from *outside* the model, misses the essential insight of Gödel's theorem. This, of course, is based on the assumption that the economist is "outside" the economy, and thus is able to develop a consistent model from the outside. But the economist, in fact, is part of the system being modeled, and interacts, directly or indirectly, with the modeled markets. So, for the economist to assume his own consistency adds another layer of inconsistent reasoning. Overall, therefore, the strategy is logically incoherent.

The incoherence of the modeling approach omits an important

aspect of economic reality. If all agents have the same model, and each agent seeks to maximize his or her payoffs, then we end up with a social dilemma. We might think that if all agents believe they have the same consistent true model of the economy, the market will achieve equilibrium quickly and smoothly, rather than oscillating (like in a cobweb model). But is it in the interest of competing agents to reach the equilibrium? There is an inherent conflict, note Roman Frydman and Michael Goldberg (2011, p. 58), between REH's presumption that people's beliefs can be adequately represented by the economist's model, and the premise that market participants are motivated by self-interest. When agents know that they have the same model, then it pays for some to deviate to make some extra gains. It is a classical social dilemma problem.

6.2 Market Efficiency

Market efficiency means that resources have been efficiently allocated, and there is no room for Pareto-improvement, at least in principle. We have already seen that the market system cannot always solve for the efficiency problem, as in the case of Prisoner's Dilemma. The interdependence of market players creates an environment that systematically fails to seize upon many Pareto-improving opportunities.

One way to see how belief in market efficiency become self-defeating was pointed out by Sanford Grossman and Joseph Stiglitz (1980, p. 404; see also Lo and MacKinlay, 1999, pp. 5-6). The argument is that information is essential for achieving efficiency. However, since information is scarce, collecting and processing information therefore is costly. If agents believe the market to be efficient, then current prices will reflect (roughly) all relevant information. Thus, there is no need to invest in information. But if market participants stop investing in information, the market becomes inefficient! At the heart of the argument is that information is essentially a public good (Romer, 1990; Stiglitz, 1995), and thus is prone to the conundrum of Social Dilemmas, which again reflect the incompleteness phenomenon. Investing in information, like cooperation in a *PD* game, becomes inferior to not investing (*Defection*).

Modern finance theory is formulated on the assumption of no-arbitrage condition. If this assumption is imposed within the pricing model used by market players, this amounts to assuming the consistency of the model within the model itself, which is contradictory.[13] Furthermore, in financial markets, players aim feverishly to make profits at the expense of other players. This can happen only if there are arbitrage opportunities. Assuming no-arbitrage therefore is self-contradictory. Assuming no-arbitrage, like assuming market efficiency, invites inefficiency, and thus bypassing arbitrage opportunities. Paul Wilmott and David Orrell (2017, p. 143) note: "Ironically, the assumption of no arbitrage creates another opportunity for arbitrage."

Market efficiency ideally is a result of the "invisible hand" of market mechanism. It is an emergent or equilibrium property that cannot be assumed or asserted at the individual player's level (see Ball, 2009, p. 10). From the perspective of Gödel's theorem, the hand is "invisible" in the sense that it cannot be explicitly asserted *within* the system, despite being "visible" from the outside. Thus, to assume market efficiency within the model means that the market's hand is not invisible anymore. Moreover, to impose an *ex post* equilibrium property onto an *ex ante* choice model induces the system to change its behavior in response. This is the basic idea behind Lucas Critique, which argues that an aggregate regularity cannot be exploited without inducing rational agents to adjust their optimal behavior accordingly.

There is no reason to confine the Critique to government actions. Any large enough entity or group of entities can produce the same effect. This has been empirically documented by Rajan, Seru, and Vig (2015), who analyze how securitization of subprime mortgages changed the underlying relations in a manner that caused the failure of the models estimating probability of default. The authors describe

---

[13]Colin Read (2013, pp. 78-80) reports that Paul Samuelson was initially reluctant to close his warrants pricing model by assuming no-arbitrage condition.

the result as "Failure of Models that Predict Failure." As Jospeh Stiglitz (2010, p. 95) points out, models based on data from pre-securitization era were used to create financial instruments, like CDOs and CDSs, that alter the data-generating processes, which makes these models inconsistent. This takes us to the next point.

6.3 The Paradox of Risk

The onset of the Global Financial Crisis, in August, 2007, was described by many as a "Minsky Moment." The basic idea of Minsky's Financial Instability Hypothesis is that, if agents perceive the economy to be stable, they are inclined to take on additional risks, leading eventually to destabilizing the economy—"stability is destabilizing" (Minsky, 1982, ch. 5; Wray, 2016). Since the period preceding the GFC was considered to be a low-risk environment, the Crisis was a manifestation of Minsky's hypothesis.

One way to understand this "paradox of risk" is to note that stability is a public good (IMF, 2009b, p. xxii): It requires the collective efforts of market players to be achieved. Since a public good is essentially a n-person Prisoner's Dilemma game, it is easy to see how believing in the stability of the system might lead to instability. If agents presume that the system is stable, each can follow his or her independent profit-maximizing strategy (i.e. *Defect* strategy), with the belief that this will not reflect negatively onto the agent's payoffs. By assuming the absence of interdependence, agents believe that their unilateral actions will have no feedback effect on their payoffs. This amounts to believing $\neg Bp$, where p is the proposition that an agent will lose upon defection. As discussed earlier, the strategy is self-defeating when all agents follow the same approach.

7 Conclusion: Wise Rationality

Gödel's theorems call for a fresh review of rationality. If the main aspect of rationality is consistency, then rational agents cannot assert their own rationality. We may call it: Wise Rationality. Wisdom is to know the limits of oneself. Gödel's theorems provide a logical framework for recognizing our limits. To be wise, therefore, we ought to acknowledge and respect our limitations.

These limitations imply that markets are inherently incompletable. Value extends beyond competitive strategies. Social dilemmas, therefore, are at the center stage of economic interactions. The implications for economic theory and policy are substantial. There is much for economics to learn from one of the greatest discoveries of the 20th century. At a time when economics is searching for an identity, the incompleteness phenomenon can be a promising starting point.

Acknowledgments

I am grateful for the continuing and fruitful discussions with Prof. Francisco A. Doria. The first time I met "Chico" (as he prefers his friends to call him) was at Workshop on Nonlinearity, Complexity and Randomness, held at University of Trento, Italy, 27–28 October 2009. Since then, we had been in constant interaction over the email and other events. Chico is a great friend and mentor, offering generous help and guidance whenever possible. He is open to discuss and investigate new ideas, regardless of how "weird" they are. His generosity and openness, though, imply no compromise on rigor and sound reasoning. His thirst for knowledge is insatiable, and while he already passed his 70th birthday, he is more active than many who are 20 years younger. I greatly value his friendship and wish him an extended and enjoyable journey in this world and beyond.

I am also thankful to Prof. Douglas Hofstadter for helpful comments and suggestions on a previous draft of this paper. Thanks to Professors Graham Priest and John Stillwell for helpful discussions. Special thanks to Prof. Freeman Dyson for helpful comments and suggestions. The author is solely responsible for the views and opinions (not to mention errors or infelicities) presented in the chapter.

References

[1] Al-Suwailem, S. (2010) "Behvioural Complexity," *Journal of Economic Surveys*, vol. 25, p. 481-526.

[2] Al-Suwailem, S., Doria, F. A. and Kamel, M. (2018). "Is Risk Quantifiable?" In *Complex Systems Modeling and Simulation in Economics and Finance*, Chen, S.-H., Kao, Y.-F., Venkatachalam, R. and Du, Y.-R. (eds.), Springer, Proceedings in Complexity, forthcoming.

[3] Aldrich, J.H. (1977) "The Dilemma of a Paretian Liberal: Some Consequences of Sen's Theorem," *Public Choice*, vol. 30, pp. 1-21.

[4] Anderson, P. (2011) *More and Different: Notes from a Thoughtful Curmudgeon*, World Scientific.

[5] Ball, R. (2009) "The Global Financial Crisis and the Efficient Market Hypothesis: What Have We Learned?" *Journal of Applied Corporate Finance*, vol. 21, pp. 8-16.

[6] Binmore, K. (1987) "Modeling Rational Players: Part I," *Economics and Philosophy*, vol. 3, pp. 179-214.

[7] Binmore, K. (1992) "Foundations of Game Theory," in J. Laffont, ed., *Advances in Economic Theory*, Sixth World Congress of the Econometric Society, Cambridge University Press.

[8] Binmore, K. (1994) *Playing Fair*, MIT Press.

[9] Binmore, K. (2015) "Rationality," in P. Young and S. Zamir (eds), *Handbook of Game Theory*, Volume 4, North Holland, pp. 1-26.

[10] Boolos, G. (1979) *The Unprovability of Consistency*, Cambridge University Press.

[11] Boolos, G. (1994) "Gödel's Second Incompleteness Theorem Explained in Words of One Syllable," *Mind*, vol. 103, pp. 1-3.

[12] Chen, Y. and J.O. Ledyard (2008) Mechanism Design Experiments, in S. Durluf and L. Blume, eds, *The New Palgrave Dictionary of Economics*, Palgrave McMillan, vol. 5, pp. 548-555.

[13] Cornes, R. and T. Sandler (1996) *The Theory of Externalities, Public Goods, and Club Goods*, second edition, Cambridge University Press.

[14] Croson, R.T.A. (2008) "Public Goods Experiments," in S. Durluf and L. Blume, eds, *The New Palgrave Dictionary of Economics*, Palgrave McMillan, vol. 6, pp. 747-753.

[15] Davis, M. (2000) *Engines of Logic*, W.W. Norton.

[16] Debreu, G. (2008) Mathematical Economics, in S. Durluf and L. Blume, eds, *The New Palgrave Dictionary of Economics*, Palgrave McMillan, vol. 5, pp. 454-458.

[17] Doria, F.A. (2017) Axiomatics, The Social Sciences, and The Gödel Phenomenon: A Toolkit, in F.A. Doria, ed, *The Limits of Mathematical Modeling in the Social Sciences*, World Scientific.

[18] Geanakoplos, J., and H. Polemarchakis (2008) "Pareto Improving Taxes," *Journal of Mathematical Economics*, vol. 44, pp. 682-96.

[19] Gödel, K. (1947) "What is Cantor's Continuum Problem?" *American Mathematical Monthly*, vol. 54, pp. 515-25.

[20] Gödel, K. (1995) Some Basic Theorems on the Foundations of Mathematics and Their Implications, in *Kurt Gödel Collected Works, Volum III: Unpublished Essays and Lectures*, edited by S. Feferman, Oxford University Press, pp. 304-323.

[21] Grossman, S. and J. Stiglitz (1980) On the Impossibility of Informationally Efficienct Markets. *American Economic Review*, vol. 70, pp. 393-408.

[22] Hofstadter, D. (1979) *Gödel, Escher, Bach: An Eternal Golden Braid*, Basic Books.

[23] Hofstadter, D. (1985) *Metamagical Themas: Questing for the Essence of Mind and Pattern*, Basic Books.

[24] Ledyard, J.O. (2008) "Market Failure," in S. Durluf and L. Blume, eds, *The New Palgrave Dictionary of Economics*, Palgrave McMillan, vol. 5, pp. 300-303.

[25] Lo, A.W. and A.C. MacKinlay (1999) *A Non-Random Walk Down Wall Street*, Princeton University Press.

[26] Luce, R. D., and H. Raiffa (1957) *Games and Decisions*, John Wiley & Sons.

[27] Minsky, H. (1982) *Can "It" Happen Again? Essays on Instability and Finance*, M.E. Sharpe.

[28] Muth, J. (1961) "Rational Expectations and the Theory of Price Movements," *Econometrica*, vol. 29, pp. 315-335.

[29] Mueller, D.C. (2003) *Public Choice III*, Cambridge University Press.

[30] Nagel, E. and J. Newman (2001) *Gödel's Proof*, edited by Douglas Hofstadter, New York University Press.

[31] Ostrom, E. (1990) *Governing the Commons: The Evolution of Institutions for Collective Action*, Cambridge University Press.

[32] Rajan, U., A. Seru, and V. Vig (2015) "The Failure of Models that Predict Failure: Distance, Incentives, and Defaults," *Journal of Financial Economics*, vol. 115, pp. 237-260.

[33] Read, C. (2013) *The Efficient Market Hypothesists: Bachelier, Samuelson, Fama, Ross, Tobin, and Shiller*, Palgrave McMillan.

[34] Romer, P.M. (1990) "Endogenous Technological Change," *Journal of Political Economy*, vol. 98, pp. S71-S102.

[35] Sargent, T. (1993) *Bounded Rationality in Macroeconomics*, Oxford University Press.

[36] Sargent, T. (2008) Rational Expectations, in S. Durluf and L. Blume, eds, *The New Palgrave Dictionary of Economics*, Palgrave McMillan, vol. 6, pp. 877-881.

[37] Sen, A. (1970) "The Impossibility of a Paretian Liberal," Journal of Political Economy, vol. 78, pp. 152-157.

[38] Sen, A. (1977) "Rational Fools: A Critique of the Behavioral Foundations of Economic Theory," *Philosophy & Public Affairs*, vol. 6, pp. 317-344.

[39] Sen, A. (2017) *Collective Choice and Social Welfare*, Expanded Edition, Penguin.

[40] Smith, P. (2013) *An Introduction to Gödel's Theorems*, second edition, Cambridge University Press.

[41] Smullyan, R. (1987) *Forever Undecided: A Puzzle Guide to Gödel*, Alfred A. Knopf.

[42] Smullyan, R. (1992) *Gödel's Incompleteness Theorems*, Oxford University Press.

[43] Smullyan, R. (2013) *The Gödelian Puzzle Book: Puzzles, Paradoxes & Proofs*, Dover Publications.

[44] Smullyan, R. (2014) *A Beginner's Guide to Mathematical Logic*, Dover Books on Mathematics. Dover Publications.

[45] Starr, R. (2011) *General Equilibrium Theory: An Introduction*, Cambridge University Press.

[46] Stiglitz, J. (1999) Knowledge as a Global Public Good, in I. Kaul, I. Grunberg, and M. Stern, *Global Public Goods*, The United Nations Development Program, Oxford University Press, pp. 308-325.

[47] Stiglitz, J. (2010) *Freefall: America, Free Markets, and the Sinking of the World Economy*, W.W. Norton.

[48] Stillwell, J. (2004) "Emil Post and His Anticipation of Gödel and Turing," *Mathematics Magazine*, vol. 77, pp. 3-14.

[49] Stillwell, J. (2010) *The Roads to Infinity: The Mathematics of Truth and Proof*, Taylor & Francis Group.

[50] Stillwell, J. (2016) *Elements of Mathematics: From Euclid to Gödel*, Princeton University Press.

[51] Tieszen, R. (2017) *Simply Gödel*, Simply Charly, Kindle Edition.

[52] Tsuji, M., N.C. da Costa, and F.A. Doria (1998) "The Incompleteness of the Theory of Games," *Journal of Philosophical Logic*, vol. 27, pp. 553-568.

[53] Weintraub, E.R. (2002) *How Economics Became a Mathematical Science*, Duke University Press.

[54] Wilmott, P. and D. Orrell (2017) *The Money Formula: Dodgy Finance, Pseudo Science, and How Mathematicians Took Over the Markets*, Wiley.

[55] Wray, L.R. (2016) *Why Minsky Matters*, Oxford University Press.

[56] Yanofsky, N. (2013) *The Outer Limits of Reason: What Science, Mathematics, and Logic Cannot Tell Us*, MIT Press.

In what sense space dimensionality can be used to cast light into cultural anthropology?

Francisco Caruso & Roberto Moreira Xavier

Centro Brasileiro de Pesquisas Físicas

> *It is a pleasure to dedicate this work to our friend Francisco Antonio Doria, scientist and philosopher, a man of multiple interests, in the hope that he will enjoy reading it.*

1 Introduction

Humans have always constructed spaces, through Mythos and Logos, as part of an aspiration to capture the essence of the changing world. This has been a permanent endeavour since the invention of language. By doing this, in fact, Humankind started constructing itself: we are beings in constant evolutionary process in real and imaginary spaces. Our concepts of Space and our anthropological ideas, specially the fundamental concepts of *subject* and *subjectivity*, are intertwined and intimately connected. We believe that the great narratives about Humanity, which ultimately define our view of ourselves, are entangled with those concepts that Cassirer identified as the cornerstones of culture: *space, time,* and *number* [1]. To explore these ideas, the authors wrote an essay, in 2017, in a book format, in which the fundamental role of real and imaginary spaces (and especially of their dimensionalities) in the History of Culture was discussed. This book, titled *O Livro, o Espaço e a Natureza: Ensaio Sobre a Leitura do Mundo, as Mutações da Cultura e do Sujeito* [2], has a preface written by Francisco Antonio Doria. As many of the issues treated there are among his multiple interests, it was decided to revisit here the problems of subjectivity and subject's relationship with the dimensionality of space including the question of the architecture of books and other writing supports.

2 A voyage through the History of Culture

In Ref. [2], the story of a navigation through the intricacies of writing (and alphabets) – in its different historical forms of registration –, by the jolts and rambles of Culture and by the extraordinary changes in Man's view of the world and of himself, is narrated. The first general idea was to highlight landmarks, traces, records of major changes in the human imaginary, in a view that can be associated with Fernand Braudel's "longue durée".

As an element of union and convergence between such apparently disparate narratives is the gradual construction of the concepts of space (real and imaginary). Special attention was given to space dimensionality and its influence on the perceptions and descriptions of the World and of Nature.

O Livro, o Espaço e a Natureza was conceived and built in order to shed some light on the confluence of the following questions:

- Is there a direct relationship between the various manifestations of Culture and the physico-philosophical conceptions of Space?

- How does the different forms of writing registration affect or were affected by the current conceptions of Space?

- To what extent can the way of thinking be conditioned by the way of writing, thus influencing Culture and our understanding of Nature?

- How is human subjectivity and the notion of subject constructed based on the characteristic cultural soil of each era?

Ultimately, the authors hope to have contributed to consolidate the acts of reading and writing in a prominent place in the History of Ideas, by underlining their links with the construction of the absolutely fundamental concepts of space and time, as well as with the humankind subjectivity own formation.

The book [2] also aims to cast new light on philosophical anthropology, linking the issue of the subject to the dimensionality of real and imaginary spaces: from the verticality of the medieval theocentric being to the postmodern multidimensional being. In fact, so far the influence of space in anthropology is considered, we share Cassirer's position *cum grano salis*. Indeed, we admit that one of the most important influences of *space* concepts on Culture occurs through the dimensionality of the writing support: two-dimensionality, in the case of *volumen*, and the three-dimensionality of *codex* [2]. On the other hand, the postmodern subject is characterized by the multidimensionality of digital networks, which tends to infinity. This particular original contribution will be the theme of our tribute to our dear friend Francisco Doria, on this very festive occasion.

3 Subjectivity and Culture: highlights

With the conquest of language, prehistoric man begins to deepen his place in the Universe, imagining and conceiving myths. Man and Nature are intertwined and gain meaning when thought as a whole. With the invention of Philosophy – the appearance of *Logos* – *Myths* are relegated to another plane and *Physis* becomes the center of reflection on Nature, with early philosophers looking for the physical (material) principle of all things: the *arkhé*. *Logos* and *Physis* are united and conceived together in the hands of the so-called physiologists. In the words of Karl Popper, "Science must begin with myths, and with criticism of myths" [3].

We cannot determine when the first man, in the face of unpredictable Nature and the starry sky, was taken by astonishment when realizing his own finiteness. Nevertheless, what we can affirm, with certainty, is that the perplexing problem of the origin and identity of Man appears in the oldest myths and has assumed a central place in Philosophy since its beginnings.

The issue of subjectivity and human identity is a task that demands a huge text which is completely out of our scope in this paper. What we intend here is to shed some light on few aspects of the subjectivity formation, especially its relation to the history of the concepts of space and time.

Indeed, at decisive moments in his own history, Humans reorganize their view of the World [4], their relationship with the Universe, through a significant change in their space-time concepts. In those moments, the perception of themselves – their identities – also changes.

In order to simplify our narrative and make it intelligible in a few pages, we will make a brief summary of the landmarks of subjectivity history, from Classical Antiquity to the end of the *Middle Ages*, then, to characterize the Modern Era, we will use the typology introduced by the sociologist Stuart Hall.

From the Greek philosophers Socrates and Plato, Man occupies the center of philosophical speculation.

Socrates is considered a landmark in the history of Greek Philosophy dividing it into two periods: the so-called pre-Socratic and the post-Socratic. Before him, physiologists were concerned with seeking the ultimate foundation of things, the primordial substances and the principles by which the plurality of *Physis* results. With Socrates, Man becomes the main focus of Philosophy. In fact, Man is thought of as a social being, somehow inserted in a society that must be just and orderly (in the sense of the Greek word *Kosmos*). For Socrates, *Truth* is inside Man, but he cannot reach it alone. It is through *dialogue*, understood as a way of exercising reason, that Truth will be discovered. Moral life is guided by thoughts which recognize the ideal values that are reached through Philosophy and Reason. Therefore, in Socratic philosophy, *Man*, *Ethics*, and *Epistémé* are linked in a stable whole through the concept of *Cosmos*.

In a very simplified way, for Socrates, *Man is his «Psyché»*. Happiness would result from the search of truth, which is sought through dialogue. Permanent questioning and inquiry are the way of building knowledge, always provisional. Thus, Man is a being who questions, dialogues, builds fleeting knowledge and concludes, with the Athenian philosopher: *I only know that I know nothing*.

According to Plato, human nature is similar to a difficult text, the meaning of which must be deciphered by Philosophy [5]. Given that, for Plato and Socrates, Man is a being who dialogues in search of truth, thus he constantly needs the other, his mirror. The mirror metaphor is recurrent in Philosophy. Gaston Bachelard, for example, referring to the human capacity of dreaming, states that *"the dreamer cannot dream in front of a mirror that is not deep"* [6], reinforcing the idea of a social being.

Aristotle starts from Plato's idea, but expands it, affirming, fundamentally, that *man is his «Nous»*. The term is commonly translated as *intellect* (or also *mind*, *intelligence* and *thought*). However, the translation, while correct in itself, leads the modern reader to a predominantly gnosiological and psychological problem, while the original semantic area in Greek includes a much broader problem: from ontology to metaphysics, from physics to cosmology, from anthropology to morality and even religion [7].

Philosophical speculation about man continues its course in the Hellenistic period. With the rise of Christianity, Eastern influence is present. A great symbolism was built around the life and death of Jesus Christ, which has influenced the human attitude for a very long time, encompassing Late Antiquity and the entire Middle Ages. It should be emphasized that we refer to the fact that Christ presented his contemporaries with a new image of Man, much more based on internal attitudes than on external appearances, in addition to having especially valued *love*, giving it at least the same weight as *justice* [8].

After a difficult period of consolidation, Christianity is adopted as the official religion of the Roman Empire by Constantine. At that time, the foundations on which the Western Cultural History will evolve are established: Greek Philosophy, Christianity and Roman Law. Saint Augustine is the philosopher who explicitly articulated this synthesis. It should be noted that this is the period in which the *codex* is consolidated as a support for writing, as will be seen in Section 3. Thus, the conditions are created for the appearance of the *medieval Man* (Section 4).

The crisis following the fall of the Roman Empire heralds a new era, a new mentality. This moment – the High Middle Ages – sees the confluence of several ideas that are articulated with the radical experience of *verticality*, in a religious society, turned to Heaven. That said, the three-dimensionality of space is associated with the concept of linear one-dimensional time, due to Saint Augustine, to form and reflect the emerging self-awareness of this *theocentric man*. Indeed, the Middle Ages comprise a long period of time when religion defines all aspects of life – family, work, the divine right to define social relations – making society essentially theocentric: Men who live verticality build cathedrals. This verticality of being is contemporary with the three-dimensionality of the *codex* as we shall see (Section 4). These interrelations, which emerge simultaneously, form the cultural soil from which a new and revolutionary mechanical vision of the world will be born, many centuries later [9]. That said, we can examine the emergence of the modern Man.

It is not difficult to realize that the multiple conceptions which Man has built about himself throughout History are so many and so varied that trying to synthesize them is practically impossible. However, there are some questions constantly revisited, since always: – Is Man marked by his own personality or is he strongly influenced by life in common, by society? – Is he essentially good, as Jean-Jacques Rousseau wants, or, as so much debated in the Christian world, does he bear the "mark of Cain" of capital sin, of wickedness?

Stuart Hall sees the history of Man in the modern period in an original way and tries to answer these questions. For him, modern

Man, who announces the Renaissance [10], looks at the world and questions it – in the manner of the ancients – and is called by him *enlightened subject* [11]. He is significantly distinguished from the *medieval man* for his openness, his new discoveries, marked by the valorization of what is *new*: New World, New Science [12], New Art, in short, Man's new relationship with the Cosmos. This subject gains, later, new meanings with the ideals of freedom and equality of the French Revolution.

As social interactions intensify, Hall identifies the emergence of a new perception of man about himself and forges the expression *sociological subject*, who has consequently his identity formed from the interaction between individual and society.

This individual has an inner core which changes through dialogue that social life and the cultural world offer. This is the characteristic subject of the 19th century and the first half of the 20th century.

If the *enlightened subject*, born in the Renaissance, according to Hall, saw the press invention and the generalization of the book, the Great Navigations, the Reformation, the Copernican and Newtonian Revolutions, the *sociological subject* experiences the era of scientific, technological and social transformations of the 20th century. Today, we are witnessing his progressive disappearance and the birth of the *postmodern subject* [13], possessing a multifaceted identity that, in reality, can be considered as a superposition of several identities which, in some cases, can even be contradictory [2].

Gradually, thinkers begin to realize that the influence of society is essential in the formation of Man. Perhaps the thinker who best expressed this view was Karl Marx. His conception was fragmentally exposed and later studied and systematized by his disciples, like Erich Fromm [14]. Supported by analysis of society after the Industrial Revolution, the Marxist conception of Man is based on two concepts: alienation and merchandise fetishism [15].

Indeed, the exacerbation of the characteristics of the *sociological subject* by the impressive increase in social relations, resulting from the advent of the train, the telegraph, the radio and television, sets the stage for a new phase of Human history, that of the *postmodern subject*. His characteristics appear little by little, but can only be clearly defined with the advent of internet, which makes men and women living beings in a multidimensional imaginary space.

In Section 7, we will examine the impact of this radical change. First, however, let us examine the role of writing in different historical moments, encompassing these significant cultural revolutions, from Antiquity to the 20th century.

4 From volume to codex

Two major phases stand out in the history of Greek civilization. The first, in which *orality* predominates, is Homer's world, and the second, marked by the *affirmation of writing*, is the universe of Philosophy, based on Plato's work. This division is not a mere academic simplification, since the alphabetic transcriptions of Homer's texts can be seen as "the beginning of a relationship between oral and written, a relationship that has proved fruitful" [16].

Such a relationship, in reality, leads us to the Platonic distinction between *doxa* and *epistémé*, which had an essential role in the dissemination of the written word, through the *volumen* [17]. Indeed, the pre-alphabetic oral and poetic tradition, characterized by *parataxis* – generically called by Havelock *oral mental state* –, constituted the main obstacle to scientific rationalism, to the use of analysis, the classification of experience, its new systematization in the sequence of cause and effect. This is why the poetic mental disposition and, therefore, poetry, is for Plato the archenemy. It is this oral mental state that Plato associates with *doxa*, considered to be an obstacle to *epistémé* – to scientific discourse – associated with *hypotaxis*, that is, to alphabetic culture.

However, it should be noted that the transformation of *orality* into *literality* in the classical period was a slow process. A beautiful work on this gradual change was made by Havelock [18]. There, the author discusses the extent to which human consciousness itself changes, when culture becomes literate. He also discusses how this new form of communication affects the content itself and the meaning of the texts.

Plato and Aristotle's legacy will open new perspectives in the way of reflecting on *Physis* and on Man, nested in a new *Zeitgeist*. These philosophers are fundamental milestones in the History of Philosophy and in the reflection on Man. It was not at all an accident that, in the middle of the Renaissance, Raffaello Sanzio represented them at the center of his famous "The School of Athens", Plato taking the *Timeo* in hand and Aristotle, his *Ethics*.

From these two notable thinkers, Philosophy starts to be written. The way is open for the establishment of what can be called a new *literary mental state*, which had significant consequences on human communication and directly reflecting on the function of *volumen*. In practice, the philosophers under the influence of Plato's severe criticism of oral culture and the impact of Aristotle's great synthesis have contributed to the changes in the status of *volumen*.

The *volumen* – and by extension, the library – becomes, then, a welcoming space for different authors and, of course, for different arguments, in a new temporality; the diachronism of oral speeches is opposed to the synchronism of written texts [19].

In spite of this achievement, another support of writing was about to change Europe: the *codex*.

Although there are records of some isolated appearances of the *codex*, already in the first century AD [20, 21], it only spreads out from the third century onwards, and has a very particular meaning for Christianity, as material support of the Bible. According to Jacques Le Goff, "Le Christianisme est une religion du Livre. Cette vérité n'a jamais été plus vraie que dans l'Occident médiéval" [22].

The *codex* is considered by some authors to be the first revolution in the history of book. Consisting of a sequential set of small groups of paper sheets sewn together, it has a format very similar to that of the modern book.

The material and practical advantages of *codex* over the *volumen* are many and significant. In general, its weight, much less than that of the parchment roll, makes it easier to handle. In addition, its shape also facilitates storage and it is no longer imperative to use both hands while reading, as required by the volume. You can also write on both sides of the sheet of paper. There is still a further but not insignificant advantage of facilitating random access to written work.

Some scholars attribute the adoption of *codex* by Christians to purely economic reasons, which seems to us a simplistic way of minimizing the impact of the change in the mentality of that time. On this matter, we agree with Úrsula Katzenstein [23] when she says that

> "Perhaps (...) some (...) prominent person in ancient Christianity (...), no matter the ultimate purpose of his inspiration, managed to (...) imagine a different format for Christians manuscripts of the scriptures, which differentiated them both from the parchment scroll of Judaism and the papyrus scroll of the pagan world (...), imposing its use on all Christianity (...). It was possibly an attempt by the ancient Christians to differentiate their writings from other literary forms, to mark them as sacred books."

Over time, *codex* takes on a very peculiar meaning, in addition to arousing superstitious respect among scholars during the Middle Ages. In short, the Book (the Bible) becomes the symbol *par excellence* of the relationship between Man and God in a new Christian *Weltanschauung*.

We have seen that Christianity's adoption of *codex* articulates with the issue of *verticality* or, in other words, with *three-dimensionality of space* [2]. It is through this third dimension that the *medieval subject* is related to God. The *codex* invention in the High Middle Ages occupies a prominent place in the formation of the emerging " three-dimensional" *theocentric subject*. Therefore, a metaphor could synthesize this long historical period: *the codex is the message,* to paraphrase McLuhan [24]. The essence of such an object which will be associated with Culture and its transmission for a long period is its three-dimensionality.

Here it is worth to emphasize that the awareness of the three-dimensionality of space and of the world, both real and imaginary, and the linearity of time pointed out by Saint Augustine are fundamental steps for the creation of Modern Science, many centuries later. These are the bases of a conceptual scenario in which movements and transformations can be studied and described by Modern Science. The greatest difficulty is to understand movement and its causes. Indeed, given the three-dimensional Euclidean space and linear time, the scenario of the World where Physics and History can be developed as possible sciences is set, but the understanding of motion will require, as we will show later, a new causal scheme, a complete revolution in Science and Culture, associated with the name of Isaac Newton, but also indebted to Copernicus and Galileo.

To conclude this Section, it is important to stress that this association between Space and God is not new. Actually, it is well known that Judeo-Christian conceptions of space were since early times identified with God. In fact, in 1st century Palestinian Judaism, the term *makom* – as the name of God – was adopted for *place*. A notable example of the confluence of stoic conceptions and Judeo-Christian beliefs around the nature of space and its identification with divine omnipresence was sustained by Iamblichus. The impact of this association is clearly seen, in a special way, in the evolution of mechanical theories in the Middle Ages, and, in the 17th and 18th centuries, culminating with the statement that space is nothing more than an attribute of God, or even identical to God. For example, for the English philosopher of the Platonic school of Cambridge, Henry More, space is the *divine extension*, while, for Isaac Newton, absolute space is *God's sensorium*.

5   The medieval subject

From a philosophical point of view, Aristotle's Cosmology is consistent with the essence of Medieval Philosophy. Indeed, in Aristotelian Cosmos, there is a clear difference between the sublunar and the supralunar worlds. The highest world, the world of celestial bodies, is made of an incorruptible substance – the *ether*, or the *quintessence*. In it, the movements of these bodies are eternal, because God moves the Universe. In our lowest world, underlining, everything is perishable, everything can decline and movements are no longer eternal. Those ideas are attractive since they are based on the judgment that the cause and origin of everything is the *One*, the Absolute. From *Unity* the variety takes place, in a type of degradation process.

The whole world is supported and hierarchically articulated by a kind of golden chain (*aurea catena*). All material or spiritual things belong to this current. In this unifying thought there are two types of hierarchy: that of *existence* and that of *value* [25]. Since they are not opposed, there is a profound unity between medieval philosophy, morals and culture. Therefore, just as the two Aristotelian worlds are not made of the same substance and do not obey the same laws of motion, so the structures of the ecclesiastical, political and social world are subject to the same principle of hierarchy [26]. The feudal system is an image of this hierarchical system. It is necessary to understand how the hierarchical order of Aristotelian Cosmos – in which the Earth occupies its center, where space is hierarchized and, in this hierarchical order, Man occupies the highest place – is convenient for the Catholic Church and medieval society. The reason is that Christian doctrine, which permeated and dominated Middle Ages in the West, is based on the assumption of the existence of a general Providence governing the world and the destinations of Man.

The adoption of the *codex* by Christianity marks the meeting of religious practice and meditation – correlated with reading the Bible and Christian comments – with the experience of Space. Perhaps the most striking example of this association is in new tendencies of Architecture towards the sky, with the construction of magnificent cathedrals [27]. As for these, it is undeniable that they introduce verticality in the medieval cities in which they were built. In a dominantly "shallow" (plane) architecture, these imposing monuments stand out to reaffirm the importance and glory of God in those theocentric societies [28].

In fact, it was Christianity that very soon adopted verticality, that is, it favored the upper-lower system, inspired by the resurrection and ascension of Christ, as well as the ascension of souls. Christian world and symbolic space thus become effectively three-dimensional. In the Middle Ages, as Jacques Le Goff teaches us [29],

> "(...) this system will guide, through the spatialization of thought, the essential dialectics of Christian values. To ride up, to rise, to go higher, this is the stimulus of spiritual and

moral life while the social norm is to continue in your place, where God placed you on earth, without having the ambition to escape your condition and taking care not to stoop, not to deprive oneself."

In this way, a *medieval subject* is shaped and trapped in his position in the theocentric society of that time, in a kind of effectively two-dimensional space, insofar he is prevented from escaping his social condition, leaving the third dimension as the only possibility of salvation.

These essential changes in the spatio-temporal categories of the Christian imaginary led, still according to Le Goff, to a slow but important process of belief in an intermediary kingdom between Paradise and Hell – the *Purgatory* –, the third place. Expanding the geography of the beyond was an immense operation for Christians, because, in the last analysis, the inclusion of the Purgatory between the lower and upper kingdoms represented the adoption of the concept of an intermediary that corresponds, from the point of view of logical structures, to profound changes in social and mental reality in the Middle Ages.

It can be said, then, metaphorically, that the Christian world is three-dimensional not only because of the importance attributed to the *upper-lower* axis in the medieval imaginary, but also because, in the Christian Middle Ages, binary logic is replaced by tertiary logic.

It is worth highlighting another general characteristic of the Late Middle Ages. The religious mental state slowly fades away and begins to give place to a new state of mind that will pave the way to the Renaissance and to the Scientific Revolution.

There is also an area of scientific knowledge, intrinsically related to the concept of Space – and also to God – which is about to change: it is the *Medieval Cosmology*. The Medieval conceptions of Cosmos, dominated by the thought of Plato and Aristotle, will not resist Copernican criticism (Section 6). The Closed World has its days numbered.

A new subject emerges from the Renaissance, who came to be called by Stuart Hall *enlightenment subject* – as already said –, which, after all, results from an inextricable entanglement of concepts which are mirrored in the crucial question of the relationship of Man with the Cosmos has in the relations between subject and object. This implies two notable capacities of this new subject: that of coming to conceive the *infinitude* of the Universe and of three-dimensional space and that of not succumbing to the amazement caused by this non-divine infinity (Section 6). The *enlightened subject* does not become small. On the contrary, as has already been seen, he grows, from the moment when he realizes how much, through his intellect, he is perfectly capable of embracing, conceiving and rationalizing all these questions that touch infinity, from the microcosm to the macrocosm. Perhaps this is why René Descartes places the soul, the intellect – *res cogitans* – at the core of his view of the subject, as opposed to *res extensa*.

6 Freud's first narcissistic wound and the new subject

The publication of *De Revolutionibus Orbium Coelestium* by Copernicus was a hard stroke to *medieval subject*, who believed himself made up in the image and likeness of God and a privileged occupant of the center of Universe. He suddenly becomes a peripheral being, as the Earth, following Copernicus, comes to be seen only as one of the planets that revolve around the Sun. This impact on the collective imagination became known as the first of the so-called three narcissistic wounds, an expression coined by Sigmund Freud. The other two wounds refer to the contributions by Charles Darwin and Freud himself.

Alexandre Koyré goes so far as to claim that the publication of Copernicus' greatest work marks the end of the Middle Ages. In his words [30],

> "The year 1543, the year of the publication of the *De Revolutionibus Orbium Coelestium* and the death of the author, Copernicus, marks an important date in the history of human thought. We are tempted to consider this date to mean the end of the Middle Ages and the beginning of modern times, because, more than

the conquest of Constantinople by the Turks or the discovery of America by Christopher Columbus, it symbolizes the end of a world and the beginning of other."

Copernicus' heliocentric system calls into question the unity of Aristotle's Physics. The unification between the physical description of motion on Earth (or in the sublunar world) and in Heaven is shaken because Aristotelian Physics, Astronomy and Cosmology depend on the hypothesis that the Earth is the center of the Universe. In the Copernican system, Astronomy becomes heliocentric, while physics remains geocentric. This fact has an immense importance in the history of ideas at the Renaissance. For instance, the Italian philosopher and historian of science Paolo Rossi, referring to the Copernican system, says that [31]

> "the admission of the terrestrial movement and the acceptance of the new system involved not only a reversal of the astronomical structure and physics, but also a modification of current ideas about the world, a new assessment of the place and meaning of man in the Universe."

There is no doubt that the Copernican Revolution scared a lot by revealing a new spatial scale. Although, in fact, the immediate result of the publication of *De Revolutionibus* has been to spread skepticism and disturbance, the theory exposed in it proved to be more effective in the long run.

However, the Copernican Cosmos is still finite. Copernicus gives movement to the Earth and claims that the visible world – the world of fixed stars – is not measurable, it is *immensum*. Koyré comments that, in a way, it is curious that Copernicus gave the first step to think of an infinite World by stopping the movement of the material orb of the fixed stars, but has hesitated to give the second, that of dissolving this sphere in boundless space.

Anyway, shortly afterwards, some Copernicans took the step that the Polish astronomer did not take, declaring that the sphere of fixed stars does not exist. "In the starry skies the stars were located at different distances from Earth and that these skies «stretched infinitely upwards»."

Inspired by Copernican ideas, Giordano Bruno stated categorically the infinity of the Universe, with the possibility of life in other planets [32].

About this possible infinite world, Blaise Pascal will say later, in the 17th century: "Le silence éternel de ces espaces infinis m'effraie" [33].

Returning to the historical perspective, we would now like to emphasize that, in our opinion, Andreas Vesalius' contribution to Anatomy goes to the opposite direction of that of Copernicus, regarding the position of Man in the Renaissance World, vis-à-vis Science.

Vesalius, in the same way as Galileo, was engaged in an ongoing struggle against philosophical authorities. In the same year that *De Revolutionibus* comes to light, his classic work *De humani corporis fabrica* is published. He did not give space in his work for metaphysical speculations. In it, to the detriment of valuing abstract theories, there is a clear valuation of observations and experiments, made from the dissection of corpses, in the best Leonardian spirit. The Belgian doctor's work has an interest that transcends the particular; its merit, in addition to being scientific, is also philosophical. Indeed, Vesalius removes Anatomy from medieval mystical speculations, while, also as an act of rupture, Galileo will seek to leave Science away from theological disputes.

On the other hand, it should be remembered that, in the Middle Ages, Anatomy was considered a "minor" activity, as it dealt with the dead and required hard (and somewhat repulsive) practical work. Vesalius' strict observance in solving his problems in perfect conformity with experience will give Anatomy a new status, and lead him to follow this new path of intellectual honesty suggested by Leonardo da Vinci, bringing Anatomy and, consequently, Man and Human body, to the core of scientific debate in the Renaissance.

Therefore, one can say that the year 1543 witnesses two antagonistic movements: one in which Man is put aside by Copernicus (as a corollary of his Astronomical Theory) and another that brings Man's body to the center of interest of Medicine.

It is worth noting that many authors consider that the Cartesian idea in which *res cogitans* is opposed to *res extensa* is an expression of these opposite movements and marks the birth of the Subject and the notion of subjectivity. We cannot forget that, if Descartes puts the intellect at the center of his reflection, Vesalius makes the human body the central object of the study of life. Thus, it is through the effort of these two thinkers that Man and Subjectivity come to occupy the center of the new Worldview.

The formation of the *enlightened subject* depends also on a central problem of Renaissance, namely that of *self-consciousness*. Thinkers, like the Italian poet and writer Francesco Petrarca, considered to be the father of Humanism, the German cardinal Nicholas of Cusa, the Italian Neoplatonic Giovanni Pico della Mirandola and the Italian philosopher, Marsilio Ficino, another exponent of Humanism, contribute to a new personality ideal of the Renaissance, to the formation of a *new subject*. Man begins to relate to the world no longer as in *Middle Ages* and this change is related to the general problem of Renaissance involving the relationship between *subject* and *object*. In Cassirer's opinion,

> "Man is for the universe, self is for the world just as the contained is for the continent. Both determinations are equally essential to express the relationship between man and cosmos. Thus, there is a constant reciprocity between them, the constant transformation of one into another."

According to Ficino, human action in all its manifestations – artistic, technical, philosophical or religious – basically expresses the divine presence of an infinite *mens* (mind) in Nature. Man is exalted as a *microcosm*, synthesis of the Universe. This view is not new, as it was already discussed by Plato and the Neoplatonists, but it will acquire new dimensions in the Renaissance and, especially, in the Enlightenment.

A beautiful example of this relationship between God and Man, different from that found in the Middle Ages, is implicit in the artistic representation of the *Vitruvian Man* by the brilliant Italian painter and scientist Leonardo da Vinci, made around 1490.

In this famous da Vinci's drawing, a naked man is seen represented simultaneously in two different positions with the arms inscribed in a circle and in a square. It is well known, since Classical Antiquity, that the *circle* is considered a symbol of perfect movements, of divine perfection or, in short, of God himself. The other geometric figure used in the representation of this Vitruvian Man is the *square* which, unlike *circle*, is not a form that appears in nature; it is, therefore, a construction of the human mind.

By simultaneously using the two geometric figures to inscribe the same man in two different positions, Leonardo is addressing the old question of *squaring the circle*, but not only from a mathematical or even a technical point of view. The Italian genius, in this drawing, is telling us two complementary things about Man: on the one hand, that the (golden) proportions of the human body reflect the divine character of the Creator; on the other, that this same Man may have something divine, since he is able to perceive in Nature the presence of a divine *mens*. God offers the circle, and Man, the square, as concatenating forms of *order*. God permeates the entire Book of Nature, from which subjects read natural relations and apprehend Nature from Geometry. It is through the intellect and a geometric framework of the world, therefore, that this new Man can rise to God. In this way, microcosm and macrocosm would be related, in the same way that Man is in harmony with God. We can foresee here traces of Baruch Spinoza's maxim *Deus sive Natura*.

7 The postmodern subject

The substitution, or rather, the tendency to replace the paper world by the digital universe, seen through the screen attached to a computer, was the starting point for the reflection of several authors on the future of the dissemination of written texts; whether or not a revolution in the art of reading and writing will result, only History will tell us. If the computer effectively replaces the book, we will have a new order of spatial imagery, with reflexes, still unknown, on the forms of contemporary thought, as the French historian Roger Chartier warns us, for example [34]:

> "the universe of electronic texts will necessarily mean a departure from mental representations and intellectual operations specifically linked to the forms that the book had in the West

seventeen or eighteen centuries ago. No order of discourse is, in fact, separable from the order of books that is contemporary with it."

Even though there are still no signs that the book will have the same fate as the *volumen* and the *codex* – on the contrary, it is known that the computer increased the amount of written texts on paper –, some significant changes in culture have undeniably been introduced by the computer.

In the mazes of networks [35], in the digital universe of the *internet*, a new type of random access is changing the relationship between Man and the text, favoring subjectivity, subjective interpretations and fragmentation of reading, which will bring unexpected consequences, as we will see. It also brings back the dream of a *bibliotheca universalis*, now without walls or borders [36]. The *book* had solved the problem of *temporality* of information; *computer* will solve the problem of *spatialization* of information. Indeed, the book allows the reader to "dialogue" with writers from different eras, but the consulted books must be in the reader's hands, whether at home, at work or in a library. Reader and printed book must be necessarily in the same space. With access to digital books stored in virtual libraries, through the computer (or modern cell phones) and the internet, the reader can be anywhere in the world.

New digital technologies also allow for a new treatment and diffusion of images, particularly notable in Medicine and Science, as well as in the fields of Art and Communication [37].

In summary, video clips in the universe of the new media – which fragments space-time – may be considered the perfect symbol of a society in which order, memory and the causal connection of facts are undervalued. To this corresponds, in Science, the interest in complex systems – chaos and fractal geometry – and the abandonment of the Cartesian program, perhaps an omen of the formation of a new mentality, of a new *postmodern mental state*.

The postmodern, contemporary subject synthesizes in his body and mind the very history of Man. He keeps the ancestral memory of all the perplexities that have built us and results, essentially, from the exacerbation of the social ties that mark men and women of the 20th century.

In reality, these social ties grew exponentially with the advent of radio and television. This fact has been technically and socially determining for the *global village* utopia and for the revolution that follows with computers and internet.

One of the most remarkable legacies of the computer and the computer age was the fact that, in 1990, the British physicist Tim Berners-Lee developed at CERN, the *World Wide Web*, or simply Web or even www, which is nothing more than a system of hypermedia documents that are interconnected and executed on the Internet, articulated in a *cyberspace*. This network, initially designed to meet specific communication needs among a large number of physicists, technicians and administrators who worked on high energy physics experiments, had a spectacular reach and is currently accessed by more than 4.5 billion people in the World.

In this scenario, however, there is a highly significant factor, which will mark a strong qualitative difference between the *modern sociological subject* and the *postmodern subject* of late 20th century and the beginning of the 21st century. It is the exaggerated *fragmentation* of everything, it is what can be called the *decentralization of the subject*. The perception that this new subject has of the World and of himself is built in an imaginary space of virtually infinite dimensionality, created by the internet and social networks.

It is the virtually unlimited www network that generates this imaginary, symbolic space, which, for all intents and purposes, has an infinite dimension. In it, Man gets lost. Pascal's perplexity in the face of an infinite (real) space is now repeated, in another context: the experience of facing an imaginary (abstract) space effectively infinite. The fractal character of the network also stands out. Space and time are fragmented. All Internet stories and narratives are incomplete instant flashes of the World, with no past or future. Man completely dissolves into what the Polish sociologist Zygmunt Bauman called *liquid modernity*.

In our opinion, the emergence of *postmodern subject*, using Stuart

Hall's expression, or *liquid subject*, in Bauman's nomenclature, is due to this almost tangible liquidity and the impact of *infinity*, with a difference: while the *enlightenment subject* was forged by facing the infinitude of the Cosmos and the real three-dimensional space, the postmodern subject, our contemporary, is characterized by the ease with which he gets lost in an infinitude of spaces in the network, with the speed of pressing an *enter* key. Or, if we prefer, virtual multidimensional spaces come to have reality in the practical life of the individual. We are referring here to the basic characteristic of several concatenated hypertexts that allow that, at each *web page*, the navigator can open another and another page and so on, practically without limits. Such text fragmentation, corresponding to a fragmentation of time and space, comes at a price. It is not hard to imagine that it can contribute to the dissolution of identity.

Indeed, it can be argued that the perception of *cyberspace* "not only develops a multiplicity of points of view, but also a set of selves". Or that "the individual self is giving way to the vague edges of identity" [38]. According to the British artist Roy Ascott, there is a glimpse of a recurring redefinition, capable of creating multiple identities that operate in different places in cyberspace. This self, at the same time, multiplied, divided and dispersed, seems to be fundamental to life in the *net*. It seems quite evident to us that these essential characteristics of this new subject force him to pulverize his own history. This kind of permanent denial of his roots and his history can lead to a corrosion of his own character, to which we are all exposed today. This is particularly true if we take into account the new labor relations in savage capitalism such as that seen in modern economy. We refer here to the point of view of the American sociologist and historian Richard Sennett [39]. Let us see, albeit very briefly, how this process can take place.

For Sennett,

> "character development depends on stable virtues such as loyalty, trust, commitment and mutual help. Features that are disappearing in the new capitalism. In some respects, the changes that mark this new system are positive and have led to a dynamic economy, but they have also eroded the idea of pur-

pose, integrity and trust in other aspects, aspects that previous generations considered essential for character formation."

Its initial premise is that the motto of survival in the modern economy can put people's emotional lives adrift. We are all exposed to a strong tendency to "decisively and irrevocably reinvent institutions, so that the present becomes discontinuous with the past". Thus, it is justified the difficulty for individuals to build their own stories, based on their professional experiences and their ties of dependence with other individuals. Not to mention the decentralization of the subject which we have just referred to. A direct consequence of this scenario is that people tend to live only in the present.

Dreaming becomes more difficult, when the uncertainties of maintaining what has been achieved professionally become significant and when there is no "deep mirror" before us. The enormous flexibility of work, on the one hand, seeks to adapt quickly to the growing social changes and the immense volatility of consumer demand and, on the other hand, implies the acceptance that "there is no long term". Both are characteristics of a society that values consumption more than citizenship. We believe, like Sennett, that this expression contains the principle of corrosion of values such as trust, loyalty and mutual commitment. In this way, the spectrum of fragmentation and volatility is widening, extending beyond the borders of Economics, and infiltrating family, social relations and, especially, the School.

This postmodern subject, who emerged in the second half of the 20th century, but who irreversibly affirms him/herself, as we stress, from the generalization of the internet and the emergence of social networks, is contemporary with radical changes in Science and Culture. We are referring to the strong presence in Physics and Chemistry – with Prigogine, for example – of the study of chaos theory and complex systems, seen no longer as reducible to a sum of simple systems, in the way imagined by Descartes. This fact has contributed to a new attitude towards the world. We do not know to where these new paths of Science will lead, by valuing indeterminacy, non-linearity and complexity. However, there is no doubt that this trend is contemporary and, in a sense, can be seen as a reflection of postmodern subject. Perhaps a broader understanding of this complex subject only results from a new psychoanalysis, built from these new scientific paradigms, as has already been suggested [40].

Finally, we must emphasize that the non-existence of an unified Physics (like Aristotle's and Newton's) can be not only an open problem in Science but a problem in which a solution would be a possible requirement for a future Enlightenment, to believe in the lessons of the past. We look forward to the day when Cosmology and Particle Physics – the old question of describing Physics in Heaven and Earth – will be unified in a new theory of the Cosmos, an achievement that would weaken the current wave of irrationalism and pseudoscience. In any case, it is difficult to imagine the impact of this new Enlightenment on man.

8   South American indian Culture: Plains and Cordilleras

> *How now, Horatio? You tremble and look pale. Is not this something more than phantasy?*
>
> Shakespeare [41]

In this brief Section, we summarize some ideas that we hope to be able to develop elsewhere. In particular, we intend to argue that many aspects of Amerindian culture can be clarified if we try to see them under the light of their space ideas, especially those concepts connected to the experience of space dimensionality, that is, space verticality, given the geographical, environmental and ecological aspects of their habitats.

Since the 19th century, under the influence of evolutionary ideas, cultural anthropologists used to refer to native and indigenous peoples of Africa, Oceania and the Americas as primitives or aborigines. In spite of isolated early criticism, this situation only began really to change after the seminal work by Levi-Strauss [42, 43, 44], in the mid

20th century, and seems to be now practically outmoded, at least in the academic world. The study of Brazilian indian Culture has played a fundamental role in these changes.

More recently, advances in Cultural Anthropology have greatly expanded the practical and empirical knowledge of Brazilian indian cultural life. See, for instance, Refs. [45, 46, 47]. New and important theoretical points of view have also appeared after 1968 events [48, 49, 50].

Perhaps the most interesting and central point of Clastres' view is his idea that there are societies not only without State but, in fact, organized against the possibility of State. Clastres studies the Guayakis, an indian people living without State of their own in the plains of Paraguay, not far from Brazil. As a matter of fact, the Guayakis recognize chiefs in their settlements. However, these chiefs [51] have no coercive power, limiting themselves to counselling, by exerting spiritual power in a conciliatory way, basically to avoid coercion and to prevent the emergence of coercive structures, especially those of organized State. Chiefs are chosen in complicated rituals and must exhibit certain leadership qualities, besides adequate kinships. In Xingu societies, similar in many aspects to the Guayakis, pubertal reclusion and excellence in *huka-huka* fights are considered to be very important points. As a consequence, there are no social hierarchies, no privileged classes accumulating wealth. Furthermore, these societies are characterized by the absence of concentration of power mechanisms. This means that separate settlements remain territorially and demographically restricted, resisting unification. Clastres' remarkable conclusion is that the Guayakis are not simply living without State or even in a pre-State formation: they are, in fact, actively engaged against State. War in this society does not aim for the expansion of Power or for the formation of an Empire. On the contrary, war is waged to guarantee the real Independence of the settlements, to prevent the formation and existence of organized State, *i.e.*, coercive Power.

Now, what calls our attention is the great contrast between the remarkable situation of these people, living in the plains, and the well known presence in the Continent of the highly organized Inca State in the mountains, the Andes Cordillera.

This situation suggests a simple interpretation: humans living in the high mountains of the Cordillera naturally develop a sense of verticality, associated to hierarchy. The great mountains, like monuments created by Nature, inspire the sense of awe and respect which we feel before a medieval cathedral. Here, it is important to stress how the cathedral constructions were used to reinforce the power of the Church and to forge the medieval subject and his relationship to a theocentric society.

This concrete and unforgettable experience of organized matter in space through its verticality is the driving force of hierarchies and their metaphors: up and down. Humans living in the Cordilleras, contemplating verticality, are prone to imagine, create, build and accept social hierarchies, organized State and Empire, which emerge as natural images in the mirror of an imposing environment. On the other hand, the Imaginary of the Guayakis flies freely in the direction of Independence, absence of hierarchies and organized State, *i.e.*, coercive power. This link between the living experience of Space and human imagination is a theme which Bachelard deeply explored in his *Poétique de l'espace* [52]. Now, our suggestion should not be considered a deterministic point of view: the experience of environment and space only opens possibilities, that is all.

A similar situation can be observed in the lowlands of Amazon. European and civilized people have constructed a metaphysical description of the world in which Nature is an invariant, it is the same for all the peoples of the world, while Culture may vary. This is, of course, *multiculturalism*. However, the point of view of the Amazonian indians is quite the opposite. For them, Nature is variable and Culture is invariant, being the same not only for all men and women, but in fact for all living animal and even spirits, which are essentially humans. From the viewpoint of the Amazonian indians, every species considers itself, and must be considered, to be human and sees the others including us as non-humans. Ethnographic evidence supports this view, for instance, the studies by Phillipe Descola [53], among others. This is called *multinaturalism*, in sharp contrast with traditional multiculturalism, and, actually, according to Viveiros de Castro [54, 55, 56, 57], corresponds to a new metaphysical approach: the Amazonian *perspectivism*. The logic underlying this view is based on the prey × predator relationship. The essential point is that, for the Amazonian indian there is no Hierarchy in Nature. Humans and animals are ontologically the same. This may possibly indicate that, for Amazonian indians, verticality does not play a fundamental role in the formation of their Imaginary. Their *Weltanschauung* is essentially two-dimensional like that of the Guayakis.

Another example may be extracted from the Bororo of Brazil, living in the state of Mato Grosso. "Although their village, patterned by cosmological concepts, has an "upper" portion and a "lower" portion this does not imply a hierarchical system. Indeed, for the Bororo, "the terms «upper» and «lower», while descriptive of an ideal topographic/cosmological feature and applicable as such to the village social units, cannot be taken as associated in any simple fashion with relative degrees of social order" [58].

Our final remark is that it should be strongly emphasized that we are not preaching geographical determinism in History. We know that, in general, ancient human groups evolved towards hierarchical societies. This represents the usual situation around the world and has been discussed and clarified by many authors. Consider, for instance, the Egyptian case. They live in a land without nearby mountains and developed a highly hierarchical society. However, their myths and legends were dominated by the presence of the Nile. Usually, they never explored the hostile desert around. The Imaginary of this extraordinary civilization was molded by the River and its cycles. Herodotus knew that. This seems to be the situation of most ancient civilizations in Mesopotamia (Tigris-Eufrates), China (Yellow and Yangtse), and possibly also India and Pakistan (Ganges and Indus) in spite of the imposing presence of the Himalaya. Our aim is very modest and simple: we are trying to understand what seems to be South American exception. We are only saying that Human Imaginary may be fed by the living conditions of the people, their natural and spiritual environment – *Physis* and *Psyche* – opening unexpected dialectic possibilities to History. This could possibly contribute to clarify unexpected situations like that of the Guaiakis or Aché. Pierre Clastres would possibly say: L'imagination au pouvoir.

We are aware of the speculative and incomplete features of the ideas exposed in this Section. There are certainly many other important factors that should be taken into account to fully understand the relation between hierarchy and space concepts in a certain society, like its cosmogony, for example. However, knowing Doria well, we are convinced that he will understand and appreciate our daring to put them on paper. Whether these conjectures will be worked out and eventually bear fruit or not, only the future will tell us.

9 Concluding remarks

The general question of the subject and his relations with Culture were dealt with here. To recap, we saw that Christianity's adoption of *codex* is linked to the issue of *verticality* or, in other words, to *space three-dimensionality*. It is through this third dimension that the *medieval subject* is related to God. The invention of this object, the *codex*, in the High Middle Ages occupies a prominent place in the formation of the emerging "three-dimensional" theocentric Man. The essence of the *codex*, which is its three-dimensionality, will be associated with culture and its transmission for a long period.

We have also argued that a new Man emerges from the Renaissance, who came to be called by Stuart Hall *enlightenment subject*, who, after all, results from an inextricable entanglement of concepts which are mirrored in the crucial question of the relationship of Man with the Cosmos. This implies two notable capacities of this subject: that of coming to conceive the *infiniteness* of the Universe and of three-dimensional space and that of not succumbing to the amazement caused by this non-divine infinity. He does not become small. On the contrary, as has already been seen, he grows, from the moment when he realizes how much, through his intellect, he is perfectly capable of embracing, conceiving and rationalizing all these questions that touch infinity, from the microcosm to the macrocosm.

We found that the great narratives about Man, which ultimately define our view of ourselves – that mirror that responds – are intertwined with those concepts that Cassirer identified as the foundation of culture: *space, time,* and *number*. The uni-dimensionality of time – associated with the Christian Worldview – is reinforced by the linearity of the text in the *codex*, which does not require repetition.

Indeed, in this essay, we suggest that the dimensionality of the writing support – *volumen, codex,* book – may have a major influence on Culture: bidimensionality, in the case of *volumen*, and the three-dimensionality of *codex*. On the other hand, the postmodern subject, as previously discussed, is characterized by the multi-dimensionality of digital networks, which tends to infinity. *Number*, on the other hand, manifests itself in the spatialization of Algebra – with the emergence of Analytical Geometry – and in the mathematization of the physical description of the World.

We are tempted to conclude that it is the dimensionality of the imaginary space in which Man lives, linked to the reading of texts and of the World, the main key to understand, in each epoch, the subject's striking features. As a new application of these ideas, we sketched, in Section 8, the possibility of understanding some cultural and social differences between two different indian groups from their different space conceptions.

At this point, we cannot resist the temptation to quote how Albert Camus refers to the problem of the dimensionality of space in the opening of his book *Le Mythe de Sisyphe*, in which he deals with the issue of the human being becoming aware of meaninglessness of his condition (absurd Man faces an absurd world and humanity):

> "There is only one really serious philosophical problem: it is suicide. To judge whether life is worth living or not is to answer the fundamental question of philosophy. The rest, if the world has three dimensions, if the spirit has nine or twelve categories, appears next."

We would now like to end the paper with an open question.

If it is true that the central issue of *subjectivity* in Post-Modernity (or Liquid Modernity) can be understood as related to the infinitude of dimensions of new virtual spaces, is it possible to establish a certain analogy with what happened in the Renaissance, when Man lived astonished by the infinitude of space and Cosmos? This leads us to an inescapable question, which we leave to the reader. Will there be a new Enlightenment in the foreseeable future?

If we had to risk an answer, we would first remember that 18th century Enlightenment was only possible after the overcoming of the crisis introduced by Copernicus in Science, by breaking the unity of Aristotelian Cosmos. Newton did it from the moment he theoretically and conceptually reunited Physics and Astronomy. That said, and reasoning by analogy, we believe that it is practically unlikely that, without a qualified theory unifying General Theory of Relativity and Quantum Theory of Matter, the credibility of Science, and the belief in Reason, which is indispensable to the new Enlightenment, will be restored.

Acknowledgment

We would like to express our thanks to José Acacio de Barros for the kind invitation to contribute to Francisco Doria's *Festschrift*, which is an honor for us. We are in debt to Hélio da Motta Filho, Felipe Silveira, and Pedro Teixeira who carefully read the manuscript, and to Mércio Pereira Gomes and Adelino de Lucena Mendes da Rocha for fruitful discussions on a few anthropology matters and for their critical remarks, although we dared to disagree in some important issues. However, some points raised by them will deserve our attention in the near future. In any case, needless to say that they are not responsible for our speculative approach.

Notes and References

[1] Ernst Cassirer. *Philosophie des formes symboliques. 2: La pensée mythique*. Paris: Édition Minuit, 1972.

[2] Francisco Caruso & Roberto Moreira Xavier. *O Livro, o Espaço e a Natureza: Ensaio Sobre a Leitura do Mundo, as Mutações da Cultura e do Sujeito*. São Paulo: Livraria da Física, 2017.

[3] Karl Popper. *Conjectures and Refutations: The Growth of Scientific Knowledge*. Chapter 1, Section VII. London: Routledge and Kegan Paul PLC, 1969.

[4] Arkan Simaan & Joëlle Fontaine. *L'image du monde. Des Babyloniens à Newton*. ADAPT Editions, 1999.

[5] *Apud* Ernst Cassirer. *An Essay on Man*. New Haven, Connecticut: Yale University Press, 1944.

[6] Gaston Bachelard. *La terre et les rêveries du repos*. Paris: Librairie José Corti, 1948.

[7] Giovanni Reale. *Storia della filosofia greca e romana*, vol IX. Milano: Bompiani, 2004.

[8] Gerhardt B. Ladner. *God, Cosmos, and Humankind: The World of Early Christian Symbolism*. English translation by Thomas Dunlap. Berkeley, Los Angeles & London: University of California Press, 1995.

[9] Eduard Jan Dijksterhuis. *The Mechanization of the World Picture: Pythagoras to Newton*. Princeton: Princeton University Press, 1986.

[10] Wallace K. Ferguson. *The Renaissance in Historical Thought: Five Centuries of Interpretation*. Toronto: University of Toronto Press, 2006.

[11] Stuart Hall. "The question of cultural identity", *in* Stuart Hall, David Held & Anthony McGrew (Eds.). *Modernity and Its Futures*. Cambridge: Polity Press, 1992, p. 274-316.

[12] *Cf.*, for example, Giambattista Vico. *La scienza nuova*. Napoli: Muziana Stamperia, 1744.

[13] An analysis of the meaning of postmodern conception in different contexts, which also seeks to identify its degree of accuracy and usefulness in describing contemporary experience, can be found in David Harvey. *Condition of Postmodernity: An Enquiry into the Origins of Cultural Change*. Cambridge, Massachusetts: Blackwell, 1992.

[14] Erich Fromm. *Marx's Concept of Man*. Continuum International Publishing Group Ltd., 1981.

[15] Karl Marx. *Economic and Philosophic Manuscripts of 1844*. International Publishers Co, 1980.

[16] Eric A. Havelock. *The Literate Revolution in Greece and its Cultural Consequences*. Princeton: University Press, 1982.

[17] Eric A. Havelock. *Preface to Plato*. Cambridge, MA: Harvard University Press, 1963, Chapter X.

[18] Eric A. Havelock. *The Muse Learns to Write. Reflections on Orality and Literacy from Antiquity to the Present*. New Haven and London: Yale University Press, 1986.

[19] Christian Jacob. "Lire pour écrire: navigations alexandrines", in Marc Baratin & Christian Jacob (Eds.). *Le Pouvoir des Bibliothèques: La Mémoire des Livres en Occident*. Paris: Albin Michel, 1996, p. 53.

[20] Colin H. Roberts & T.C. Skeat. *The Birth of the Codex*. London: Oxford University Press, 1983.

[21] Guglielmo Cavallo. "Testo, libro, lettura", *in* Gugliemo Cavallo, Paolo Fedeli & Andrea Giardina (Eds.). *Lo spazio letterario di Roma antica*, v. II, *La circolazione del testo*. Roma: Salerno Editrice, 1989, pp. 307-341; "Libro e cultura scritta", *in Storia di Roma*, v. 4, *Caratteri e Morfologie*. Torino: Einaudi, 1989, pp. 693-734.

[22] From Jacques Le Goff's Preface to the book of Philippe Bruc. L'Ambiguïté du livre: Prince, pouvoir, et peuples dans les commentaires de la Bible au Moyen Age. Paris: Beauchesne Éditeur, 1994.

[23] Úrsula E. Katzenstein. A origem do Livro: da Idade da Pedra ao Advento da Impressão Tipográfica no Ocidente. São Paulo: Editora HUCITEC e Instituto Nacional do Livro, 1986, p. 37.

[24] Marshall McLuhan. Understanding Media: The Extensions of Man. McGraw-Hill Paperbacks, 1965.

[25] Ernst Cassirer. "The place of Vesalius in the Culture of Renaissance". Yale Journal of Biology and Medicine, v. 16, n. 2, pp. 109-120 (1943).

[26] On the hierarchical conception of medieval society see Chapter 3 of the classic book of Johan Huizinga. The Waning of the Middle Ages: A Study of the Forms of Life, Thought and Art in France and the Netherlands in the XIVth and XVth Centuries. New York: St. Martin's Press, 1984.

[27] Keith D. Lilley. City and Cosmos: The Medieval World in Urban Form. UK: Reaktion Books, 2009.

[28] Regarding the analogies between art, philosophy and theology in the architecture of the Middle Ages, we refer the reader to the classic of the critic and historian of German art Erwin Panofsky. Gothic Architecture and Scholasticism. New York: The World Publishing Company, 1968.

[29] Jacques Le Goff. La Naissance du Purgatoire. Paris: Éditions Gallimard, 1981.

[30] Alexandre Koyré. La Révolution Astronomique: Copernic, Kepler, Borelli. Paris: Hermann Impr. Union, 1961.

[31] Paolo Rossi (Ed.). La Rivoluzione Scientifica: da Copernico a Newton. Torino: Loescher Editore, 1973, p. 125.

[32] For a history of the theories of the infinite and the plurality of worlds see: Pierre Duhem. *Medieval Cosmology: Theories of Infinity, Place, Void, and the Plurality of Worlds*. Chicago and London: The University of Chicago Press, 1985.

[33] Blaise Pascal. *Pensées*. Paris: Guillaume Desprez, 1670.

[34] Roger Chartier. *L'Ordre des Libres: Lecteurs, Auteurs, Bibliothèques en Europe entre XIVe et XVIIIe siècle*. Aix-en-Provence: Alinéa, 1992.

[35] An extensive analysis of the social and economic impact of the networks was done by Manuel Castells. *The Rise of the Network Society*. Malden, MA, and Oxford: Blackwell Publishers, 1996.

[36] R. Howard Bloch & Carla Hesse. *Future Libraries*. Berkeley and Los Angeles: University of California Press, 1995.

[37] Martin Lister (Ed.). *The Photographic Image in Digital Culture*. London: Routledge, 1995.

[38] Roy Ascott. "The Architecture of cyberception". In M. Toy (Ed.). *Architectural design. Architects in cyberspace*, v. 65, n. 11/12. London: Academy Editions, 1995, pp. 38-41.

[39] Richard Sennett. *The Corrosion of Character: The Personal Consequences of Work in the New Capitalism*. New York: W.W. Norton & Company, 1998.

[40] Stanley R. Palombo. *The Emergent Ego: Complexity and Coevolution in the Psychoanalytic Process*. Madison, Connecticut: International Universities Press, 1999.

[41] William Shakespeare. *Hamlet*, act 1, scene 1.

[42] Claude Levi-Strauss. *Tristes tropiques*. Paris: Librairie Plon, 1955.

[43] Claude Levi-Strauss. *Anthropologie Structurale*. Paris: Librairie Plon, 1958.

[44] Claude Levi-Strauss. *Pensée Sauvage*. Paris: Librairie Plon, 1958.

[45] Aparecida Vilaça. *Paletó e eu: memórias de meu pai indígena*. São Paulo: Editora Todavia, 2018.

[46] Carlos Fausto & L.A. Lino da Costa. "Recent Studies of Amazonian Ontologies". *Religion and Society*, v. 1, n. 1, pp. 89-109, 2011.

[47] Tânia Stolze Lima. *Um peixe olhou para mim: o povo Yudjá e a perspectiva*. São Paulo: Editora UNESP, 2011.

[48] Pierre Clastres. *Chronicle of the Guayaki indians*. New York: Zone Press, 1998.

[49] Pierre Clastres. *La societé contre l'État: recherches d'Anthropologie Politique*. Paris, Éditions de Minuit, 1974.

[50] Pierre Clastres. *Archeology of Violence*. New York: Semiotext(e), 1994.

[51] Pierre Clastres. Échange et pouvoir: philosophie de la chefferie indienne. *L'Homme*, v. 2, n. 1, pp. 51-65, 1962.

[52] Gaston Bachelard. *Poétique de l'espace*. Paris: PUF, 1957.

[53] Philippe Descola. *Beyond Nature and Culture*. Chicago: University of Chicago Press, 2005.

[54] Eduardo Viveiros de Castro. "O mármore e a murta: sobre a inconstância da alma selvagem". *Revista de Antropologia*, v. 35 (dezembro) p. 21, 1992.

[55] Eduardo Viveiros de Castro. *A inconstância da alma selvagem*. São Paulo: Cosac & Naify, 2002.

[56] Eduardo Viveiros de Castro. *Metafísicas canibais: elementos para uma antropologia pós-estrutural*. São Paulo: Ubu Editora, 2009.

[57] Eduardo Viveiros de Castro. *Perspectivismo e multinaturalismo na América Indígena*. São Paulo: Ubu Editora, 2018.

[58] Stephen M. Fabien. *Space-time of the Bororo of Brazil*. Gainesville: University of Florida, 1992.

Revisiting the systemic golden years from a contemporary organisations perspective

Maurício Vieira Kritz*

Abstract

In the wake of Bertalanffy's General System Theory, established in the middle of last century, there was a far-reaching quest for transdisciplinary studies and unified treatment of sciences that lasted for decades and was supported by achievements in general systems theory. Yet, the development of the foundations of GST started to fade in the second half of the 1980's. As a consequence, the lure of an unified treatment for the sciences, that has been boosted by the ubiquity of the system idea, came into an idle steady state, thinning into *ad hoc* case studies.

This writing addresses questions related to the reasons why the impetus of unification within the sciences centred on the system concept faded and to the ostensible insufficiency of the system concept for handling problems in Weaver's *organised complexity* class. It also argues about how 'organisation', a straightforward generalisation of the 'system' concept, may disturb this state of affairs, reinvigorating systemic foundational developments, centred now on organisations, and the quest towards unified treatment of several scientific domains.

1 Introduction

The quest for unified, cross-disciplinary, descriptions and explanations of natural phenomena is long-standing. It is subjacent to the Scientific Revolution, in centuries XVI and XVII. It sustains the search for an unified field theory and a theory of everything in the reach of physical sciences [35]. It supports the view of physics as "the" exemplar science and the seductive trend of reducing every scientific explanation to physics-like explanations (the effective meaning of "reductionism"). Yet, biological phenomena, complex phenomena, (self-)organisation and chaos have severely quivered this tradition in the last century. Since the middle of last century, the pervasiveness of the system notion across scientific disciplines suggested the foundational portion of system

---

*National Laboratory for Scientific Computation, Petrópolis, RJ, Brazil and School of Biological Sciences, FBMH, University of Manchester, UK

science and its abstract system's framework [26; 17; 5] as a basis for possible unified treatments across distinct sciences.

The establishment General System Theory (GST) in the middle of last century alongside a collection of intertwined sibling disciplines, theories, and methodologies — cybernetics, robotics, computer science, information theory, automata theory, behaviour theory, game theory, dynamical systems, asymptotic analysis, computational modelling and so on — paved the way for a more systematic study of problems in Weaver's "organised complexity" class [41]; which contains all biological, ecological and environmental phenomena.

Strictly speaking, organised complexity embraces phenomena that have the following characteristics among others: interdependence of a huge number of variable aspects, hierarchy, and changes in the pattern of possibilities (channels) of interaction. Any of these characteristics affects by itself propensities in the system's behaviour and, hence, need to be considered as part of the system-state [18]. The interdependence in organised-complexity phenomena and their relation to phenomena studied by more basic sciences is summarised in Figure 2 of the next section with the concur of organisation and information as guidelines; that for the moment can be considered to have their informal meaning.

Since its consolidation, GST aimed at an isomorphism of concepts, non-reductionism, the unity of sciences, novel epistemological values and a thinking reorientation [39]. It was pivotal in amalgamating its sibling disciplines to handle problems in various disciplines [25; 45; 17; 42]. Within the reach of this efforts, cross-disciplinary analogies and comparisons appeared everywhere. This unleashed a far-reaching quest for trans-disciplinary studies and an unified treatment of sciences that lasted for decades; an effort based on the ubiquity of the system idea, on the achievements of GST, and on the seamless and natural integration of GST with other disciplines.

As an intellectual discipline, system science has two major parts: the foundational and the phenomenological [45, Preface]. The foundational is primarily concerned with assigning precise meaning to concepts and with the development of an abstract framework for representing phenomena as systems and handling them. This part is centred on the exploration of properties of GST and sibling disciplines foundational concepts. The phenomenological part develops methods, techniques, and algorithms [32] for studying the behaviour of systems arising while describing natural or artificial phenomena by means of elements produced in the foundational part.

A system's *behaviour* may be summoned by changes in the system's state along time and space. Common kinds of systems studied in this domain are: differential systems, finite-state systems, modular systems, stochastic systems, learning systems, distributed systems etc. Another aspect of the phenomenological part is the ontological version of the above strategy. Namely, the application of the methods, techniques and algorithms above to systems that represent phenomena in many "domains of enquiry" to understand their nature or to solve specific questions. It is worth noting that inferring phenomena properties and characteristics through the system's behaviour is largely independent of the enquiry domain although is interpretation may not be, which adds to the fascination of finding unified treatments across scientific disciplines.

These two parts have progressed side by side until the middle of the 1980's, when the foundational part started fading. The phenomenological branch of systems science, though, has maintained a continued impetus, successfully flourishing everywhere as an enormous number of publications testify[1]. At a first sight, system sciences have revived as systems biology in the beginnings of the 2nd milenium [43]. However, a deeper inspection of its productions reveals that systems biology is primarily phenomenological remaining centred on *ad hoc* case-studies, while lacking philosophical foundations and theoretical-foundational developments [4]. In terms of using the system idea to understand natural phenomena, the quest for unified treatments across the sciences has slowly drifted since the 1990s towards taming complex phenomena and investigating self-organisation though dynamics.

The system idea is ubiquitous and occurs in practically all domains of thought [39; 27]. As a consequence of the innovative strength of systems science and its interaction with scientific disciplines meagre in mathematics, a *systems thinking* developed collaterally, spreading beyond the scientific milieu. Until now, systems thinking sustains investigations in the most unexpected domains [42], despite the enormous difficulty of inserting the systemic perspective in the educational process [29].

This paper addresses questions about the loss of impetus in the foundational aspects of system science. Questions related to "Why the efforts concerning the unification of sciences grounded on the system concept faded?" and "Why is the system concept apparently ill-suited to handle W. Weavers *problems of organised complexity*?". On the top of that, I shall start questioning what can be done about this distressing situation and to discuss how organisations, a straightforward generalisation of systems recently introduced, and its framework [19] can help in warming-up developments along these directions.

2   Phenomena and the Ubiquity of Systems

When we observe with scientific interest a portion of Nature and nothing changes, we tend to say that what we observe is an *object*. That is, something we can name and describe but has a stable and well-defined identity in space-time, remaining invariable, as far as one can see. On the contrary, if there

---

[1] An interesting investigation is to peruse the many volumes of the *International Journal of General Systems* [1] counting the frequency of appearance of foundational and phenomenological articles.

are observations — of attributes, properties or structures — that change, we tend to say it is a phenomenon. Therefore, it is wise to remark since the beginning the fundamental role of observers for the existence proper of phenomena [9].

Hence, no change, no phenomenon. As well, no observer, no phenomenon. An observer is required to state that something has changed and, eventually, what has changed and how. Furthermore, experience tells us that changes are a consequence of interactions or exchanges of any sort. Figure 1 schematically represents the main elements of a phenomenon: borders, things, interactions, exchanges, and observation (of changes). Note that some of these elements are only implicitly represented in the picture and that attached to each *thing* there is always a collection of aspects that identify and classifies it within a study.

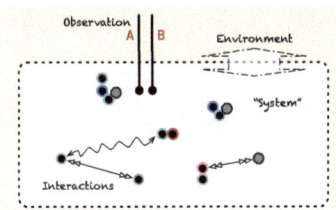

Figure 1: Things and Interactions: a Scheme for Phenomena.

It is important to remark that:

- Beyond the recognition and distinction of changes, observers are required to identify the totality of the interesting aspects (changing or not) as intertwined into a single phenomenon. That is, the weaving between aspects that turn them into something unique;

- The frontiers of a phenomenon, represented in Figure 1 by a dashed line, are arbitrary and a *choice* of the observer even if the perception subjacent to the *selection* between what is or is not part of the phenomenon is not conscious;

- Things belonging to a phenomenon may always interact with things that are not considered part of the phenomenon, this kind of interaction should be studied in a differentiated manner to keep what should be out, out.

- Even if these in-out interactions are disregarded and not studied, their documentation is vital when the phenomenon is delimited.

Considering the fuzziness at the border of ecological systems, this last care is *sine qua non* in environmental sciences and other studies of complex phenomena where the identity of a phenomenon itself strongly depends on the identification of interactions and their strength. The characterisation of an environmental phenomenon or thing may be easily disturbed by small changes in these interactions [18]. This fact is an important source of complexity in this category of phenomena [44].

As suggested by Figure 1, making explicit what is exchanged through interactions among the things composing a phenomenon and considering the remarks in [44], we can organise interesting scientific phenomena into an hierarchy having as guidelines the complexity of its interacting entities and the complexity of what is exchanged in each occurring interaction; where complexity relates to organisation and ability to decide. The latter complexity is usually associated with the quantity and quality of information that is exchanged in interactions and with the degree of difficulty in determining the phenomenon's boundary (see the arrows in Figure 2 and [18]).

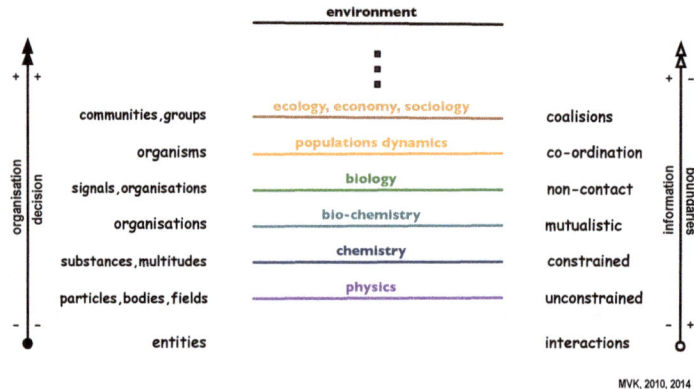

Figure 2: An hierarchy of phenomena, guided by their relations with organisation, information, sharpness of boundaries and determinism. Improved from [18].

It is straightforward to conclude that entities in Figure 2 are the "things" that form a phenomenon. An important detail to note in Figure 2 is the short number of typical entities enrolled. Indeed, when we observe and study phenomena we consider the interacting things as elements of certain classes: particles, bodies, fields, substances, waves, individuals, firms, populations, fluids, organisations, solid matter, organelles, organisms etc. Each of these classes has a collection of characteristic attributes that frame our perceptions about what goes on within the phenomenon with this type of things. For instance, particles have no volume. They may possess mass, color, spin and other lumped attributes but it makes no sense to vary any of of these attributes in a space-domain around the location of a particle.

In classical descriptions, it makes no sense to talk about a particle's frequency, although it makes sense to talk about the frequency of its oscillations around a point. Quite the contrary, it does make sense to refer to vibrations of a body or molecule and its frequency, since bodies have volume and molecules a spatial organisation. Discussing thing-types, or arguing in terms of thing-types, is outside the scope of this writing. However, as a hint of its charm, we may remark that individuals in a population are particles that doubt and make decisions.

A system may be defined in a simple manner as a collection of interrelated things [26; 17; 5; 20]. As such, systems immediately and straightforwardly match the scheme in Figure 1 by equating interrelations to interactions and possibilities of interactions, independently of the nature of these interactions. Hence, the ubiquity of systems result from the possibility to straightforwardly associate any phenomenon with a system.

3 Systems, Models, and the Scientific Process

Scientific explanations [6; 9] about empirical phenomena are developed over an abstract representation of theirs [31]. These abstract descriptions are iteratively and incrementally constructed in a cycle running through steps of observation accumulation, model construction and theory development [20] (refer to the left part of Figure 3). In this cycle, fundamental and more immediate properties of models provide a basis to enunciate axioms supporting the development of theories that unveil hidden, often unexpected, properties of the phenomenon and its behaviour [21], changing our perspective of the world.

To reach this abstract description we employ the concepts of systems and of models. A bit more formal than the previous section but still of great generality, systems (S) are sets of things T associated by means of one or more relations R. That is, they are pairs $S = \{T, R\}$ pairs of things and (collections of) relations [17]. Establishing the pair $S = \{T, R\}$ makes explicit and almost fixes what belongs to a system S, and to the phenomenon it represents.

This characterisation of a system is atemporal making no reference to space as well and focusing on things and their possibilities of interaction or association. It worth noting, though, that it encompasses several other possibilities, including very complete definitions dynamical systems that include time, space, control and observation operators and other matters [14; 26].

This definition contains, nevertheless, the quintessence of the system concept because it makes evident two fundamental concepts that reside at the core of the system idea [34]: *thing-hood* (T) and *system-hood* (R). The first part concerns the attributes and properties belonging to things composing the phenomenon, while the second to whatever keep these things together among the changes to which they are submitted by the system's behaviour, providing their identity.

Obviously, there are properties, attributes and characteristics that belong to these two realms and cannot be assigned to one or the other. Consider, for example, collective aspects like temperature, density, concentration, saturation etc, which cannot be ascribed to one thing, since they reflect a collective state. They are ascribed to locations but result from the interaction of many things nearby.

Models are *abstract representations* of something in a context different from the context where this something lives, that are useful to know and understand a real entity, the modeled entity. Models are *expressed* by means of figures, diagrams, sketches, symbols, maquettes, propositions, mathematical construct and so on. Models are always models **of** something or models **for** something and, thus, make reference to things observed beyond the model's context proper [3]. Moreover, models are never unique. This bring us to the necessity of justifying models and to semantic values and an ontology associated with the process of constructing models [3; 21]. In the scientific process, justifications are elaborated during the building of models and, sometimes, theories.

A first step in constructing models is to handle the complexity of things interacting to instantiate a phenomenon. As hinted in the previous section, to a great extend, things are considered as exemplar elements of certain well-known classes: particles, waves, bodies, substances, individuals, and so on; which possess certain sell-established and specific characteristics and properties. This is indeed our first step in the modelling process (Figure 3). These thing-

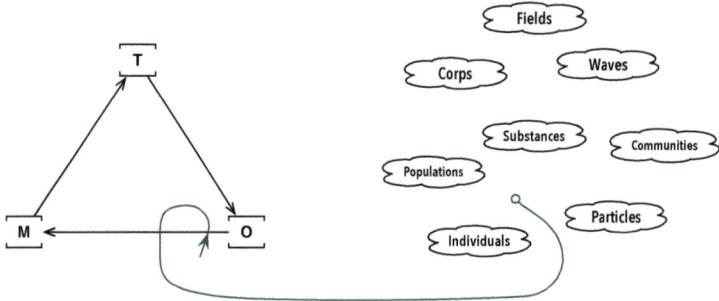

Figure 3: The Scientific Process: a (magic) cycle around observation, model construction, and elaboration of theories.

classes are characterised by aspects and properties important to what is being questioned.

Their choice though cast the subsequent modelling shaping what can be described and explained about the phenomenon. An effortless example emerge from a billiard or snoker table. When modelling interactions among balls in 2 dimensions, we may represent balls as particles or disks. Despite that most interactions can be represented with either choice, the chock between two spinning balls cannot be described in the particle representation, even if we consider the rotation velocity of the balls as an attribute. Next, a collection of aspects attached to things is selected as the system-state and the interrelations and interactions creating the phenomenon are described having this collection as a ground. The system concept is cardinal in this last step.

4 Interaction Graphs and Dynamics

As remarked above, the 'static' definition of systems may be enlarged with dynamics and companion procedures like control and observation. The formal exercise in this section provides a closer look in associating systems to dynamical behaviour and will be useful in clarifying some insufficiencies of the system concept and to introduce the idea of organisation. Let us start with interaction graphs.

Interaction graphs are direct outcomes, almost a rewriting, of the system definition (Figure 4). An *interaction graph* is a directed or undirected graph, $\mathbf{g} = \{N, A\}$, where the nodes are the things in a phenomenon, that is, $N = T$ and the arcs of $\mathbf{g}$, elements of the set $A(\mathbf{g})$, are given by the relation(s) of S in the following way. Let $\tau \colon N \longrightarrow T$ be a bijection between N and T and let

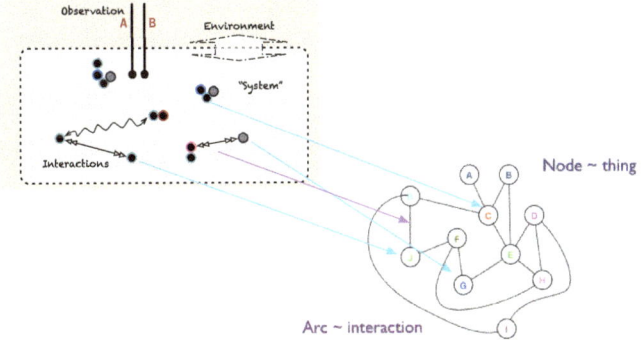

Figure 4: Phenomena and interaction graphs. Aspects are attached to either nodes (things) or interactions (arcs).

$A_{i,j} = (n_i, n_j)$ denote a node pair, then $[A_{i,j} \in A(\mathbf{g})]$ **iff** $\tau(n_i) \, R \, \tau(n_j)$, for a relation R in S.

In physical phenomena, where any thing can interact with any other thing as long as they come close together, the interaction graph of any motion phenomenon (Figure 2) is the complete graph K_n, where $n = \#(N)$ is the number of things interacting in the physical phenomenon. Straightforwardly, any two nodes are adjacent. For the solar system, for instance, the interaction graph is K$_{10}$ (Figure 5(a)). In chemical phenomena, g arises from the list of chemical reactions by considering substrates as nodes and reactions as arcs. Simple reactions give rise to graphs but, in general, (bio-)chemical networks are hypergraphs [16]. Hypergraphs may be represented as bi-partite graphs where contains substrates and the other the reactions; thus, any two reactions may be connected by more than one substrate. In this case, the representation graph of the hypergraph [2] is the interaction graph, which is usually not complete. Another example of interaction graphs commonly used are the trophic webs of ecological systems [12]. A virtual one is shown in Figure 5(b).

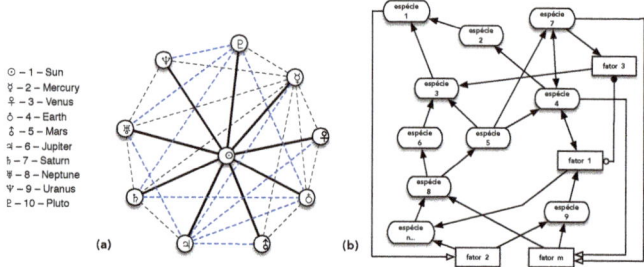

Figure 5: Two examples of usual interaction graphs in physical and ecological systems. Note: Not all K$_{10}$ arcs are depicted.

There are many ways of associating dynamics with an interaction graph (g(S)). Motivated by the simple physical and chemical phenomena described above, I focus below on phenomena which aspects vary continuously and possess a reactive-diffusive behaviour. More details and a deeper discussion may be found in [22].

Continuous reaction-diffusion phenomena (R−D) are generally described by variations of the following mathematical model. Let $\{c_1,\ldots,c_p\} = \vec{c}$ be a set of continuous variables depending in $(\vec{x},t)$ associated with R−D relevant aspects that form the state of the R−D system. The time behaviour of a R − D system are the solutions of

$$\frac{\partial \vec{c}}{\partial t} = \mathcal{D}(\vec{c}) + \mathcal{R}(\vec{c}), \qquad (1)$$

where $\mathcal{D}$ describe diffusion in the space domain and $\mathcal{R}$ the interactions among the variables $\{c_i\}$. Typically, in R − D systems the diffusion $\mathcal{D}$ does not affect interactions between variables, nor is it affected by these interactions. If $\mathcal{R}$ is a linear operator, it may be written as a matrix $\mathcal{R}$ and $\mathcal{R}(\vec{c}) = \mathcal{R} \cdot \vec{c}$. In the linear case, the interaction graph g(R−D) = $\{N, A\}$ is such that $N = \{c_1,\ldots,c_p\}$ and $[A_{i,j} \in A(\mathsf{g})]$ iff $[\mathcal{R}_{i,j} \neq 0]$, where $A_{i,j} = (c_i, c_j)$. The same association can be done for quite general $\mathcal{R}$ [22]. This allows us to establish a bridge between dynamical systems and interaction graphs by the following transformation. Let $\mathbf{G}_n$ denote the class of all graphs $\mathbf{g}$ which node set $N(\mathbf{g})$ has n elements, and let $\mathcal{C}^r(\mathbb{R}^n; \boldsymbol{T}(\mathbb{R}^n)) = \mathcal{C}^r(\mathbb{R}^n)$ be the class of all vector-fields $\vec{F} : \mathbb{R}^n \longrightarrow \boldsymbol{T}(\mathbb{R}^n) = \mathbb{R}^n$ that are $r \in \mathbb{Z}, r \geq 0$ times differentiable and satisfy Equation 1. The minimum value of r depends on properties of operators $\mathcal{D}$ and $\mathcal{R}$.

Each R−D dynamical system is thus attached to a graph by the mapping:

$$\begin{aligned} \mathcal{G} : \mathcal{C}^r(\mathbb{R}^n) &\longrightarrow \mathbf{G}_n \\ \vec{F} &\mapsto \mathbf{g}, \end{aligned} \quad (2)$$

where $G = \mathcal{G}(\vec{F})$ is given by the procedure introduced above. The mapping $\mathcal{G}$ is a well defined function and $dom(\mathcal{G}) = \mathcal{C}^r(\mathbb{R}^n)$. That is, any dynamical systems is associated to a single interaction graph [22].

5 Inadequacy of Systems

As said in the introduction, problems in Weavers organised complexity class [41] have the following characteristics: interdependence of a huge number of variable aspects, hierarchy, and changes in the pattern of possibilities (channels) of interaction, that is, in the systems structure or its interaction graph. All these characteristics are present in any biological, ecological and environmental phenomena. However, despite the imperative necessity of having *simple* models, equations and mathematics as brilliantly explained in [32], discussing the challenges imposed by a "huge number of variable aspects" nowadays is pointless due to the ever bigger computational power available each year. This section addresses the challenges of phenomena which *things* result from phenomena at other scales and of phenomena where the possibilities of interaction between things vary.

Hierarchy is already present in simple and well-known phenomena, running unnoticed. It is present in planetary systems; since the planet-moon systems are analogous to the larger one. Nevertheless, the planet-moon systems affect the sun-planet system only through its mass which remains constant irrespective of behavioural changes in the planet-moon or planet-sun systems. Therefore, the two phenomena can be decoupled and studied separately. A second instance of hierarchy in physics occurs wherever "collective" variables intervene, since these attributes are indeed statistics over samples from a pool of repetitive events and phenomena whose number approaches infinity. Variables like temperature, pressure and concentration are in this class that leads to constitutive relations and axiomatic theories.

This is not the case in biological phenomena where the cell behaviour depends on the behaviour of its organelles that may contain other organelles or stable molecular assemblages joining different scales and performing distinct "biological functions" (see nevertheless [37]). Upwards, one can always identify sub-systems with the same main 19 functions replicated at each organisational scale [27]. All living systems being open this means in reality that behaviours associated with units and sub-unities are what is really interacting [10]. This means that a biological entity, and not just aspects of it, may alternate between *thinghood* and *systemhood* in different moments and from different perspectives. Moreover, the state of the system must contain information about hierarchies, hierarchy dependencies and multi-level interactions since they change [11] and affect the propensities [38] of future outcomes.

System scientists use the expression *system of systems* [42] to refer to situations of this kind. However, it is clear that a "system of systems" does not satisfy the definition in section 3 above; nor is it natural to handle their dynamics in the framework sketched in section 4, despite the huge and beautiful knowledge accumulated recently about dynamical systems [32]. We do not know how the behaviour of a system that is a "thing" in another can be easily represented in terms of its own behaviour, since behaviour is dependent on the larger phenomenon's evolution and cannot be handled like a parameter. Moreover, treating such systems as a *flat* bigger system leads immediately to stiffness, due to largely different characteristic length and time scales; and to a high level of dynamical complexity [36; 32].

Phenomena where the interaction possibilities and constrains among its components vary are called systems with variable structure since long [28]. In the nomenclature just introduced (section 4), these are systems S_v which $g(S_v)$ changes. A graph may change in two complementary ways: either its node-set varies, augmenting, diminishing, or maintaining the number of nodes, or the arc-set changes, deleting or adding arcs.

Changes in systems structure may be driven from inside, by the system's behaviour, from outside, due to changes in the context or environment where the system lays, or both. Changes from outside are easier to explain and handle and will be used as examples in the sequel. An important collection of systems of this kind are the ecosystems in the Amazon Floodplains [23; 13].

Due to the Amazonian climate, there are annual floods in the main rivers of the Amazon basin that covers a non-trivial portion of the forest surface. At some points, the centennial mean difference in the rivers surface height between low and high waters seasons reach around 18–20 meters. Parts of the forest remain flooded for more than 2/3 of the year. As a consequence of biological adaptation, the waters trophic network and the forest trophic network merge into a mixed trophic network during flood. For instance, fish, whose food becomes scarce due to abrupt changes in concentration, eat the fruit of trees spreading seeds as a consequence. Furthermore, the flood may bring populations of foreign species into the system imposing other structural changes to the ecosystem (Figure 6).

From the stated, it is clear to see that nor the general, nor the dynamical, characterisations of systems can cope with the hierarchical or variable-structure challenges inherent in "problems of organised complexity".

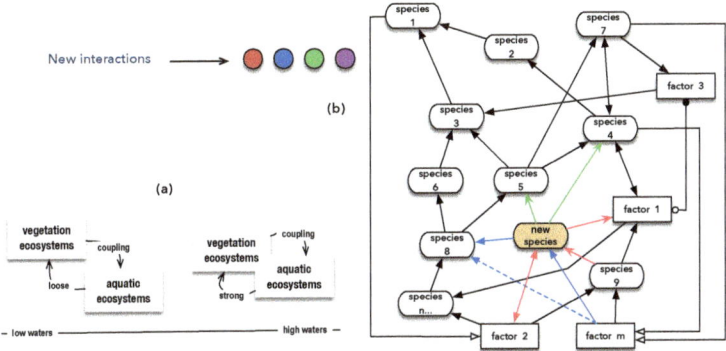

Figure 6: Natural variable-structure systems: (a) Trophic variation in Amazon Floodplains ecosystem; (b) Trophic pressures caused by invading populations.

6 Organisations in a Nutshell

Organisation is consensually considered the most distinctive aspect of living systems and entities [40; 33; 10; 24], what makes it a strong candidate to support the construction of a theoretical basis for biology. Notwithstanding, the vast majority of these efforts try to characterise organisation by means of biological function, what brings in all problems inherent to biological function [37]. The more severe being its *a posteriori* and environment-interaction dependent identification.

The concept of organisation introduced below does not rely on biological function, hierarchy or relationships with environmental entities. It is more basic from the epistemological stand, reporting to the intrinsic nature of things: how are they assembled and what is the interaction structure of phenomena.

Like energy, organisation is something immaterial. Like energy, it depends on relations and relative arrangements between the elements that compose the organisation, or between collections of these elements. However, the idea of relations and relativeness is far more general for organisations and do not depend on distance between its components or portions thereof, or any type of topology. It doesn't depend also on external contextes, like the presence of force or action fields. To put it simply, organisation is what distinguish a pile of bricks from the walls of a house that are made with the bricks of the pile. The pile and the walls are two different organisations of the very same bricks.

Figure 7 show two wall blueprints that may be made with the same pile of bricks. The difference is purposefully small: a door (same area) that moved

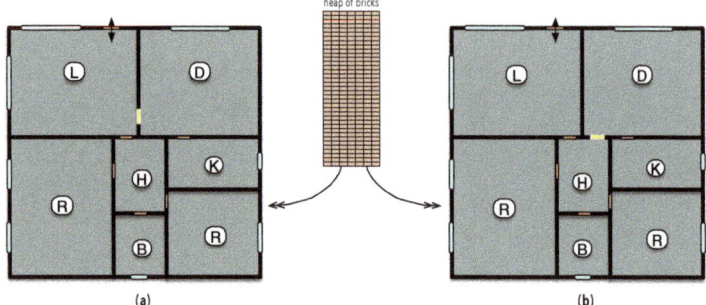

Figure 7: From a pile of bricks into houses. The (small) difference appears in yellow.

from a wall to a nearby wall. The energy required to build either is the same. Nevertheless, the circulation in both changed in important ways; for one, the access to rooms in version (a) can be accomplished only from the living room. This suggests that several organisation-transformations, resulting in distinct organisations, may require the same amount of work. Therefore, quantitative characterisations of organisations are undetermined. An operational definition for organisation is the following recursive scheme:

Definition 6.1 *An Organisation is one of the following:*

- *an atom;*
- *a set of organisations;*
- *a group of organisations put somehow in relation to one another;*
- *nothing else.*

This definition purposefully lacks a better clarification of the expression *somehow in relation with one another*, since this may be instantiated in several ways. Notwithstanding, it includes as organisations sets of atoms, organisations, or both. In the case of the above example, for instance, the bricks are atoms.

Definition 6.1 makes no reference to hierarchy, environment dependence, or purpose (function in biology). It can be made mathematically formal by using graph theory, elements of metamathematics, and methods common in computer science [19]. A detailed description of the constructs necessary to represent organisations mathematicaly is outside the scope of the present work but the main idea of their construction can be presented pictorially.

Hypergraphs (Figure 8(a)) will be our way of expressing *somehow in relation with one another* without further specification and often depicted as bi-partite graphs. To be able to consider an organisation as an element of another (system of systems...) and to allow for transformations of organisations and "dynamic" creation and destruction of organisations, we will need meta-elements: a hook and as many meta-variables as needed; and to extend hypergraphs with these elements (Figure 8(b)). Finally, we need to "assign", with an interpretation

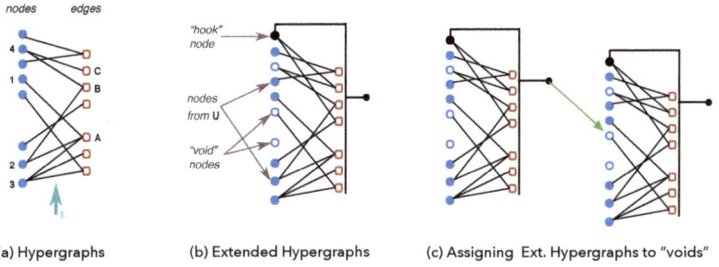

Figure 8: Whole-part graphs I: basic elements of the construct

similar to those of pointers in computer science, a hypergraph nicknamed *hook* to a meta-variable (a "void") in another extended hypergraph (Figure 8(c)).

This settled, the mathematical representation of organisations —whole-part graphs— is recursively defined by:

Definition 6.2 *An object γ is a* wp*-graph, that is, $\gamma \in \Gamma$ if and only if:*

1. $\gamma \in \mathcal{H}$,

2. $\gamma \doteq {}_m h^\star \hookleftarrow\, < \gamma_1^\circ, \ldots, \gamma_n^\circ >$,

3. *nothing else.*

Where γ° means that its "upmost" hyper-graph ${}_m h^\star$ of γ has a 'hook' as node. And the symbol $\doteq$ reads is build as or is given by and has a double interpretation: as a programming assignment during construction of wp*-graphs and as a mathematical equality in Γ [19].*

In Definition 6.2, $\mathcal{H}$ is the class of all extended hypergraphs and ${}_m h^\star$ is an element of $\mathcal{H}$ that has m meta-variables as nodes.

It is important to note that Definition 6.2 does not define one structure, as usual in the sciences of programming but, instead, defines a space Γ of all possible whole-part graphs. Each element of Γ looks like the sketches in Figure 9. both sketches are the same, Figure 9(a) with more details than Figure 9(b).

Interaction graphs are a particular kind of organisations, as any graph is, and hence belong to Γ. Organisations, as seen from the systemhood side, also

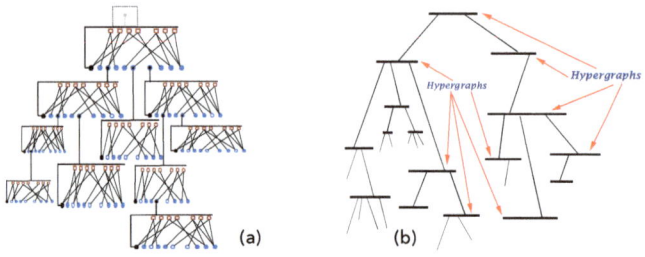

Figure 9: Whole-part graphs II: points in space Γ

reflect possibilities of interaction among its component elements and parts. The association between interaction graphs and dynamics described in section 4 can be extended from $\mathbf{G}_n$ to Γ with minimum effort by considering the set of all possible solutions of a given dynamical system, that is, the collection of orbits or surfaces defined by $\vec{F}$ irrespective of initial conditions [32].

Let $\mathcal{P}$ be a space of attributes, containing the space-time $\{(\vec{x},t)\}$. The *synexion* space, $\mathcal{B}$, is the class of all associations between $\gamma \in \Gamma$ and subsets of $\mathcal{P}$, for each $\gamma \in \Gamma$; that satisfy the following constrains:

$$\left.\begin{array}{rl} \mathcal{V}(\gamma) \subset \mathcal{P} & \text{if } \gamma \in \mathcal{H}, \\ \mathcal{V}(\gamma) = \bigcup_{i=1}^{k} \mathcal{V}(\gamma_i) & \text{if } \gamma = \{\gamma_1,\ldots,\gamma_k\}, \\ \mathcal{V}(\gamma) \supset \bigcup_{i=1}^{n} \mathcal{V}(\gamma_i) & \text{if } \gamma \cong {}_m h^\star \hookleftarrow s, \\ & s = <\gamma_1^\circ,\ldots,\gamma_n^\circ>, \end{array}\right\} \quad (3)$$

where $n \leq m$. That is,

$$\begin{array}{rcl} \mathcal{B}: \Gamma & \longleftrightarrow & \wp(\mathcal{P}) \\ \gamma & \longleftrightarrow & \mathcal{V} \end{array} \quad (4)$$

such that $\mathcal{V}(\gamma)$ satisfies the hierarchical constrain 3.

Synexions are not sets in the usual sense (Figure 10). Subsets of $\mathcal{V}(\gamma)$ must also conform to conditions (3) and be formed from subsets of $\mathcal{V}(\beta)$, where β is a part of γ. Thus, we may have $\mathcal{V}(\gamma)_1 \cap \mathcal{V}(\gamma)_2 = \emptyset$ as organised volumes, even though $\mathcal{V}_1 \cap \mathcal{V}_2 \neq \emptyset$ as usual subsets of $\mathcal{P}$.

7 Portal towards Fenceless Creativeness

In Γ, a wealth of transformations, operations, relations, structures and finite automata can be defined. There is as well a complexity measure that attributes a complexity value to each element of Γ, i.e., to each organisation. A small sample of these is depicted in [19]. This, together with the association with dynamics and "concrete" (physical) behaviour inherent in synexions provide a

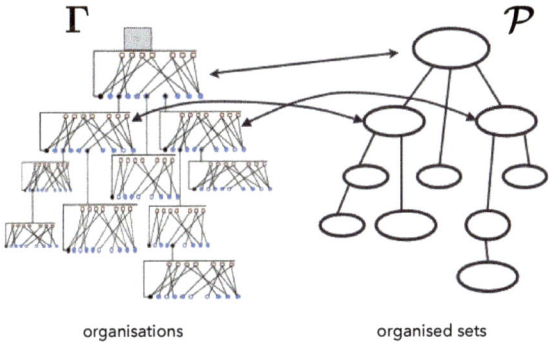

organisations   organised sets

Figure 10: Synexions or organised sets

lot of freedom in modelling the many variable-structure systems and hierarchies arising in living phenomena.

It should be emphasised, though, that the complexity measures defined over Γ and inhered by elements of $\mathcal{B}$ (synexions) are not the same as the complexity related to dynamical systems, entropy, information or self-organisation, currently available in the literature (see [30; 46; 15; 36; 24; 32; 5] for an overview of the many senses of complexity). Therefore, an interesting open problem is to search answers for questions like *"What are the similarities and distinctions among these various concepts of complexity?"* and yet, since the complexity in Γ may be directly connected to the interaction structure of any dynamical system [22], *"Is the Γ-complexity monotonically related to dynamical complexity?"*.

Chaos exist in low dimension systems. We learn from [7; 8] that one cannot decide simply, i.e. algorithmically, whether a dynamical system may present chaotic behaviour or not and that systems presenting chaotic behaviour are everywhere (they are dense in the space of dynamical systems), being easy to select one when modelling whatever phenomena. From another stand, biological entities and phenomena are organised and anything but chaotic. Organisms are even plenty of organelles and systems that prevent the expression of fluctuation and noise. Guided by these remarks, it is possible to extend the previous questions into two bolder ones: *"Does organisation in life phenomena tame complexity and chaos in same sense?"*; if it does, *"Which sense is this?"*.

However, the real breakthrough enabled by generalising systems into organisations has an epistemological character and is illustrated in Figure 11. Organisations blur the separation between *thinghood* and *systemhood*, since they can at times be things that interact and/or a representation of the interaction-channels that bind the phenomenon together. This allows for considering a

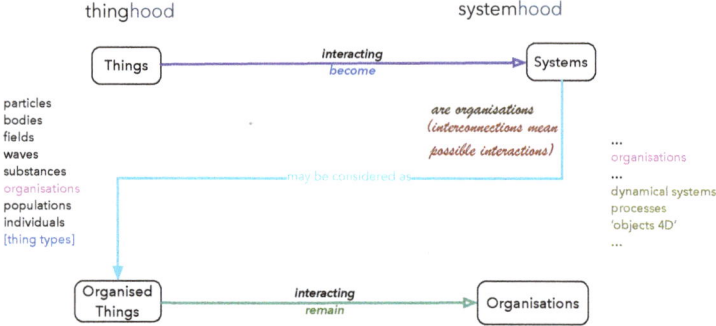

Figure 11: Thinghood, Systemhood, Totalhood

phenomena, however complex, as a collection of interacting things and 'simpler' phenomena. The number of possible ways of doing this 'decomposition' is bounded from below by the partition number [19, Sec.4.2, Eq. 25]. Hence, the fenceless creativity.

It have not escaped my attention how integrative and unifying this characteristic of organisations is.

References

[1] R. Belohlavek, G. J. Klir, and E. Bartl. "contents" of the International Journal of General Systems. Taylor & Francis Group, London, 1974–current. URL https://www.tandfonline.com/loi/ctas20.

[2] C. Berge. *Graphs and Hypergraphs*. North-Holland, Amsterdam, 1973.

[3] J.-Y. Béziau and M. V. Kritz. Théorie et modèle I: Point de vue général et abstrait. *Cadernos UFS Filosofia*, 8(Fasc. XIII):9–17, ago-dez 2010.

[4] F. C. Boogerd, F. J. Bruggeman, J.-H. S. Hofmeyr, and H. V. Westerhoff, editors. *Systems Biology: Philosophical Foundations*. Elsevier B.V., Amsterdam, 2007.

[5] E. Bresciani and I. M. L. D'Ottaviano. Basic concepts of systemics. In A. P. Júnior, W. A. Pickering, and R. R. Gudwin, editors, *Systems, Self-Organisation and Information: an interdisciplinary perspective*, volume 1, pages 47–64. Routledge, London, UK, 2018.

[6] K. J. W. Craik. *The Nature of Explanation*. Cambridge University Press, London, 1943.

[7] N. C. A. da Costa and F. A. Doria. Undecidability and Incompleteness in Classical Mechanics. *Int. Journal of Theorietical Physics*, 30(8):1041–1054, 1991.

[8] N. C. A. da Costa, F. A. Doria, and A. F. Furtado do Amaral. Dynamical System Where Proving Chaos Is Equivalent to Proving Fermat's Conjecture. *Int. Journal of Theorietical Physics*, 32(11):2187–2206, 1993.

[9] N. R. Hanson. *Patterns of Discovery: An Inquiry into the Conceptual Foundations of Science.* Cambridge University Press, Cambridge, 1958.

[10] F. M. Harold. *The Way of the Cell: Molecules, Organisms and the Order of Life.* Oxford University Press, Oxford, UK, 2001.

[11] F. M. Harold. Molecules into Cells: Specifying Spatial Architecture. *Microbiology and Molecular Biology Reviews*, 69(4):544–564, Dec. 2005.

[12] A. M. Jones. *Environmental Biology.* Routledge Introductions to Environment Series. Routledge, London, 1997.

[13] W. J. Junk, M. T. F. Piedade, F. Wittmann, J. Schöngart, and P. Parolin, editors. *Amazonian Floodplain Forests: Ecophysiology, Biodiversity and Sustainable Management*, volume 210, Dordrecht, 2010. Springer.

[14] R. E. Kalman, P. L. Falb, and M. A. Arbib. *Topics in Mathematical System Theory.* McGraw-Hill Book Co., Inc., New York, NY, 1969.

[15] S. A. Kauffman. *At Home in the Universe: The Search for Laws of Self-Organization and Complexity.* Oxford University Press, New York, NY; Oxford, UK, 1995.

[16] S. Klamt, U.-U. Haus, and F. Theis. Hypergraphs and Cellular Networks. *PLoS Computational Biology*, 5(5):e1000385, May 2009.

[17] G. J. Klir. *Facets of Systems Science.* Plenum Press, New York, NY, 2nd edition, 2001.

[18] M. V. Kritz. Boundaries, interactions and environmental systems. *Mecánica Computacional*, XXIX: 2673–2687, Noviembre 2010. Downloadable from http://www.cimec.org.ar/ojs/index.php/mc/article/viewFile/3183/3110.

[19] M. V. Kritz. From Systems to Organisations. *Systems*, 5(1):23, Mar. 2017.

[20] M. V. Kritz. De la modélisation à la créativité mathématique. In J.-Y. Beziau and D. Schulthess, editors, *L'Imagination. Actes du 37e Congrès de l'ASPLF (Rio de Janeiro, 26-31 mars 2018)*, volume 1 of *Academia Brasileira de Filosofia*, pages 267–289, Londres, 2020. ASPLF, College Publications.

[21] M. V. Kritz and J.-Y. Béziau. Théorie et modéle II. *Cadernos UFS de Filosofia*, 10(Fasc. XIV):7–16, jul-dez 2011.

[22] M. V. Kritz and M. T. dos Santos. Dynamics, systems, dynamical systems and interaction graphs. In M. M. Peixoto, D. Rand, and A. A. Pinto, editors, *Dynamics, Games and Science II*, volume 2 of *Springer Proceedings in Mathematics*, pages 507–541, Berlin, 2011. Springer-Verlag.

[23] M. V. Kritz, C. M. Dias, and J. M. da Silva. *Modelos e Sustentabilidade nas Paisagens Alagáveis Amazônicas*. Notas em Matemática Aplicada. SBMAC – Sociedade Brasileira de Matemática Aplicada e Computacional, São Carlos, SP, 2008.

[24] G. Longo, P. A. Miquel, C. Sonnenschein, and A. M. Soto. Is information a proper observable for biological organization? *Progress in biophysics and molecular biology*, 109(3):108–114, Aug. 2012.

[25] M. D. Mesarović, editor. *System Theory and Biology*. Springer-Verlag, New York, NY, 1968.

[26] M. D. Mesarovic and Y. Takahara. *General Systems Theory: Mathematical Foundations*, volume 113 of *Mathematics in science and engineering*. Academic Press, New York, NY, 1975.

[27] J. G. Miller. *Living Systems*. McGraw-Hill Book Co., Inc., N. York, NY, 1978.

[28] R. R. Mohler and A. Ruberti, editors. *Recent Developments in Variable Structure Systems, Economics and Biology*, volume 162 of *Lecture Notes in Economics and Mathematical Systems*. Springer-Verlag, Berlin Heidelberg, 1978.

[29] E. Morin. *Les sept savoirs nécessaires à l'éducation du futur*. Seuil, 2000.

[30] G. Nicolis and I. Prigogine. *Exploring Complexity: An Introduction*. W.H. Freeman and Company, New York, NY, 1989.

[31] A. Plagnol. Psychologie, épistémologie et˜théorie de˜la˜représentation: fondation analogique et˜données séminales. *Psychologie Française*, 52 (3):327–339, sep 2007. doi: 10.1016/j.psfr.2007.02.003. URL http://linkinghub.elsevier.com/retrieve/pii/S0033298407000489.

[32] A. J. Roberts. *Model Emergent Dynamics in Complex Systems*. Mathematical Modeling and Computation. SIAM, Society for Industrial and Applied Mathematics, Philadelphia, Pennsylvania, 2015.

[33] R. Rosen. Biological Systems as Organizational Paradigms. *International Journal of General Systems*, 1(3):165–174, 1974.

[34] R. Rosen. Some comments on systems and system theory. *International Journal of General Systems*, 13(1):1–3, 1986.

[35] C. Schiller. *The Motion Mountain*, volume 1–6. CreateSpace Independent Publishing Platform, March 2013–2016. URL http://www.motionmountain.net/index.html.

[36] F. Schweitzer, editor. *Self-Organization of Complex Structures: From Individual to Collective Dynamics*, Boca Raton, 1997. CRC Press, Taylor and Francis Group, LLC.

[37] J. Shrager. The fiction of function. *Bioinformatics*, 19(15):1934–1936, Oct. 2003.

[38] R. E. Ulanowicz. *A Third Window: Natural Life beyond Newton and Darwin*. Templeton Press, West Conshohocken, PA, 2009.

[39] L. von Bertalanffy. *General Systems Theory*. Allen Lane The Penguin Press, London, 1968.

[40] C. H. Waddington, editor. *Biological organization, cellular and subcellular*, London, 1959. UNESCO, Pergamon Press.

[41] W. Weaver. Science and Complexity. *American Scientist*, 36:536–544, 1948.

[42] G. M. Weinberg. *An Introduction to General Systems Thinking*. Dorset Hause Publishing, N. York, NY, silver anniversary edition, 2001 (1st ed. 1975).

[43] O. Wolkenhauer. Systems biology: The reincarnation of systems theory applied in biology? *Briefings in Bioinformatics*, 2(3):258–270, Sept. 2001.

[44] J. Wu. Landscape ecology, cross-disciplinarity, and sustainability science. *Landscape Ecology*, 21:1–4, 2006.

[45] L. A. Zadeh and E. Polak, editors. *System Theory*, volume 8 of *Inter-University Electronics Series*. TATA McGraw-Hill Publishing Co. Ltd., Bombay, New Delhi, 1969.

[46] W. H. Zurek, editor. *Complexity, Entropy, and the Physics of Information*, volume VIII of *Santa Fe Institute Studies in the Sciences of Complexity*. Perseus Publishing, Cambridge, MA, 1990.

Randomization and Fair Judgment in Law and Science

Julio Michael Stern[*],
Marcos Antonio Simplicio[†],
Marcos Vinicius M. Silva[‡],
Roberto A. Castellanos Pfeiffer[§]

Abstract

Randomization procedures are used in legal and statistical applications, aiming to shield important decisions from spurious influences. This article gives an intuitive introduction to randomization and examines some intended consequences of its use related to truthful statistical inference and fair legal judgment. This article also presents an open-code Java implementation for a cryptographically secure, statistically reliable, transparent, traceable, and fully auditable randomization tool.
Keywords: Randomization; Truthful inference; Fair judgment; Judicial autonomy and independence.

---

[*]Institute of Mathematics and Statistics of the University of Sao Paulo jstern@ime.usp.br
[†]Polytechnic School of the University of Sao Paulo mjunior@larc.usp.br
[‡]Polytechnic School of the University of Sao Paulo mvsilva@larc.usp.br
[§]Law School of the University of Sao Paulo, roberto.pfeiffer@usp.br

κληρω νυν πεπαλασθε διαμπερες ος κε λαχησιν:
*Let the lot be shaken for all of you,
and see who is chosen.* Iliad, VII, 171.

מִדְיָנִים יַשְׁבִּית הַגּוֹרָל וּבֵין עֲצוּמִים יַפְרִיד
*Casting the dice puts judgment quarrels to rest and keeps
distinct essential powers duly separated.* Proverbs 18:18.

1 Introduction

Francisco Antonio Dória has had a consistent interest in randomness and chaos and, together with his collaborators, has investigated fundamental aspects of such phenomena. This article is our contribution to the Festschrift celebrating Doria's 75th birthday.

This article analyses some pragmatical aspects of applying randomization in empirical science and law, considers some philosophical implications or premises justifying or motivating these applications, and offers some tools that promote good randomization practices. *The Cardsharps* (1594) marks the beginning of the independent career of the great Italian master Michelangelo Merisi da Caravaggio (1571-1610). This painting displays a wealthy but innocent looking boy playing cards with his opponent, a cardsharp, that cheats in two ways: On the one hand, the cardsharp hides in his belt spurious cards that he intends to use in illegitimate ways; on the other hand, a sinister looking and strategically positioned accomplice gives him access to privileged and undue information. Finally, the cardsharp carries a dagger, hinting at the dangers lurking in this environment of misrepresentation and deception.

Caravaggio gives a beautiful depiction of some themes discussed in this article. First, the social importance of activities involving randomization, that is, the random setting of some variable of interest, like the drawing of dice or, in this painting, the distribution of playing cards. Second, it suggests the question – Why to randomize?

Figure 1: *The Cardsharps* (1594), by Michelangelo Caravaggio

that is – Why should a rational agent abdicate the opportunity of making a deterministic choice introducing, on purpose, a random component in making a decision? If so – What is the role played by randomization? Finally – How to randomize? that is – What dangers could jeopardize a randomization process? and, if necessary – How to shield or immunize the process against these dangers?

In order to answer these questions, we have to pay attention to some topics in Statistics, Computer Science and Cryptography; in addition, we have to examine some details concerning the design of empirical trials or the operation of legal systems. In this article, we investigate each one of the questions just raised, looking for an intuitive understanding of the role(s) played by randomization.

In the final sections of the paper, we present an easy to use, open-code, traceable, auditable, secure, and statistically sound randomization toll that is ready for use in empirical trials and legal applications. This kind of secure randomization tool can prevent the possibility of misrepresentation and deception, as depicted in the painting by Caravaggio. Moreover, even in situations where no misdeeds actually occur, the use of such a tool can be beneficial by fostering public confidence in the soundness of important decisions, by strengthening the resilience of public institutions, and by favoring the peaceful resolution of conflicts.

2 Social Importance of Randomization

Gambling and lotteries exchange billions of Dollars every day worldwide. Hence, ensuring honesty and transparency in these activities should already be considered a meritory task. However, since ancient times, sortition (i.e., selection by lottery) is used for many other purposes. In the Iliad, one of the oldest texts of western culture (aprox. 1200BC), the Argonauts (crew of the ship Argo) selected a man to execute a dangerous task by sortition – see this paper's first opening quotation. In the same manner, modern societies often resort to sortition for drafting. Figure 2 displays some photographs related to compulsory enlistment for service in the USA, namely, military drafting during the Civil (left) and the Vietnam (center) wars, and selection for jury duty (right).

In order to gain public trust, the sortitions for the Vietnam war were conducted in public view: Balls with calendar dates were placed in a transparent urn and some anniversary dates were then picked, giving the (un)lucky winners the opportunity to serve their country in the battlefield. A post hoc statistical analysis of these drawings revealed a significant bias favoring latter days of the calendar, corresponding to the last balls placed inside the urn, an unexpected effect of an ill-conceived randomization process that generated misunderstanding, frustration and conflict.

Figure 2 (right) shows a letter calling a citizen for jury duty in the USA. In this process, an eligible citizen was chosen at random by running a computer program. Post hoc analyses of this randomization process revealed no significant bias or any other statistical anomaly. Nevertheless, the code of these computer programs were never made public, making the randomization process opaque and

Figure 2: Draft lotteries in war and peace

non-verifiable, thus generating mistrust and resentment.

Finally, in the world of science, good clinical trials are conducted by (double) blind and random attribution of patients to two or more distinct treatments. The objective of such a trial is to find out if a new or alternative treatment is significantly better than the old or standard one, according to well-established statistical criteria. In this situation, some frequently asked questions are: Why should a patient's treatment be selected at random? Why not give him or her the freedom to chose his or her proffered treatment? Why hide from a patient information about his or her own treatment?

3 Why to Randomize?

Imagine a clinical trial where patients are free to choose a treatment according to their own will. Among patients, there will be rich and poor, people with different degrees of instruction, people with better or worst networks for support, etc. Obviously, rich, well educated, and well connected patients will have better access to good information and advice and, therefore, will be prone to make better decisions. Moreover, these same patients likely have better overall living conditions and, therefore, even with the same treatment, might have a better chance of recovery. Hence, this freedom of choice would automatically introduce *confounding effects*: After the trial is over, we would not be able to (completely) discern the beneficial and adverse consequences of distinct treatments from consequences of preexisting conditions.

Similar unwanted interference is generated by the *placebo* and *nocebo* confounding effects. If a patient knows to be receiving either a new, experimental and possibly wonderful drug, or else an old and possibly not very effective drug, his or her moral may be, respectively, lifted or depressed. That, in turn, may affect his or her overall health and chance of recovery. This is why, in a good clinical trial, treatment information is denied (blinding or censorship) to patients, and commonly also to their direct caretakers (double-blinding).

There are in the medical literature plenty of examples of clinical trials that came to wrong conclusions in consequence of such confounding effects. The best known antidotes against these confounding effects rely in some form of randomization. The idea is to chose a patient's treatment based on a random variable that is independent of any potentially confounding variable. In so doing, the random element in the choice of treatment has the effect of breaking causal links that should not interfere with the experiment, allowing the trial to adequately focus on the causal links of interest – see Stern (2008) and Pearl (2009) for further details.

Finally, let us consider the use of randomization in the legal system, like the selection of jurors or judge(s) for a given case. Figure 3 displays two pictures from ancient Egypt. On the left, a stone carving of approx. 2400BC shows two merchants using a two-pan balance to correctly measure amounts of goods for a fair commercial transaction. On the right, the Hunefer papyrus, of approx. 1275BC, shows the scale used by Maat, the goddess of justice, where the heart and the (de)merits of a man are measured. It should be clear that these two scales are essentially distinct – they belong to distinct contexts. The figure at the center suggests the possibility of "mixing" these two essences: Perhaps Maat could make a more benevolent assessment in the scale of justice if she, or her priests, received goods of commercial valuable... There we have, once more, a confounding effect, characterized by spurious influences between powers belonging to essentially distinct systems:

Figure 3: Essentially different powers: Economy and Justice

in this case, the economic system and the justice system.

How to avoid such confounding effects caused by spurious influences, fostering autonomous decisions in a strong and independent justice system? Surprisingly, the Hebrew bible already offers very good advice at this respect, as stated in the second opening quotation of this paper. Interestingly, the Hebrew root עצם, *etzem*, whose literal meaning is bone, also generates words meaning essence (the etymological origin of the English word), strength, power and the modern Hebrew word for independence.

Judges, even if perfectly honest, do not come to court as a blank slate, nor should they. Every judge has his or her own history of decisions and opinions. Hence, if the selection of judges could be influenced by the litigants or other interested parties, the richer, better informed, well connected, or otherwise more powerful parties would likely have an advantage in directing the case to a judge sympathetic to their arguments. For this reason, in many modern democracies, the *distribution* of a new judicial case must take the form of a random choice among the available judges or courts qualified to judge it.

4 How to Randomize?

Previous sections discussed several applications of randomization and explanations of why to use it. This section describes some desirable characteristics of such a randomization process, including:

Statistical honesty: In a set of sortitions (random selections) in the system, the probability of any group of outcomes should be exactly as prescribed by the established rules.

Cryptographic security: The outcome of the sortition should be unpredictable; moreover, no external agent should be able to influence the randomization process, even if the agent knows in detail the randomization mechanism being used and has state of the art knowledge of all relevant technologies involved.

Transparency: All relevant information about the sortition process must be of public knowledge, including any pertinent detail about the randomization mechanisms being used.

Auditability: All relevant occurrences of an actual randomization process must be traceable and auditable. Furthermore, it should not be possible to conceal any improper use of the randomization system.

The first two requirements are of technical nature, stipulating that, in the randomization process, we should use "honest dice that cannot be tampered with" – or else a more convenient device, like a computational algorithm that adequately mimics all the relevant characteristics of "honest dice". For more technical discussions on theses characteristics, see Marcondes et al (2019), Saa and Stern (2019) and Silva et al. (2020).

In order to emphasize the importance of the last two requirements, let us discuss a form of cheating known as *rerandomization*. In this kind of cheating, the agent responsible for a sortition has the privilege of using the randomization mechanism out of public scrutiny, examine the outcome, and chose at will either to make this process and its outcome public, or to hide this first try and randomize a second time, as if the first try never happened. Imagine for example the classic process of picking a ball from a transparent urn. However, instead of making a live presentation, the sortition ceremony is recorded for broadcasting at a later time. A dishonest agent could repeat and record the process twice, and only release the recording that best fits his or her goals, as if it were the only recording ever made. It should be clear that the repeated use of this subterfuge gives the agent in charge some latitude to pick and choose, biasing the final outcome according to his or her convenience.

Authority, Transparency and Understanding

Why is transparency even required in a randomization process? Would it no be possible, or even easier, to anchor the credibility of the process on a *principle of authority*? If a given authority is responsible for a randomization process, doesn't the requirement of transparency imply an implicit doubt? If so, doesn't the requirement of transparency imply disrespect for the same authority?

These are basic questions in philosophy of law, that can only be answered in a context that specifies the fundamental values and goals chosen by a given society. Niklas Luhmann (1985, 1989), a celebrated scholar in philosophy of law, postulates that the fundamental goal of the justice system is – "the congruent generalization of normative expectations". That is, the final objective of the legal system is the construction of a harmonious society, where citizens have a coherent view of what constitutes a good set of rules for social behavior (normative expectations). Moreover, a legal system should provide mechanisms that stimulate citizens to conform to normative expectations and inhibit their transgression.

This conception of law requires from every citizen a well founded trust that the justice system is efficient and fair, preferably obtained by conscious understanding of laws and regulations and their forms of implementation. Moreover, a justice system conceived according to such principles is weak or fragile if sustained on blind faith on ad hoc authority, but strong and resilient if sustained by a conscious, engaged, and participative community. The articles of Silva et al. (2020) and Stern (2018) expand these ideas.

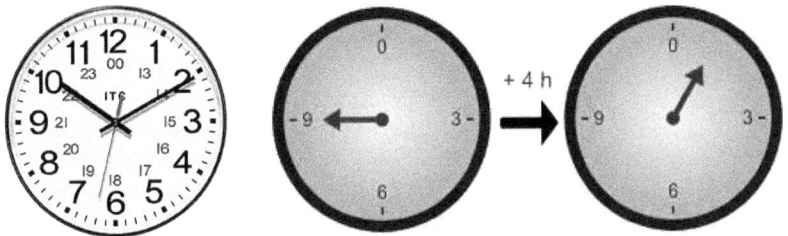

Figure 4: Modulo 12 or Clock Arithmetic

5 Modular Arithmetic and Trusted Roulettes

In this section we discuss intuitive ideas for how to implement an honest randomization device that all interested parties can trust. In following sections of this article we offer a viable technological solution to the problem of randomization that satisfies all requirements stipulated in previous sections, following the general ideas hereby discussed.

Modular arithmetic is an integer arithmetic system in which numbers "wrap around" after reaching a maximum value, m, called the modulus. A familiar example is the standard reading of a clock. After noon (12 o'clock), we restart counting from $1, 2, \ldots$ (p.m.). Notice that, in Figure 5, the position corresponding to noon (or midnight) is marked either by the modulus value, $m = 12$, or by the value zero – that is mathematically more convenient. In general, for positive integers, n and $m > 1$, we define $n \bmod m$ (read n modulo m) as the remainder of the division of n by m. For example, see Figure 5 (left): $13 \bmod 12 = 1$, $14 \bmod 12 = 2$, ... $23 \bmod 12 = 11$, and $24 \bmod 12 = 12 \bmod 12 = 0$. Figure 5 (right) illustrates the modular arithmetic operation $(9 + 4) \bmod 12 = 1$.

Now imagine we have a game using a roulette or wheel of fortune, see Figure 5, with k participants, also known as the *stakeholders*, all of them wanting the privilege of spinning the roulette, and not trusting anyone else to do the job. How can we break this deadlock?

Figure 5: Rolling Dice and Spinning Roulettes

We can solve the aforementioned impasse using the following protocol:

1. Provide each stakeholder with a well-balanced roulette, marked according to the numbers set $\{0, 1, 2 \ldots (m-1)\}$;

2. Ask each stakeholder to spin his roulette *honestly*, that is, with a not fully controlled and strong enough initial impulse so to produce any of the possible outcomes, $\{0, 1, 2 \ldots (m-1)\}$, with the same probability, $(1/m)$. Moreover ask each stakeholder to use his roulette *independently*, that is, to do so without sharing any information with other stakeholders or interested parties;

3. Collect and add, using modulo m arithmetic, the results produced by each one of the k stakeholders in order to produce the final result: $n_f = (n_1 + n_2, \ldots + n_k) \bmod m$.

We can guarantee that the final result, n_f, produced by this protocol is equivalent to an "honest roulette", as long as at least one (any one) of the k stakeholders does his job as required. This guarantee is a corollary of the following theorem: Let x and y be independent random variables in $\{0, 1, 2 \ldots (m-1)\}$. Then, if any one of these random variables, x or y, is uniformly distributed, so is $z = (x + y) \bmod m$. Imagine, for example, that variable x is not random at all, but rather a known constant, c, namely, the initial state of the roulette. Furthermore, imagine that y is independent of x and uniformly distributed in $\{0, 1, 2 \ldots (m-1)\}$. Under these conditions, the theorem states that the final state of the roulette, $z = c + y$ is uniformly distributed, corresponding to the intuitive idea that, when using a well-constructed roulette, a strong enough impulse will produce a final outcome that "forgets" the initial state of the roulette. For further details and formal mathematical analyses, see Scozzafava (1993).

In many applications in statistics, clinical trials, and complex sortitions, we need a random variable x uniformly distributed in the interval $[0, 1[$ of the real line. In computational procedures, this continuous variable can be approximated by a fraction n/m, where m in a large integer, and $n \in \{0, 1, 2, \ldots (m-1)\}$. This fraction can be translated to standard floating point notation, and then be further transformed into random variables with several probability distributions of interest in statistical modeling, see Hamersley (1964), Ripley (1987). Such uniform or non-uniform random variables can, in turn, be used in dynamic clinical trials, haphazard intentional sampling, adaptive sampling procedures, and other complex applications of interest in statistical modeling and decision science, see for example Fossaluza (2015) and Lauretto et al. (2012, 2017) and the bibliography therein.

In the sequel, we describe a software implementing the protocol outlined in this section, including all necessary precautions in order to guarantee cryptographical security. In this software, every stakeholder is required to input a random number n between 0 and $m = 9,999,999$. Each stakeholder has the responsibility of producing his or hers random 7-digit decimal number in a way he or she finds appropriate.

Random decimal digits can be easily produced, for example, by rolling 10-faced dice, available in several shapes with the required symmetry conditions, see Stern (2011). Figure 6 exhibits 10-faced dice shaped as a pentagonal trapezohedron (left) and a pentagonal antiprism (center), and also 20-faced dice shaped as an icosahedron (right); in order to produce a random decimal digit using icosahedral dice, the user should read only the last digit for outcomes ranging

Figure 6: Polyhedral dice for decimal digits

from 10 to 20.

6 Intuitive Cryptology

One crucial requirement of the random drawing approach described in Section 5 is that the the roulettes are run independently. Otherwise, a dishonest stakeholder S_i could wait until the results of all roulettes are revealed, and then run his/her own roulette for manipulating the final outcome of the drawing: for example, suppose that the sum of the contributions from all stakeholders except S_i is $n = 3$ for, say, $m = 12$; after learning this value of n, S_i could force his/her own roulette to give $n_i = 2$, thus obtaining $n_f = n + n_i \bmod m = 5$ as the final (manipulated) outcome of the drawing.

To ensure this independence property, Silva et al. (2020) builds upon the properties of hash-based bit-commitment mechanisms. Intuitively, a hash function H is a cryptography construct analogous to fingerprinting for humans, as illustrated in Figure 7 – we refer the reader to Beutelspacher (1994), Bultel et al. (2017) and Fellows and Koblitz (1994) for intuitive introductions on key ideas of cryptology, an to Rogaway and Shrimpton (2004) for a concise but formal explanation of properties of cryptographic hash functions. Specifically, given the fingerprint for an unknown human being, it takes a lot of computational power to look all around the world for the owner of that fingerprint; on the other hand, given a fingerprint and the corresponding human, it is quite easy to check whether or not they match, and it is hard (almost impossible) to find two different people with the same fingerprint (even considering identical twins).

Similarly, suppose that someone computes the hash of a number n, i.e., a value $h = H(n)$, which acts as a "fingerprint" for n; then, if only h is revealed, but n is kept secret, there is no simple mechanism for finding the value of n. Of course, one could test every possible value of n, checking if a guess n_i is such that $H(n_i) = h$, just like finding the owner of a fingerprint given only the fingerprint itself. However, the computational effort for performing such brute-force attack would be very large. Actually, in practice the cost for hash functions would be even larger than searching for a human who owns a fingerprint: while there are a few billions of humans in the world, a hash function can be used in such a manner that the number of tests would be as large as the number of atoms in the whole planet Earth! For this purpose, it suffices to combine the value of n with a large and unpredictable (e.g., random) mask r when computing the hash, i.e., to make $h = H(r, n)$. As a result, even if there are only a few possible values of n to be tested, determining whether or not a guess n_i is correct would require testing all possible values of r too. Therefore, it suffices to use a large-enough mask r (e.g., 256-bits) to ensure that any attempt of determining n via brute force would be computationally infeasible.

When both n and h are revealed, on the other hand, it is easy to verify whether or not they match: it suffices to compute $H(n)$ directly, and check if the output of this computation is identical to the provided value of h. However, like different humans should not have the same fingerprint, it is computationally hard to find two distinct values of n (say, n_1 and n_2) that have the same hash h. Hence, once $h_1 = H(n_1)$ is revealed, one can say that the person who revealed it is "committed" to revealing n_1, i.e., it would be hard to trick someone into believing that h_1 was computed from any other input $n_2 \neq n_1$.

Such properties are used by Silva et al. (2020) to build a two-phase procedure for ensuring the fairness of random draws:

Figure 7: Hash functions and their similarity with human fingerprinting.

1. Commitment phase: first, each stakeholder S_i runs a roulette (honestly or not), getting a value n_i as result. Then, S_i computes the hash of n_i, denoted h_i, and reveals only h_i to the other stakeholders, keeping n_i itself secret. This prevents S_i from learning the roulette results from his/her peers, and vice-versa.

2. Reveal phase: only after all hashes are received, every stakeholder S_i reveals its own n_i. The outcome of the drawing is then computed locally by S_i by adding every n_i together using modular arithmetic as explained in Section 5. In this case, even if S_i is malicious and tries to delay the revelation of n_i until it learns the partial outcome of the drawing from the values revealed by his/her peers, it would be already too late: after revealing h_i in the commitment phase, S_i has no choice but to reveal the already chosen n_i, rather than some other value that might lead to a more desirable (but unfair) drawing outcome.

7 *Java* Implementation

We developed a simple Java library for implementing the protocol described in Silva et. al (2020), and made it available under the MIT License at https://doi.org/10.24433/CO.6108166.v1 This library can, thus, be freely adapted for the needs of specific application scenario. To help in this task, we also provide a simple proof-of-concept graphical interface for testing purposes, which is depicted in Figure 8. More precisely, this figure shows:

(a) A simple configuration interface for drawing a number among $m = 10{,}000{,}000$ candidates, i.e., from 0 to 9,999,999. The number of stakeholders participating in the drawing and additional metadata related to it can also be defined.

(b) A snapshot of the Commitment phase, as seen by Stakeholder S_0 in a drawing involving 5 stakeholders. In this snapshot, S_0 is then free to choose a number n_0 to commit, which is combined with a random mask for better security against brute force attacks. Meanwhile, S_1, S_3 and S_4 have already sent the hashes h_1, h_3 and h_4 of their own commitments, n_1, n_3 and n_4, respectively; as a result, these stakeholders cannot modify the chosen values n_1, n_3 and n_4 anymore.

(c) A snapshot of the Reveal phase, as seen by Stakeholder S_3. The figure shows that S_3 is the only one who has not yet revealed the chosen value for n_3, while all of his/her peers have already revealed n_0, n_1, n_2 and n_4. Nevertheless, S_3 can only reveal the correct n_3 (and corresponding mask), since the revealed value must match the committed value h_3.

(d) The completion of the protocol, when one of the eligible numbers (namely, 6,932,980) is picked with uniform probability based on all stakeholders' contributions n_0, n_1, n_2, n_3 and n_4.

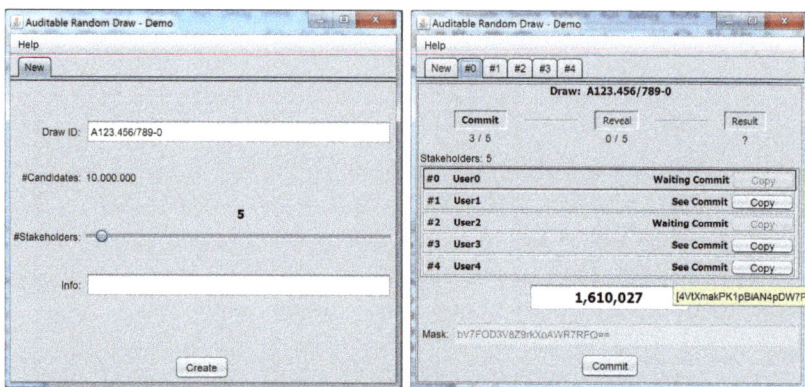

(a) Random drawing with 5 stakeholders (S_0, S_1, S_2, S_3 and S_4) and modulus value $m = 10,000,000$.

(b) S_1, S_3 and S_4 after commitment, as seen by S_0. Value committed by S_4 is $h_4 = \mathtt{4VtCmakPK1pBiAN4pDW7Pj}\ldots$

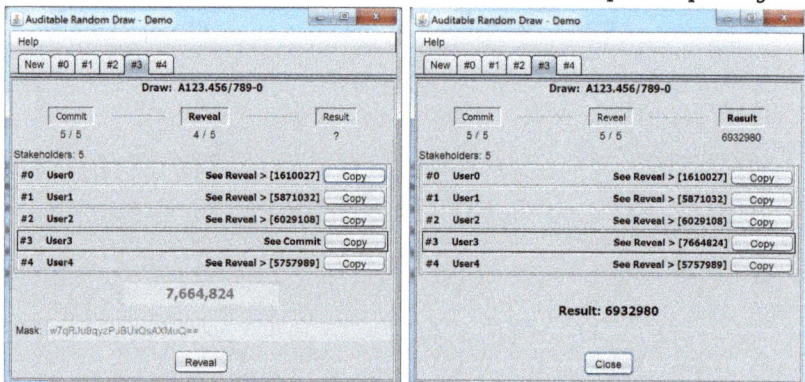

(c) S_0, S_1, S_2 and S_4 in reveal phase. Partial result as seen by S_3.

(d) Drawing result: $1{,}610{,}027 + 5{,}871{,}032 + 6{,}029{,}108 + 7{,}664{,}824 + 5{,}757{,}989 \mod 10{,}000{,}000 = 6{,}932{,}980$.

Figure 8: Proof-of-concept Java implementation: screenshots

8 Final Remarks

Previous articles of this research group have explored the need of randomization procedures in legal systems, like the random assignment (distribution) of legal cases to individual judges or courts, the sortition of jurors for a given case, etc., see Marcondes et al. (2019), Saa and Stern (2019), Silva et al. (2020). Moreover, these papers provide extensive discussions on how to build honest (statistically non-biased) and cryptographically secure procedures and protocols, on the sociological and political importance of using fully transparent and auditable procedures, and on the positive effects of using procedures fully compliant with the aforementioned desiderata in the constitution of strong and autonomous legal institutions.

Finally, breaking away from vicious old habits can always be stimulated by respectful criticism, by firm encouragement, and by making available user friendly tools that facilitate the adoption of virtuous new habits without the imposition of additional difficulties beyond the already heavy load of overcoming corporate inertia. This paper provides such a tool, fully compliant with all technical desiderata, user friendly, written in freely available and open source code. The authors hope it will be soon put to use by Brazilian legal institutions and, if necessary, stand ready to help in this endeavor.

Acknowledgments

This work was supported by: Ripple's University Blockchain Research Initiative; CNPq (Brazilian National Council for Scientific and Technological Development – grants PQ 307648/2018-4 and 301198/2017-9); and FAPESP (São Paulo Research Foundation, grants CEPID-CeMEAI 2013/07375-0 and CEPID-Shell-RCGI 2014/50279-4). The authors are grateful for suggestions received from participants of the Interdisciplinary Colloquium on Probability Theory, held on October 10, 2019 at IEA-USP (Institute of Advanced Studies of the University of Sap Paulo), for early conversations with Julio Adolfo Zucon Trecenti from ABJ (Brazilian Jurimetrics Association), and for the mobile interface design conceived by Giovanni A. dos Santos and Joao Paulo A. S. E. Lins. The authors are grateful for the invitation of Jean-Yves Beziau, from ABF (Brazilian Academy of Philosophy), and for the effort of José Acácio de Barros and Décio Krause, organizers of the Festschrift celebrating Doria's 75th birthday.

References

- A. Beutelspacher (1994). *Cryptology*. Mathematical Association of America.
- X. Bultel, J. Dreier, P. Lafourcade, M. More (2017). How to explain modern security concepts to your children. *Cryptologia*, 41, 5, 422–447. 10.1080/01611194.2016.1238422
- M. R. Fellows, N. Koblitz (1994). Kid Crypto. *Congressus Numerantium*, 99, 9–41.
- J.M. Hammersley and D.C. Handscomb (1964). *Monte Carlo Methods*. Methuen.
- V. Fossaluza, M.S. Lauretto, C.A.B. Pereira, and J.M. Stern (2015). Combining optimization and randomization approaches for the design of clinical trials. *Springer Proceedings in Mathematics and Statistics*, 118, 173–184. doi:10.1007/978-3-319-12454-4_14
- M.S. Lauretto, F. Nakano, C.A.B. Pereira, J.M. Stern (2012). Intentional Sampling by Goal Optimization with Decoupling by Stochastic Perturbation. *American Institute of Physics Conference Proceedings*, 1490, 189-201. doi:10.1063/1.4759603
- M.S. Lauretto, R.B. Stern, K.L. Morgam, M.H. Clark, J.M. Stern (2017). Haphazard Intentional Allocation and Rerandomization to Improve Covariate Balance in Experiments. *American Institute of Physics Conference Proceedings*, 1853, 050003, 1–8. doi:10.1063/1.4985356
- N. Luhmann (1985). *A Sociological Theory of Law*. Routledge.
- N. Luhmann (1989). *Ecological Communication*. The Univ. of Chicago Press.
- M. Naor, M. Yung (1990). Public-key cryptosystems provably secure against chosen ciphertext attacks. *Proceedings of the 22nd Annual ACM Symposium on Theory of Computing*, May 13-17, Baltimore, MD, p. 427-37.
- D. Marcondes, C. Peixoto, and J.M. Stern (2019). Assessing randomness in case assignment: The case study of the Brazilian Supreme Court. *Law,*

Probability and Risk, 18, 2/3, 97–114. doi:10.1093/lpr/mgz006
- J. Pearl (2009). *Causality: Models, Reasoning, and Inference*. Cambridge Univ. Press.
- B. Ripley (1987). *Stochastic Simulation*. Wiley.
- P. Rogaway, T. Shrimpton (2004). Cryptographic hash-function basics: Definitions, implications, and separations for preimage resistance, second-preimage resistance, and collision resistance. *Lecture Notes in Computer Science*, v.3017 p.371-88.
- O.T. Saa, J.M. Stern (2019). Auditable Blockchain Randomization Tool. *Proceedings*, 33, 1,, 17.1–17.6. doi:10.3390/proceedings2019033017
- P. Scozzafava (1993). Uniform distribution and sum modulo m of independent random variables. *Statistics and Probability Letters*, 18, 4, 313-314.
- M.V.M. Silva, M.A. Simplicio, R.A.C. Pfeiffer, J.M. Stern (2020). *A Fair, Traceable, Auditable and Participatory Randomization Tool for Legal Systems*. arXiv:2006.02956
- J.M. Stern (2008). Decoupling, Sparsity, Randomization, and Objective Bayesian Inference. *Cybernetics and Human Knowing*, 15, 2, 49-68.
- J.M. Stern (2011). Symmetry, Invariance and Ontology in Physics and Statistics. *Symmetry*, 3, 3, 611-635. doi:10.3390/sym3030611
- J.M. Stern (2018). Verstehen (causal/interpretative understanding), Erklären (law-governed description/prediction), and Empirical Legal Studies. *JITE - Journal of Institutional and Theoretical Economics*, 174, 1, 105-114. doi:10.1628/093245617X15120238641866

www.ingramcontent.com/pod-product-compliance
Ingram Content Group UK Ltd.
Pitfield, Milton Keynes, MK11 3LW, UK
UKHW061222180426
11947UKWH00026B/1979